Synthetic Nitrogen Products

Synthetic Nitrogen Products

A Practical Guide to the Products and Processes

Gary R. Maxwell
DuPont Chemical Solutions Enterprise
Memphis, Tennessee

Kluwer Academic / Plenum Publishers
New York, Boston, Dordrecht, London, Moscow

Library of Congress Cataloging-in-Publication Data

Maxwell, Gary R.
Synthetic nitrogen products: a practical guide to the products and processes/by Gary R. Maxwell.
p. cm.
Includes bibliographical references and index.
ISBN 0-306-48225-8
1. Nitrogen compounds. 2. Nitrogen compounds—synthesis. I. Title.

TP245.N8M378 2004
661′.0711—dc22

2003062098

ISBN: 0-306-48225-8

233 Spring Street, New York, New York 10013

http://www.wkap.nl/

10 9 8 7 6 5 4 3 2 1

A C.I.P. record for this book is available from the Library of Congress

Printed in the United States of America

PREFACE

Industrial products that are made from, or contain, nitrogen are described in parts of some encyclopedias and standard reference works. However it is not always simple to determine from these varied sources the present status of the technology and markets for various nitrogen products.

We therefore perceived a need for a text that provides a comprehensive description of: 1) products that are made from or that contain nitrogen; 2) the processes that produce these products; and 3) the markets that consume these products. I have attempted to present the material in a standardized format that should make this book easy to use and helpful to the readers. The standard format for each product is: Introduction, Process, Production, and Uses, with some variations in different chapters.

This book provides information that could be used by a wide range of readers:

- Fertilizer companies—to evaluate different production processes and review general trends in the market.
- Basic chemical companies—to evaluate different production processes and review general trends in the market.
- Specialty chemical companies—to investigate new chemical production and/or sales opportunities and the processes that could make those sales a possibility.
- Chemical distributors—to obtain a feel for the general market size for some chemicals and the basic handling and distribution procedures for various chemicals.
- Engineering Companies—to evaluate different production processes and review general trends in the market.
- Engineering and Chemistry Students—to learn more about practical applications of the principals that they have experienced in their classrooms and laboratories.

- Business Students—to obtain information so they can evaluate industries related to the nitrogen or nitrogen products.
- University professors—to obtain practical examples of how different processes and products are used.
- Government Employees—to obtain information for elected officials to support their legislative proposals, to help develop market-based programs within agencies and to provide guidelines in the preparation of regulations that affect nitrogen products.
- Investment Analyst and Venture Capital Companies—to gain an understanding of the technology that is used to produce various chemicals and to review past trends in various markets.
- This information could be useful to companies in most regions of the world: North American companies, South American companies, Asian Pacific Companies, Middle East companies, European companies, Central American and Caribbean companies, and South African companies. However, developing or third-world countries will probably benefit the most from the comprehensive information in this book.

The volume is arranged into chapters for each of the nitrogen products that are discussed. The first four chapters provide basic information about nitrogen and nitrogen products and processes. Chapters 5 through 22 provide detailed descriptions of various nitrogen or nitrogen-containing products.

I have provided numerous tables, figures, and graphs to provide a visual summary and description of much of the information that is discussed. In addition, I have used an extensive number of references that allow the reader to obtain additional information about the subject of interest to them. I have also attempted to provide as many Internet references as possible. The Internet provides a vast amount of resources and references that are readily available and allow individuals to quickly obtain information that they need so they can be more productive in their endeavors.

The index, containing numerous key words, enables the reader to rapidly find the required information.

I would like to express my thanks to Jim Kent for his encouragement and for suggesting to the publisher that this book should be written. Many colleagues inside the DuPont Company supported the development of this book by reading parts of the manuscript and providing me with reference material through our library and information services. I would also like to thank my uncle, Charles D. Scott, PhD for taking the time to review some of the key chapters in this book. Several companies and organizations graciously gave us permission to use their copyrighted information in this book and made suggestions on how to improve the description of their technologies. Without their cooperation, this book would not have been very informative.

I would like to thank Kluwer for their confidence in me in the development of this book as well as Ken Howell, Senior Editor, Brian Halm, Production

Editor, and Catherine Wilson, Editorial Assistant, for their assistance and encouragement in preparing this book.

Gary R. Maxwell

DuPont Company

CONTENTS

1

INTRODUCTION TO NITROGEN

Nitrogen is one of the most familiar substances on earth. It makes up 78% of the air that surrounds us. Nitrogen is found in a combined form in a few mineral compounds, such as potassium nitrate and sodium nitrate. It is also in a host of complex organic compounds, i.e., proteins that all animals use as food to build up tissues.

Nitrogen products are among the most important chemicals produced in the world today. The largest quantities are used as fertilizers, but nitrogen products also find very important uses in the manufacture of nylon and acrylic fibers, methacrylate and other plastics, foamed insulation and plastics, metal plating, gold mining, animal feed supplements, herbicides, and many pharmaceuticals.

1.1. NITROGEN CHARACTERISTICS

Nitrogen is a non-toxic, non-flammable, colorless, tasteless and odorless gas. It is slightly lighter than air with a specific density of 0.967 (air = 1.0). It is non-detectable by any of the human senses. It is considered non-hazardous and is listed along with other utilities on some sites. Thus we can often be lulled into a false sense of security and complacency concerning nitrogen. As a gas, nitrogen is a "silent-killer", and some people consider it to be one of the most dangerous gases. This is because a person can suffocate in an enclosure that contains a nitrogen atmosphere or at an opening on a vessel being purged with nitrogen and never know that they were deprived of oxygen – until it is too late.

Inhalation of a nitrogen-enriched atmosphere may cause dizziness, drowsiness, nausea, vomiting, excess salivation, diminished mental alertness, loss of consciousness, and death. If a person enters an atmosphere of nitrogen, he or she can lose consciousness without any warning symptoms in as little as 20 seconds. Death can follow in 3 to 4 minutes. A person falls as if struck by a blow on the head. One deep breath of 100% nitrogen can be fatal because nitrogen will displace carbon dioxide in the body completely.

Breathing is stimulated and controlled by carbon dioxide (CO_2) present in the lungs. As the CO_2 level increases, the brain sends a message to increase respiration. When the CO_2 level drops, the rate of respiration will also decrease in order to maintain the proper balance. In the absence of a carbon dioxide signal to the brain, breathing stops. While some chemicals or substances may effect some people to a greater degree than they will others - based on our tolerance of the substance, nitrogen does not discriminate. It affects every individual the same way. It displaces oxygen. If we do not have enough oxygen, we will die.

No one should work in or enter atmospheres containing less than 19.5% oxygen - unless they are equipped with a self-contained breathing apparatus or a breathing air mask. This is also true of rescue personnel who can be overcome by the same oxygen-deficient atmosphere as the initial victim[39, 146]. While it is widely known that exposure to excessive amounts of nitrogen inside equipment can result in swift death, the fact that people can be equally affected while standing near openings of nitrogen-purged equipment may be less understood. This situation must not be underestimated! Fatalities have occurred when personnel were working *outside* an open manhole on a vessel that was being purged with nitrogen[146].

Nitrogen gas is classified as a "simple asphyxiant". This means nitrogen will displace oxygen in high concentrations and create oxygen deficient (<19.5%) atmospheres without significant physiological effects. Exposure limits are not normally given to "simple asphyxiants" because the limiting factor is the available oxygen. Therefore, nitrogen has no exposure limit.

In addition to being an inhalation hazard, Liquid Nitrogen is a colorless, odorless, and extremely cold under pressure. Contact with this liquid or the cold vapors can cause severe frostbite.

The properties of nitrogen are listed in Table 1.1.

TABLE 1.1 Physical Properties of Nitrogen

Property		Value
Molecular Weight		14.0067
Boiling Point, °C (°F)		-195.8 (-320.5)
Freezing Point, °C (°F)		-209.86 (-346.0)
Critical Temperature, °C (°F)		-146.95 (-232.4)
Critical Pressure, kPa (psi)		3,393 (492.3)
Critical Volume, cc/g-mol		89.5
Specific Heat at 20°C and 147 psia		
$c_{p.}$ (BTU/lb °F)		0.247
c_v (BTU/lb °F)		0.176
Specific Heat at 21°C	c_p(J/(kg-K)	1,046

TABLE 1.1 Physical Properties of Nitrogen (continued)

Property	Value
$k = c_{p.}/c_v$	1.41
Heat of Formation of gas, ΔH (kJ/mole)	0.3604
Heat of Fusion at Melting Point (J/g)	25.6
Heat of Vaporization at Boiling Point	
J/g	199
BTU/lb	85.5
Solubility in Water, g N_2 per 100 g H_2O at 760 mm	
0°C	0.00239
20°C	0.00189
40°C	0.00139
60°C	0.00105
Specific Gravity	
Relative to Air (at 1.0)	0.967
-195.8°C (Liquid)	0.808
-252°C (Solid)	1.026
Specific Volume (at 70°F and 1 atm), ft^3/lb	13.80
Density, kg/m^3	
At 20°C	1.16
Liquid at Boiling Point	808.6
Solid at Melting/Freezing Point	1028
Thermal Conductivity, W/m-K	
Gas @ 21°C	0.026
Liquid @ -200°C	0.14456

1.2. NITROGEN SEPARATION AND PURIFICATION

Industrial demand for high purity nitrogen is met by producing most of the nitrogen from liquid air. After air has been liquefied by pressure and cooling, it consists almost entirely of a mixture of liquid oxygen (which boils at –183°C) and liquid nitrogen (which boils at –195°C). By careful control it is possible to separate these two gases by fractional distillation and produce nitrogen with a purity of 99.9% or greater[1].

Nitrogen production via PSA (pressure swing absorption) is based on the principle that nitrogen and oxygen have different absorption rates on carbon molecular sieves (CMS). This principle was first demonstrated commercially in the Bergbau-Forschung (or BF) process that separates nitrogen from air by selective adsorption over activated carbon in a PSA system[3].

The KURASEP (**Kur**aray **Sep**aration) Process for separating nitrogen from air is a variant of the PSA process in which a carbon molecular sieve is used as the adsorbent[1]. The Kuraray Chemical Company developed this process[3]. This process can produce "nitrogen gas" containing less than 100 ppm oxygen (99.99% N_2) and five-nines nitrogen (99.999% N_2, 10 ppm O_2) can be produced if nitrogen yield is sacrificed slightly.

The MOL-nitrogen (**Mol**ecular Sieve **P**ressure **S**wing **A**dsorption) process is a version of the PSA process for separating nitrogen from air. It was developed by Kobe Steel[1]. Most PSA processes for nitrogen production use molecular sieve carbon as the adsorbent, but this one uses Zeolite X. Water and carbon dioxide are first removed in a two-bed PSA system, and then the nitrogen is concentrated and purified in a three-bed system[4]. The purity of the product is 99.999% N_2, 10 ppm O_2.

The NitroGEN (**Nitro**gen **Gen**erator) process is a version of the VPSA (Vacuum Pressure Swing Adsorption) process for separating nitrogen from air by vacuum pressure swing adsorption[1]. It was developed by the Linde Division of Union Carbide Corporation. The name has also been used for two membrane systems for extracting pure nitrogen from air[5, 6].

Depending on volume and purity requirements, nitrogen can be supplied by several different production systems or delivery systems. Figure 1.1 illustrates some of these systems[159]. Liquid refers to delivery by trucks or in cylinders; Permeation refers to a membrane system; Adsorption refers to a PSA system; and Distillation refers to a cryogenic air separation plant.

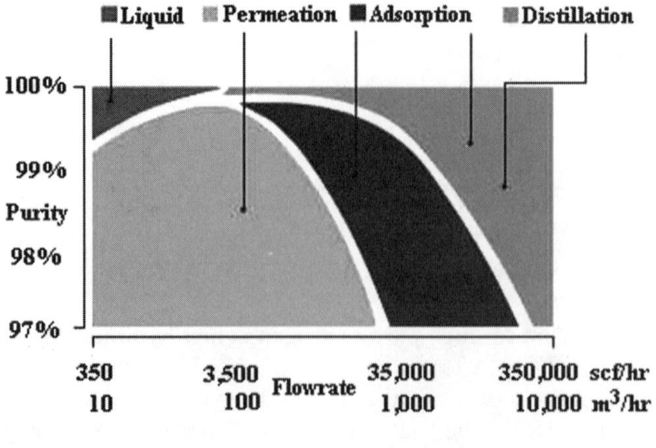

Figure 1.1. Nitrogen supply options based on flowrate and nitrogen purity. (Copyright 2002 Air Products and Chemicals, Inc.)

A flowsheet for a plant that produces nitrogen in a cryogenic air separation plant is shown in Figure 1.2[150].

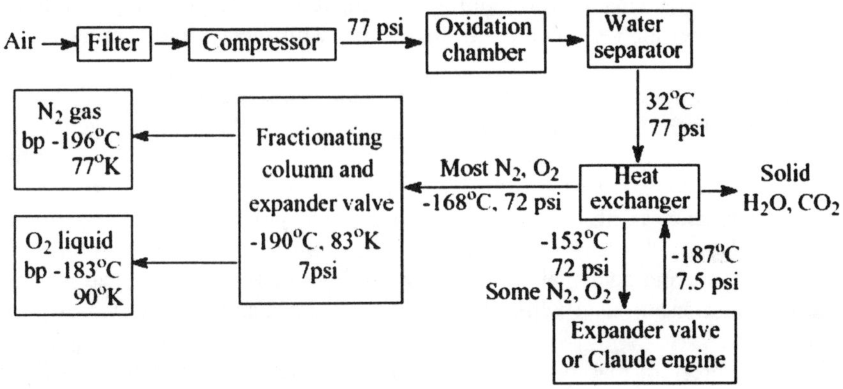

Figure 1.2. Nitrogen production flowsheet in a cryogenic separation plant.

In a cryogenic nitrogen production process the following steps occur:

1. Air, which contains 78% Nitrogen, 21% Oxygen and 0.9% Argon, is fed to the process.
2. The air is filtered to remove particulates.
3. The air is then compressed to 77 psi.
4. Next the air enters an Oxidation Chamber that removes hydrocarbons by converting them to carbon dioxide (CO_2) and water (H_2O).
5. The stream then passes through a water separator to remove most of the water.
6. The process stream then enters a heat exchanger that a) cools the stream to -168°C and b) removes the CO_2 and H_2O as solids.
7. The cooling for the heat exchanger is provided by a small amount of nitrogen and oxygen that are fed to an expander valve.
8. The stream that leaves the heat exchanger is primarily nitrogen and oxygen.
9. This stream goes through another expander valve that cools it to -190°C, and then it enters the fractionating column.
10. The nitrogen and oxygen are then separated based on boiling point (N_2 = -196°C and O_2 = -183°C).
11. Further fractionation of the oxygen stream removes the argon and gives a stream that is 99.5% O_2.
12. Argon, with a boiling point of -183°C, can be obtained as a middle fraction and distilled further.

1.3. PRODUCTS UTILIZING NITROGEN

Nitrogen does not unite easily with other elements to form compounds. Therefore economical fixation of nitrogen from the atmosphere has been a never-ending quest. At high temperatures and pressures nitrogen combines with most of the elements to form nitrogen compounds although many of these compounds are not very stable. It also reacts with metals and semimetals to form nitrides.

When nitrogen combines with certain elements, it produces dyes that rival the rainbow in the brilliance of their colors. With other elements nitrogen forms drugs such as morphine, quinine and acetanilid that ease pain and combat diseases. Nitrogen-bearing compounds are used to make some very deadly compounds such as nitroglycerin, guncotton, and ptomaines. Some nitrogen compounds provide pleasant-tasting flavors and sweet perfumes while others are so vile in taste and odor that they are beyond description. In nitric acid we find nitrogen in a strong acid while in ammonium hydroxide we have it in a well-known base[7].

In the rest of the book various synthetic nitrogen compounds will be discussed. The production processes for these compounds will be described along with the way the compound is used by society.

2

NITROGEN FIXATION

Nitrogen Fixation involves the combination of nitrogen from the air with other elements to form other compounds. On the roots of beans, peas, clover and other leguminous plants are nodules. When these nodules are examined under a powerful microscope, they are found to be the homes of great masses of 'nitrogen-fixing" bacteria. The bacteria have the ability to extract nitrogen from the air and convert it into soluble nitrates. These nitrates renew the fertility of soil since plants use them for food. Every plant must have nitrogen to live, but no plant or animal can use free nitrogen for nourishment. Soluble compounds of nitrogen must be present to make the soil fertile. Such compounds can then be used by plants to produce proteins that in turn can then be used by animals to produce tissues and enzymes. The physical basis of life is the plant or animal cell. The content of such cells includes protoplasm and various organelles that contain nitrogen-bearing compounds such as proteins and enzymes[7]. The Nitrogen Cycle is shown in Figure 2.1 along with some examples of different fixation processes [143].

2.1. NITROGEN OXIDES

Most nitrogen products have been based on nitrogen oxides and ammonia; so the history of fixation of nitrogen as nitrogen oxides and ammonia is of interest.

The five oxides of nitrogen are[7]:

- N_2O Nitrous Oxide
- NO Nitric Oxide
- N_2O_3 Nitrogen Trioxide
- NO_2 Nitrogen Dioxide
- N_2O_5 Nitrogen Pentoxide

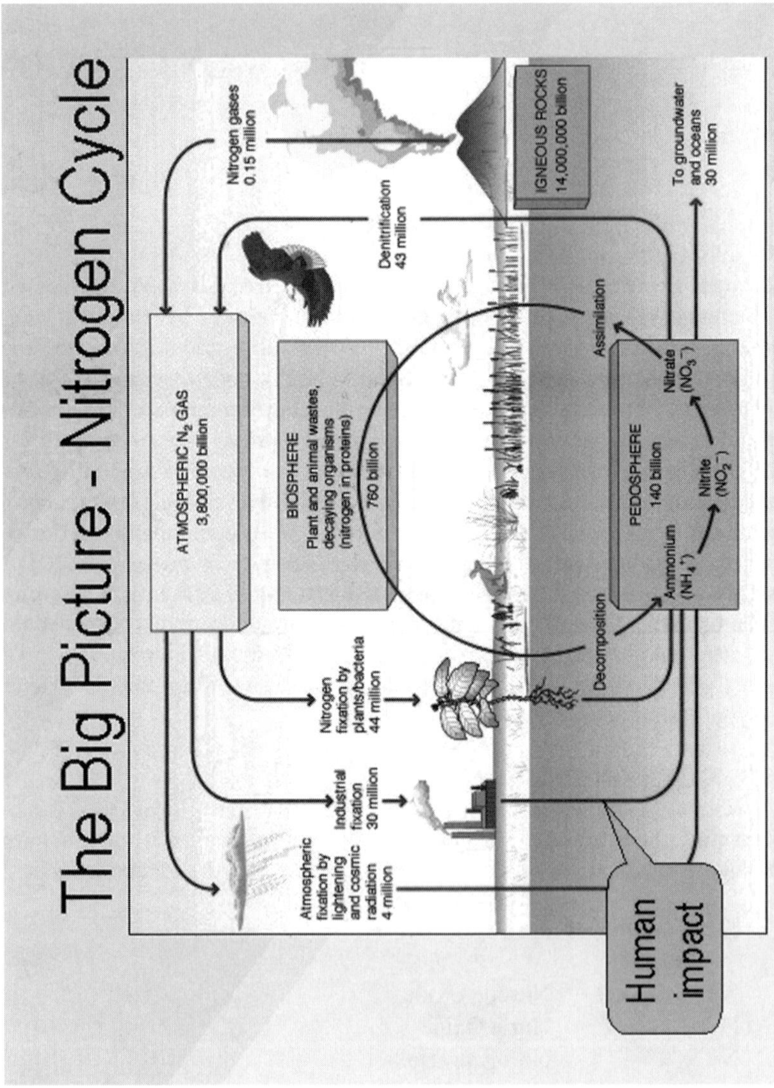

Figure 2.1. The Nitrogen cycle.
(Reproduced by permission of W.H. Freeman & Co.)

Nitrogen trioxide and nitrogen pentoxide are the anhydrides of nitrous acid and of nitric acid. They are difficult to prepare and are unstable.

Nitrous oxide (laughing gas) is a colorless gas. It has a slightly sweet odor and taste and is somewhat soluble in water. It supports combustion of most substances almost as well as oxygen. Since the 1840's the major use of nitrous oxide has been as an anesthetic – especially by dentists. It is also used as a propellant in some aerosol cans, in atomic absorption spectrophotometry, in cryosurgery and in racecar engines to provide extra power and acceleration.

Nitric oxide is a colorless gas that is insoluble in water. It does not support the combustion of most substances. The most noticeable chemical characteristic of nitric oxide is the ease with which it combines with oxygen to form nitrogen dioxide. Nitric oxide is produced in considerable quantities in the industrial preparation of nitric acid. In these reactions the gas is not isolated. It is immediately oxidized to nitrogen dioxide.

In the late 1980's nitric oxide (NO) was discovered as a product of enzymatic synthesis in mammals. And in 1998 the Nobel Prize in Physiology or Medicine was awarded to the scientists (Louis J. Ignarro, Robert F. Furchgott, and Ferid Murad) that discovered the role of NO as a biological messenger. Although one of the simplest biological molecules in nature, nitric oxide has found its way into nearly every phase of biology and medicine. Its uses range from its role as a critical endogenous regulator of blood flow and thrombosis to a principal neurotransmitter mediating erectile function to a major pathophysiological mediator of inflammation and host defense. These major discoveries have stimulated intense and extensive research into a vast array of fields including chemistry, molecular biology, and gene therapy. The Nitric Oxide Society has been formed to coordinate meetings, and they also publish a research journal. More information about the Nitric Oxide Society is available at www.apnet.com/no.

Nitrogen dioxide is a red-brown gas that is very soluble in water. It has an unpleasant odor and is quite poisonous. When nitrogen dioxide is cooled, its brown color fades to a pale yellow. Molecular weight determinations show that this yellow substance has a formula of N_2O_4 (nitrogen tetroxide):

$$\underset{\text{Nitrogen Dioxide (Brown)}}{2\,NO_2} \quad \leftrightarrow \quad \underset{\text{Nitrogen Tetroxide (Yellow)}}{N_2O_4} \qquad (2.1)$$

The importance of nitrogen dioxide lies in the fact that it combines readily with water to form nitric acid.

$$3\,NO_2 + H_2O \quad \rightarrow \quad 2\,HNO_3 + NO \qquad (2.2)$$

This is the final step in the synthetic production of nitric acid. Nitrogen dioxide also acts as an oxidizing agent in the manufacture of sulfuric acid.

Nitrogen dioxide causes the sulfur dioxide to be oxidized to sulfur trioxide, which then unites with water to form sulfuric acid.

In 1780 Henry Cavendish, an English scientist, used an electric spark to combine atmospheric nitrogen and oxygen to make nitrogen oxides. In 1902, Bradley and Lovejoy performed experiments at Niagara Falls aimed at developing a commercial process using an electric arc to produce nitrogen oxides. The experiments failed, apparently because the electric arc "flame" area was too small, and product gases were not removed from the reaction zone fast enough. In 1905, with the help of these results, Birkeland and Eyde, in Nottoden, Norway, developed the first large-scale manufacturing process for making nitrogen oxides from air by using an arc spread out to form a flame.

The English scientist Joseph Priestley first discovered nitrous oxide in 1793. He made N_2O by heating ammonium nitrate in the presence of iron filings, and then passing the resulting gas through water to remove toxic by-products.

Nitrous oxide production technology can be licensed from the Sanghi Organization in Worli, India (www.sanghioverseas.com). Their technology is based on heating ammonium nitrate to 250°C and then purifying the resulting gas stream.

Nitrogen oxides can be produced from the nitrogen and oxygen in air by raising the air to very high temperatures. Much experimental work has been done to develop a commercial process based on this concept, but to date the concentration of the nitrogen oxides has been too low for economic success. Also, the high temperatures in this process make commercial application of this process unattractive. A typical high temperature process operates at 2000°C; therefore, the choice of and the maintenance of materials of construction are critical.

Other processes have been proposed for nitrogen fixation, but none of these have approached a commercially attractive stage. For example, it was shown that low energy ultraviolet photons can form nitric oxide at room temperature under oxidizing conditions on the surface of a titanium dioxide-containing material[8].

Research continues on the formation of nitrogen oxides. The high temperature fixation of nitrogen as oxides remains an environmental problem rather than a commercially attractive process, as large quantities of nitrogen oxides are produced by fixation in high temperature combustion processes such as power plants, automobile engines, and home furnaces. The US EPA promulgated regulations in 2000 (NOx SIP Call) that require several states in the Eastern United States to reduce their NOx emissions by a combined total of 35% to 40% by 2005 (see www.epa.gov/ttn/otag).

2.2. NITRIDES

In some of the early nitrogen processes, atomic nitrogen, which is formed by ionizing molecular nitrogen with electrical discharges, was used. Atomic nitrogen reacts much more readily with the elements than molecular nitrogen does. High melting points, hardness, and resistance to corrosion characterize the nitrides of high melting transition metals (TiN, ZrN and TaN). Some of the nonmetallic nitrogen compounds (BN, Si_3N_4, and AlN) are corrosion- and heat-resistant, ceramic-like materials with semiconductor properties.

An example of high temperature nitrogen reactions is the Serpek process that produces aluminum nitride[1]. In this process bauxite and coke are mixed and heated in a nitrogen atmosphere at 1,800°C to produce aluminum nitride. Boiling the aluminum nitride with potassium aluminate then produces ammonia:

$$Al_2O_3 + 3\,C + N_2 \rightarrow 2\,AlN + 3\,CO \qquad (2.3)$$

$$2\,AlN + 3\,H_2O \rightarrow 2\,NH_3 + Al_2O_3 \qquad (2.4)$$

This process was invented by O. Serpek in Germany in 1909 and is summarized in British patents 15,996 and 15,997.

The van Arkel or Iodide Process was first used in The Netherlands in 1925 to make especially pure nitrides[1]. This process can make several metal nitrides (TiN, ZrN, HfN, VN, NbN, BN and AlN) by passing a mixture of the metal halide with nitrogen and hydrogen over a hot tungsten wire. The metal halide decomposes and the resulting nitride deposits on the wire. The process is described in U.S. Patent 1,671,213.

2.3. AMMONIA

The fixation of nitrogen as ammonia is also important. The alchemists prepared ammonia by distilling leather scraps, hoof, horns and other animal refuse. Since they found that the best ammonia was obtained by distilling deer antlers, the alchemists called ammonia, "Spirits of Hartshorn[7]." Nature forms ammonia by decomposition of proteins that contain approximately 16% of their weight in nitrogen. The odor of ammonia can often be detected around manure piles and other decaying organic matter.

A process for synthesizing ammonia from nitrogen and hydrogen, using high temperatures and pressures and an iron-containing catalyst, was invented by Fritz Haber at BASF in 1908[1]. In 1909, C. Bosch of BASF built a pilot plant using an osmium-based catalyst, and in 1913 a larger plant was built at Oppau, Germany. The Haber-Bosch process has been continuously improved and is still of major importance worldwide[9, 10, 11, 12, 13, 14].

The chemistry for this process is relatively simple:

$$N_2 + 3\,H_2 \leftrightarrow 2\,NH_3 \qquad (2.5)$$

A fairly high temperature of ~500°C has been employed to give a compromise between acceptable reaction rate and ammonia yield. High pressure favors the forward reaction and forces the equilibrium to the right, since fewer moles are on the right side of the above chemical reaction.

The Mond gas process produced ammonia by gasifying coal at a relatively low temperature by using a mixture of air and steam[1]. The steam was used to increase the yield of ammonia. L. Mond developed the process at Brunner Mond in Winnington, Cheshire, England in 1883. It was invented primarily to produce ammonia that was needed for the Solvay process that makes sodium carbonate. The South Staffordshire Mond Gas Corporation near Birmingham, England commercialized the process. This company distributed the ammonia gas to local industry through the world's first gas distribution grid[29, 30].

Some other early processes for making ammonia are listed below along with some ammonia process developments from the 1980's and 1990's:

- The Casale ammonia process, which was designed by L. Casale. He founded Ammonia Casale of Lugano, Switzerland in 1921[1, 12, 19].
- The Claude-Casale process, which is a high-pressure ammonia synthesis process that was developed by G. Claude in the 1920's[1, 12, 20, 21].
- The Fauser process, which was developed by G. Fauser in Italy in 1924[12].
- The Mont Cenis process was an early ammonia synthesis process that was similar to the Haber-Bosch process except that it used coke-oven gas[1]. It was started up by The Royal Dutch Group at Ymuiden, The Netherlands in 1929[31, 32].
- The Braun process is a variation on the Haber-Bosch process ammonia synthesis process in which the synthesis gas is purified cryogenically[1]. It has been widely used since the mid-1960's[18]. (Synthesis gas is a mixture of hydrogen, carbon monoxide and carbon dioxide - see Table 5.9 for more details).
- The AMV process for making ammonia was invented by ICI and announced in 1982[1]. It uses a new catalyst and operates at a pressure close to that at which the synthesis gas has been generated, thereby saving energy. Construction licenses have been granted to Chiyoda Corporation, Kvaerner and Mannesman. In 1990 the process was being used in the CIL plant in Ontario, Canada and also in Henan Province, China[15, 16, 17]. The European patent for this process is 49,967.
- The LCA (Leading Concept Ammonia) process makes ammonia from air and natural gas[1]. It is essentially a simplified form of the standard ammonia synthesis process that is more suitable for smaller plants. Thermal economics are achieved in the steam reforming section. The process was developed in the mid-1980's by ICI, and two units began operating at the ICI plant in Severnside in 1988. The first non-ICI installation was designed by KTI for Mississippi Chemicals in Yazoo City, MS[26, 27, 29].

- The BYAS (**By**pass **A**mmonia **S**ynthesis) process can be used for economical expansion of existing ammonia synthesis plants. This process introduces additional natural gas at an intermediate stage in the process. The additional nitrogen in the air, which also has to be introduced, is removed by PSA (pressure swing absorption). The process was developed and offered by Humphreys and Glasgow in the UK[1].
- The KAAP (**K**ellogg **A**dvanced **A**mmonia **P**rocess) process is the first high-pressure ammonia synthesis process that makes ammonia from nitrogen and hydrogen without the aid of an iron-containing catalyst[1]. The catalyst was developed by British Petroleum (BP) and contains ruthenium supported on carbon. The KRES (Kellogg Reforming Exchanger System) process provides the reformer gas for this process. The process was developed by MW Kellogg in 1990 and first installed by the Ocelot Ammonia Company (now Pacific Ammonia Inc. - PAI) at Kitimat, British Columbia between 1991 and 1992. Another plant was installed at Ampro Fertilizers in Donaldsonville, LA in 1996[22, 23, 24, 25].

In 2002 the Haber process was the most commercially attractive ammonia process even though it had high compression costs, and a large expenditure of energy was required to produce the feed hydrogen. Improvements such as the AMV process and the KAAP process may provide attractive cost reduction opportunities in ammonia production.

The more important ammonia production processes (including some of those mentioned above) will be discussed in more detail in subsequent chapters.

Ammonia and some of its salts and derivatives are used as fertilizers. Ammonia also is a very important building block for many other chemical commodities as shown in Figure 2.2.

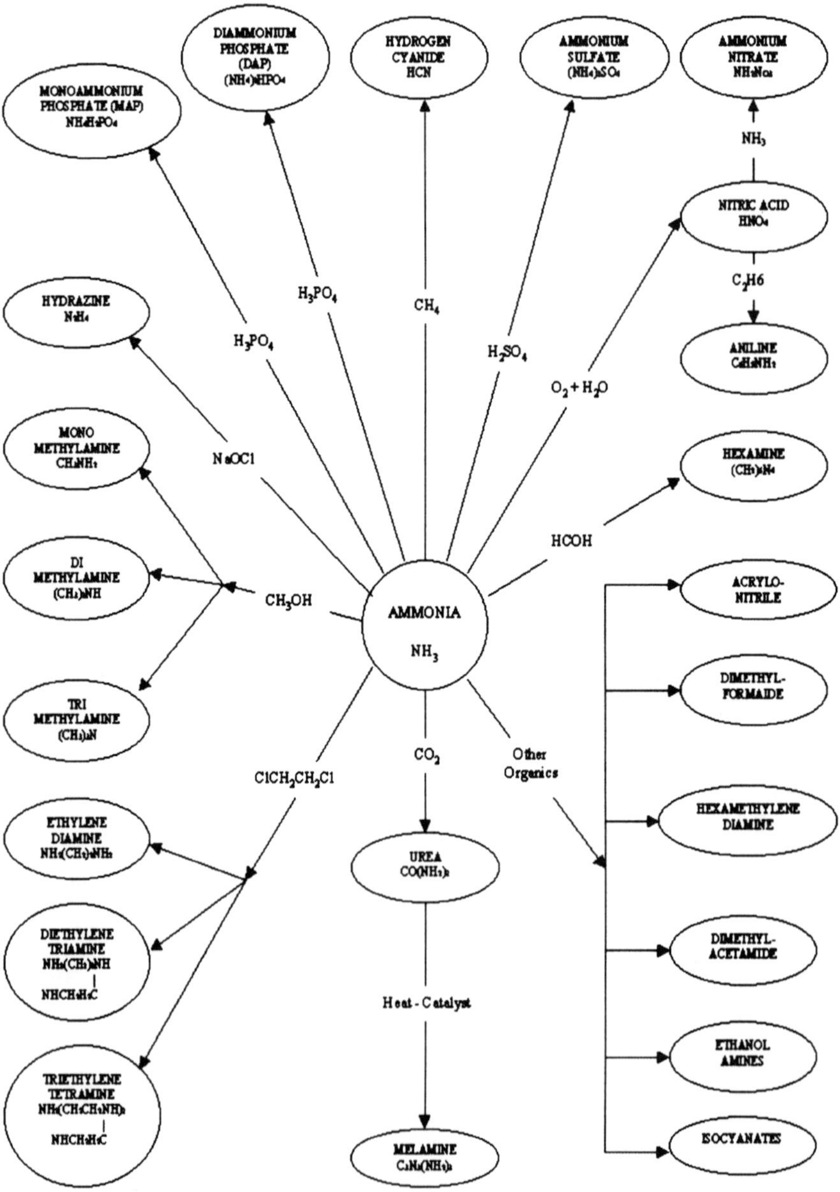

Figure 2.2. Synthetic nitrogen products.

3

NITROGEN PRODUCTS: PRODUCTION AND CONSUMPTION

The demand for nitrogen in a chemically fixed form (as opposed to elemental nitrogen gas) drives a huge international industry that encompasses the production of seven key chemical nitrogen products: ammonia, urea, nitric acid, ammonium nitrate, nitrogen solutions, ammonium sulfate and ammonium phosphates. Such nitrogen products had a total worldwide annual commercial value of about $US 50 billion in 1996. The cornerstone of this industry is ammonia. Virtually all ammonia is produced in anhydrous form via the Haber process (as described in Chapter 2). Anhydrous ammonia is the basic raw material in a host of applications and in the manufacture of fertilizers, livestock feeds, commercial and military explosives, polymer intermediates, and miscellaneous chemicals[35].

On a worldwide basis, nitrogen fertilizer consumption accounts for more than 85% of the total nitrogen industry. In the United States, with its significant industrial markets, fertilizer use accounts for only approximately 80% of total nitrogen consumption[35]. About 97% of nitrogen fertilizers are derived from synthetically produced ammonia. The rest are produced as by-product ammonium sulphate from caprolactam processes, acrylonitrile processes, metal refining operations and small quantities of natural nitrates, especially from Chile[35]. Ammonia is the starting material for six primary nitrogen products: urea, ammonium nitrate, ammonium phosphates, ammonium sulfate, nitric acid and nitrogen solutions (i.e., aqueous solutions that are mixtures of selected nitrogen chemicals)[36].

3.1 NITROGEN INDUSTRY: CAPACITY AND PRODUCTION

Up through the 1960's, the development of the nitrogen industry took place in the developed countries of Western Europe, North America and Japan. However, in the 1970's and early 1980's, the construction of new plants shifted to the natural gas-rich countries of the Caribbean and Middle East. Additional plants were also built in some large consuming countries such as China, India, Indonesia and Pakistan. At the same time, many plant closures occurred in Western Europe and Japan. Western European share fell from 20% in 1980/81 to 11% in 1997/98. In 1980/81 the developing countries accounted for 31% of nitrogen fertilizer production. By 1997/98 their share had increased to 55% (see Figure 3.1). The main regions that produced nitrogen fertilizers in 1997/98 are summarized in Figure 3.2. Although at least 60 countries produce urea, over one-third of the world's production of nitrogen fertilizer is in just two countries, China and India[36].

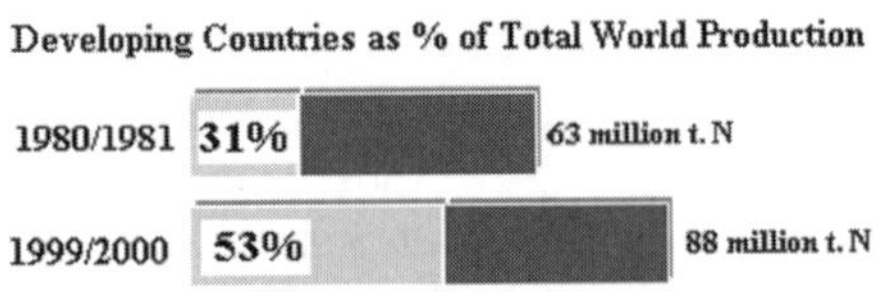

Figure 3.1. Nitrogen fertilizer production - 1980/81 & 1999/2000.
(Reproduced by permission of International Fertilizer Industry Association)

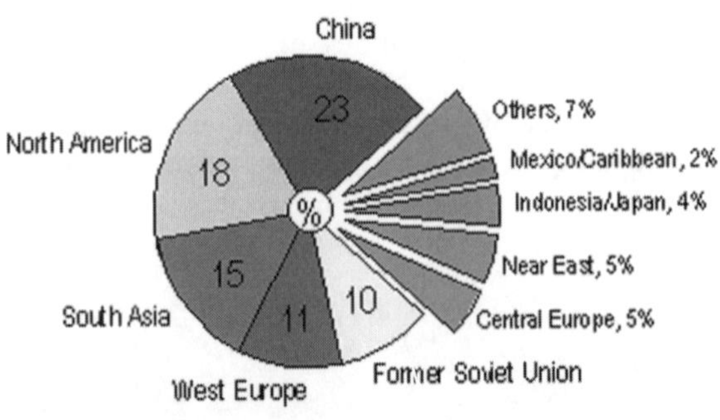

Figure 3.2. Distribution of nitrogen fertilizer production, 1997/98.
(Reproduced by permission of International Fertilizer Industry Association)

World ammonia capacity increased by nearly 14% from 1984 to 1996 while capacity for urea, the primary downstream nitrogen product, increased by 45%. The increases were due primarily to 1) a desire by some major importing countries to become more self-sufficient and 2) the construction of export-oriented capacity in the Middle East and in the former Soviet Union (prior to its breakup). Ammonium phosphate capacity increased by 9% between 1984 and 1996. Ammonium nitrate capacity declined by 2% from 1984 to 1996 while ammonium sulfate capacity declined by 8%[35].

In the future, developing nations are expected to continue to account for most of the increases in ammonia and urea capacity. Ammonia capacity is expected to increase by about 20 million tonnes and urea capacity by about 12 million tonnes of nitrogen between 1996 and 2002. The availability of relatively low-cost feedstock (usually natural gas) will be a major determinant as to where this new capacity is installed. Ammonium nitrate and ammonium phosphate capacity are also expected to rise[35]. The following tables summarize anticipated world capacity for nitrogen products by year (Table 3.1) and by major regions or countries (Table 3.2)[148].

Table 3.1. Anticipated World Capacity for Nitrogen Products - by Year[148]

Year	Ammonia	Urea	Ammonium Nitrate	Ammonium Phosphates (as P_2O_5)
1999	157,539	125,080	58,123	29,160
2000	161,165	130,812	57,717	30,014
2001	161,849	133,607	57,596	31,209
2002	165,175	139,683	57,596	31,451
2003	169,135	143,410	57,596	31,451
2004	170,903	139,886	57,596	31,451
	Average Annual Growth Rate (percent)			
	0.2%	0.2%	0.0%	0.2%

(Reproduced by permission of International Fertilizer Industry Association)

The trend in nitrogen product production generally parallels the changes in product capacity. However, ammonium sulfate production is expected to rise as world industrial production increases, even though a significant increase in capacity is not likely. Also, ammonium nitrate production is projected to be flat even though some new capacity will be built[35].

Table 3.2. 1997/98 World Production of Nitrogen Fertilizer - by Country[36]

Country / Region	Nitrogen Fertilizer Production (thousands of tonnes of N)
China	19.94
India	10.08
North America	14.97
Western & Central Europe	13.57
Former Soviet Union	7.99
Middle East	5.61
Others	12.77
Total	84.93

(Reproduced by permission of International Fertilizer Industry Association)

3.2. NITROGEN PRODUCT CONSUMPTION

The apparent consumption of ammonia increased by about 15% between 1984 and 1996, while the apparent consumption of urea increased by about 54%. The developing countries are largely responsible for the increased consumption (see Figure 3.3 and Figure 3.4)[36]. The industrialized regions of the United States, Western Europe and Japan have experienced a flat to slightly downward trend in the consumption of nitrogen products[35].

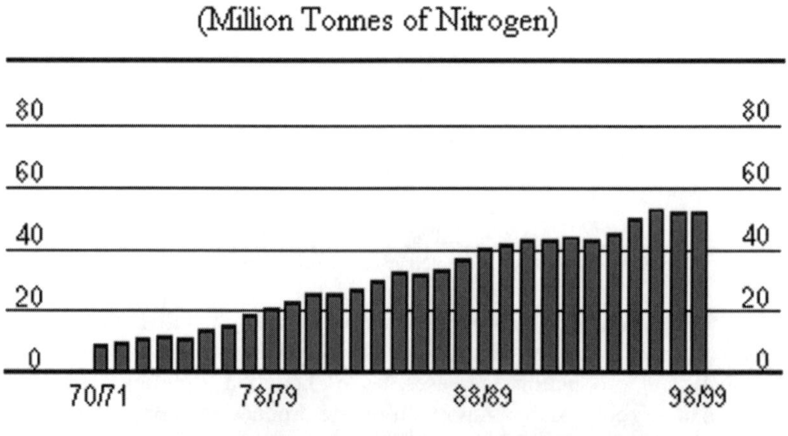

Figure 3.3. Nitrogen fertilizer consumption in developing countries. (Reproduced by permission of International Fertilizer Industry Association)

By 2002, urea was expected to account for about 61% of worldwide consumption of the four major downstream nitrogen products. World urea consumption is forecast to increase at an average annual rate of 2.4% between 1999 and 2004[36]. By definition, apparent consumption is equal to production plus imports minus exports. Thus apparent world consumption equals reported world production[35].

The International Fertilizer Industry Association (IFA) presents worldwide nitrogen fertilizer consumption figures that are lower than the figures developed by SRI. However both organizations present growth rates that are similar for the nitrogen products. The statistics from IFA are shown in Figure 3.5, Table 3.3 and Figure 3.6[36].

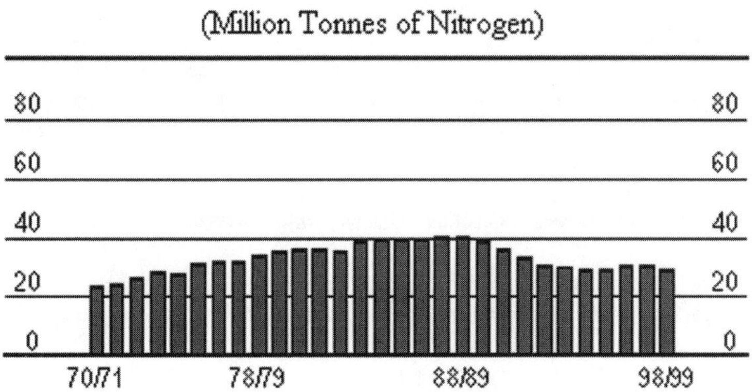

Figure 3.4. Nitrogen fertilizer consumption in developed countries.
(Reproduced by permission of International Fertilizer Industry Association)

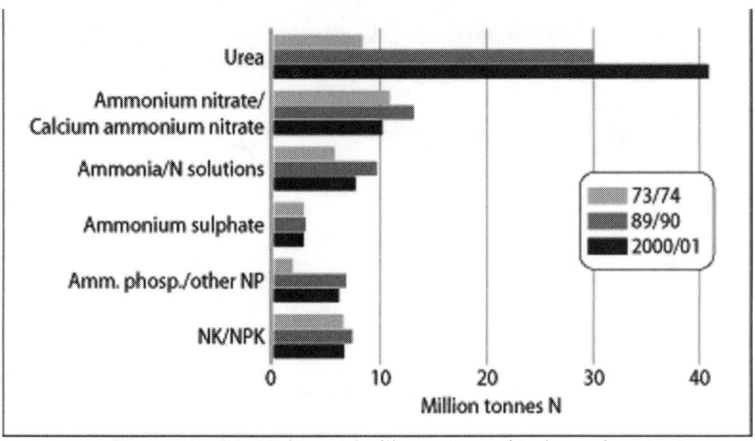

Figure 3.5. World nitrogen fertilizer consumption: by product.
(Reproduced by permission of International Fertilizer Industry Association)

Table 3.3 World Nitrogen Fertilizer Consumption

Nutrient	Million Tonnes of Nutrients 1973/74	1987/88	1997/98
Ammonium Sulphate	2.74	3.03	2.61
Urea	8.29	27.17	37.57
Ammonium Nitrate	7.30	9.88	7.10
Calcium Ammonium Nitrate	3.54	4.45	3.58
Ammonia – Direct Application	3.57	5.29	4.18
Nitrogen Solutions	2.12	4.16	3.99
Mono-Ammonium Phosphate/ Di-Ammonium Phosphate & Other NP compounds	1.71	5.62	6.22
NK / NPK compounds	5.91	7.46	6.51
Others[a]	3.94	8.72	9.19
Total nitrogen	39.12	75.78	80.95

[a] Others : Calcium nitrate, sodium nitrate, ammonium chloride, calcium cyanamide, ammonium bicarbonate

(Reproduced by permission of International Fertilizer Industry Association)

As can be see from Table 3.3 and the Figure 3.4, the consumption increase has been particularly strong in the case of urea. Between 1973/74 and 1997/98, urea consumption increased from 8.3 million tonnes Nitrogen to 37.6 million tonnes Nitrogen. This is equivalent to an increase from 22% to 46% of total world nitrogen consumption[36].

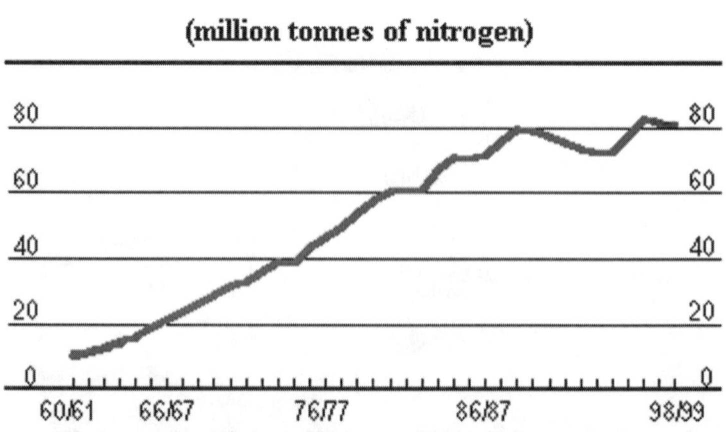

Figure 3.6. World nitrogen fertilizer consumption: 1960 to 1999.
(Reproduced by permission of International Fertilizer Industry Association)

As shown in Figure 3.6, from 1960 to 1989, there was a sustained increase of almost 6% per year in world nitrogen fertilizer consumption - with temporary set-backs due to the oil crises in the 1970's. From 1989 to 1994, world nitrogen fertilizer consumption fell. This was due mainly to the decline in use in the countries of Central Europe and the former Soviet Union. Consumption in Western Europe also fell, but all of these decreases were partially offset by increased consumption in Asia[36].

From 1994 to 1998 world nitrogen fertilizer consumption increased at an average rate of 3% per year. Socialist Asia, South Asia and Latin American led the consumption increases. Consumption in Western Europe stabilized while demand in the former Soviet Union continued to fall. From 1998 to 2000, world nitrogen fertilizer consumption increased more slowly than it had in the previous period[36].

From 2000 to 2005, world nitrogen fertilizer consumption is forecast to increase by 2.3% per year. A gradual decline is expected in Western Europe and Japan. The consumption increases are forecast to occur primarily in the regions listed in Table 3.4:

Table 3.4. Distribution of Forecast Increases in Nitrogen Fertilizer Consumption

Region	% of Consumption Increase
Socialist Asia (China & South Asia)	43
Latin America	15
Central Europe & Former Soviet Union	14
Southeast Asia	14
North America	7

These forecasts assume a relatively stable global economy. In the past, major unpredictable shocks to the global economic system have had an important impact on fertilizer demand[36].

The world nitrogen fertilizer market is summarized by region in Table 3.5[36]. Although world consumption grew substantially between 1984 and 1996, it actually peaked at 79.6 million tonnes of nitrogen in 1989 and declined through 1994 because of a sharp drop in fertilizer consumption in Eastern Europe, the Former Soviet Union (FSU) and Western Europe. The decline in the Eastern Block countries was the result of severe economic strife as that region converted from a centrally planned to a market-driven economy. The consumption decrease in Western Europe was related mainly to lower farm subsidies and concern about groundwater contamination by nitrates. Total world nitrogen fertilizer consumption resumed its long-term upward growth trend in 1995, and substantial growth is expected by 2002[35]. It should be noted that Japan is one of the few countries where the use of slow-release fertilizers is being developed for field crops, especially crops grown in paddies[36].

Table 3.5. World Consumption of Nitrogen Products

(thousands of tonnes of N)

	1984		1996		2002 (Estimated)	
Country	Quantity	Share (Percent)	Quantity	Share (Percent)	Quantity	Share (Percent)
Socialist Asia	15.85	22.3	26.00	31.4	25.25	29.7
South Asia	6.99	9.8	13.46	16.2	15.06	17.8
North America	11.76	16.6	12.87	15.5	12.85	15.2
Western Europe	10.80	15.2	10.12	12.2	10.21	12.0
East Asia	3.40	4.8	5.07	6.1	5.41	6.4
Latin America	3.21	4.5	4.41	5.3	4.64	5.5
Near East	2.59	3.6	3.65	4.4	4.21	5.0
FSU	10.28	14.5	2.72	3.3	2.42	2.9
Central Europe	4.56	6.4	2.31	2.8	2.01	2.4
Africa	1.18	1.7	1.34	1.6	1.39	1.6
Oceania	0.40	0.6	0.99	1.2	1.24	1.5
Totals	71.02	100.0%	82.94	100.0%	84.69	100.0%
Developed Countries	38.95	54.8%	29.99	36.2%	29.67	35.0%
Developing Countries	32.07	45.2%	52.95	63.8%	55.02	65.0%

(Reproduced by permission of International Fertilizer Industry Association)

In Western Europe a large number of fertilizer producers want to exit the market, but few operators are willing to expand or move into the business[149]. This sentiment has been reinforced by the European Commission's 2002 proposals to reform the European Union's (EU) system of farm subsidies. The Commission wants to pay farmers less for the amount they produce and more for their efforts. The Commission's goal is to 1) improve the quality and safety of food and 2) conserve the countryside[149].

In early 2002 the European Fertilizer Manufacturers' Association (EFMA) estimated that between 2002 and 2012, nitrogen fertilizer use would fall by 6 percent in the EU. With the Commission's proposals mentioned above, the decline could be slightly more than forecast. Executives at some fertilizer producers have also said that the Commission's proposals will result in even more stagnation in the European fertilizer market than had previously been expected[149].

In early 2002 the European Fertilizer Manufacturers' Association (EFMA) estimated that between 2002 and 2012, nitrogen fertilizer use would fall by 6 percent in the EU. With the Commission's proposals mentioned above, the decline could be slightly more than forecast. Executives at some fertilizer producers have also said that the Commission's proposals will result in even more stagnation in the European fertilizer market than had previously been expected[149].

In 2002 the major Western European producers of bulk fertilizers are Norsk Hydro, Kemira, Atofina, Cargill and Kali und Salz (K+S). Norsk Hydro is now targeting its fertilizer investment in areas outside Europe. They see North and South America as offering the best opportunities for global growth. They reached this conclusion after deciding that the prospects for the fertilizer market in Asia were less bright than expected. They discovered that a lot of low-quality assets in Asia have a high level of state ownership. And this leads to products being sold at uneconomic prices[149].

The nitrogen fertilizer sector in Western Europe suffers from 1) overcapacity and 2) a high level of imports from Eastern Europe and elsewhere. Atofina is planning to withdraw from the fertilizer business altogether - once they have sorted out the consequences of the September, 2001 explosion at Grande Paroisse's AZF ammonium nitrate plant at Toulouse, France (which killed 30 people) [149].

In an effort to stabilize earnings in the cyclical nitrogen products market, Potash Corp has decided to emphasize industrial sales (rather than agricultural sales) from their North American production base and use low-cost production in Trinidad to supply the North American fertilizer market. Since initiating this strategy in 1997, they have increased industrial sales of nitrogen products from ~25% of production to 55% of nitrogen sales in 2001[153].

Figure 3.7 illustrates the importance of North America and Asia in the production and consumption of nitrogenous fertilizers. The surplus of production over consumption in the FSU and the Middle East illustrates their importance as exporters[36].

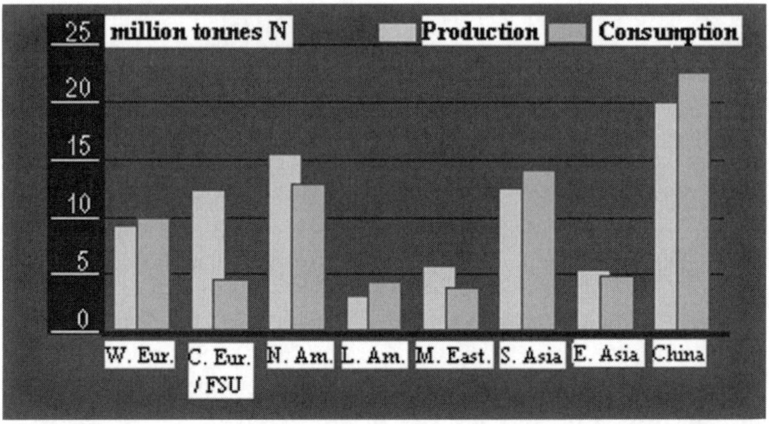

Figure 3.7. Compare production and consumption of nitrogen fertilizers by region for 1997 /1998. (Reproduced by permission of International Fertilizer Industry Association)

The values of nitrogen fertilizer shipments in the United States are summarized in Table 3.6. Both volume and price can affect these values, but these values give an indication of the economic impact of this industry.

Table 3.6. Value of Nitrogen Fertilizer Shipments in the United States

	(Millions of Dollars)				
Year	NH_3, Nitric Acid & Ammonium Compounds	Urea	Nitrogen Fertilizers of Organic Origin	Other Nitrogen Fertilizers	Total Value
1992	2,691.4	708.8	155.5	33.3	3,589.0
1993	2,862.9	641.6	147.6	85.1	3,737.2
1994	3,443.4	735.9	170.7	78.0	4,428.0
1995	3,699.4	769.8	196.9	95.1	4,761.2
1996	3,675.2	915.0	162.9	110.4	4,863.5
1997	3,511.9	784.8	70.4	140.9	4,507.9
1998	2,967.1	674.6	23.6	181.0	3,846.2
1999	2,263.3	600.9	63.5	142.3	3,070.0
2000	2,557.6	653.7	96.2	150.2	3,457.7

3.3. AMMONIA PRODUCTION, CONSUMPTION AND FORECASTS

Since no economical nitrogen fixation process that starts with nitrogen oxides has been discovered, ammonia has developed into the most important building block for synthetic nitrogen products worldwide. Prior to World War II, ammonia production capacity remained relatively stable. But during the war the need for explosives caused an increase in the production of ammonia for nitric acid manufacture. Then, after the war, the ammonia plants were used to manufacture fertilizers. As a result, there was a rapid increase in fertilizer consumption. The advantages of fertilizers were emphasized, and production capacity increased by leaps and bounds.

From 1940 to 1950 the number of ammonia plants doubled; then from 1950 to 1960 the number more than doubled again. Since 1963, there has been a revolution in ammonia-manufacturing technology. The advent of large single-train plants has resulted in a large increase in production capacity, the shutdown of a number of smaller plants, and a reduction in manufacturing costs. Capacity tripled in the period from about 1958 to 1968.

Worldwide, ammonia capacity grew from 119 million tonnes in 1980 to a peak of 141 million tonnes in 1989. Virtually all the growth in capacity occurred in the Former Soviet Union (FSU) and Asia. Between 1989 and 1995 capacity remained relatively flat, with increases in Asia being offset by closures in Europe and the FSU[36]. Ammonia demand grew at a rate of 3.2% per year from 1993 to 1998, but the growth rate is expected to slow to 0.5% per year from 1998 through 2002[37].

U.S. capacity declined in the 1970s, and it continued to decline during the 1990's by 5% to 10% to about 13 million tonnes per year. During the 1990's U.S. ammonia capacity was between 13 million and 14 million tonnes per year with operating rates over 95%. The U.S. capacity is forecast to reach 17 million tonnes per year by 2005. Worldwide ammonia capacity is forecast to grow at over 2% per year and approach 165 million tonnes per year by 2005[57].

In 2002 it was estimated that anhydrous ammonia is produced in about 80 countries, and about 85% of all ammonia production is used to make nitrogen fertilizers. About 6% of the ammonia production is applied directly to the fields[36]. In the United States the distribution of ammonia use differs slightly from the worldwide uses. In the USA only 80% of the ammonia is used to make fertilizers, but 20% is used as a direct application fertilizer. Chemical intermediates (such as acrylonitrile and caprolactam) account for 19% of the ammonia use while the remaining 1% is used in pulp and paper, metals and refrigeration applications[37].

In 2001 several major North American ammonia producers unexpectedly curtailed some of their ammonia and urea production. These producers cited over-capacity and low demand that were caused by adverse weather conditions and depressed crop prices. Although many plants take periodic shutdowns, the combination of import competition that depressed prices and very high feedstock costs in many industrialized countries during 2000 resulted in shutdowns of some high-cost plants which actually resulted in a decline in world capacity in 2000[57]. The combination of poor demand and increased competition continued into 2002 and resulted in the shutdown of additional ammonia capacity in North America. By July 2002, about 30% of North American ammonia capacity had been curtailed. And more shutdowns may occur when a new 640,000 ton-per-year, low-cost ammonia plant is started up in Trinidad during the summer of 2002[58].

Table 3.7 and Figure 3.8 show how ammonia capacity grew in different regions from 1974 to 1998. Western Europe's share fell from 13% in 1988 to only 9% in 1998. In 1974 the developing countries accounted for only 27% of ammonia capacity. But by 1998 their share had increased to 51%[36,57].

Table 3.7. Ammonia Capacity by Region

(thousands of metric tons of N)

Region	Capacity	1987 Share (%)	1998 Share (%)	1999 Capacity	1999 Share (%)	2005 Capacity	2005 Share (%)
China	18,675	16.9	22	30,450	23.6	33,460	24.6
Former Soviet Union	21,725	19.7	17	19,340	15.0	18,455	13.6
North America	16,390	14.8	15	18,955	14.7	18,410	13.6
South Asia	8,935	8.1	12	15,750	12.2	16,705	12.3
Western Europe	15,635	14.1	9	11,870	9.2	11,255	8.3
Middle East	4,100	3.7	7	5,950	4.6	7,795	5.7
Central Europe	9,830	8.9	6	7,560	5.9	6,820	5.0
Indonesia & Japan	5,800	5.3	4	7,725	6.0	8,340	6.1
Mexico & the Caribbean (Including Venezuela)	5,705	5.2	3	6,415	5.0	8,265	6.1
Other Countries	3,700	3.4	5	4,790	3.7	6,330	4.7
Totals	110,495	100.0	100	128,805	100.0	135,835	100.0

The capacity at the major producers in the Western Hemisphere is summarized in Figure 3.9. The reference to 'Offshore Capacity' refers to capacity that is in the Western Hemisphere, but is outside North America[153].

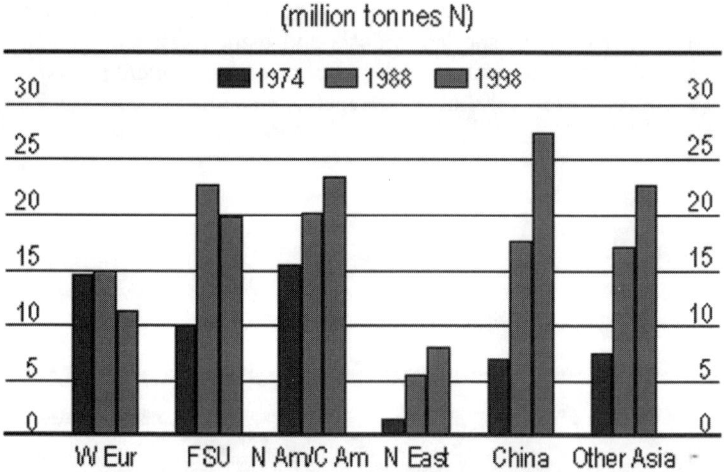

Figure 3.8. Ammonia capacity in different regions.
(Reproduced by permission of International Fertilizer Industry Association)

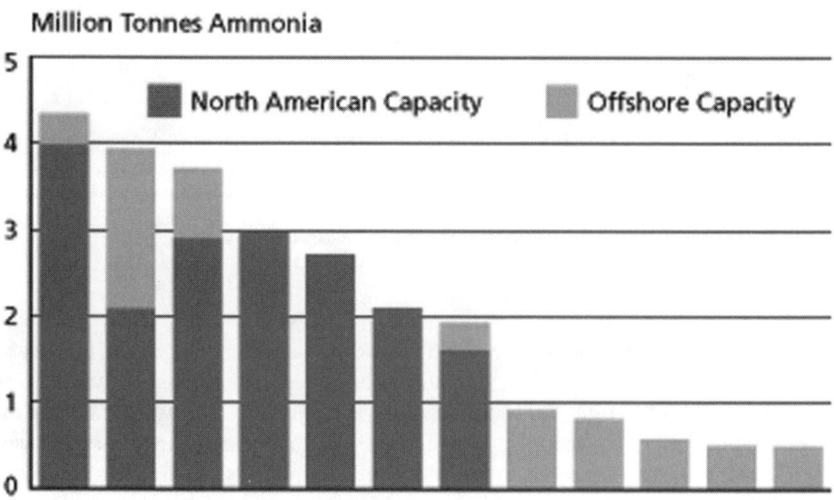

Figure 3.9. Ammonia capacity at major producers in the Western Hemisphere.
(Reprinted by permission of Fertecon)

Worldwide operating rates at ammonia plants fell from over 85% in the late 1990's to as low as 80% in 2001. As shown in Figure 3.10, the operating rate is expected to reach ~85% in 2003 and remain around this level through 2010[153].

A summary of ammonia production and idle capacity for various regions is shown in Figure 3.11. Asia is the largest producer of ammonia and had the largest amount of idle capacity in 2001. The FSU, the next largest producer, idled capacity due to transportation costs and plant inefficiencies. In Central Europe, capacity was idle due to high gas costs and inefficient plants. In North America, approximately 20% of plant capacity was idled in 2001, and this grew to 30% in 2002[153, 158].

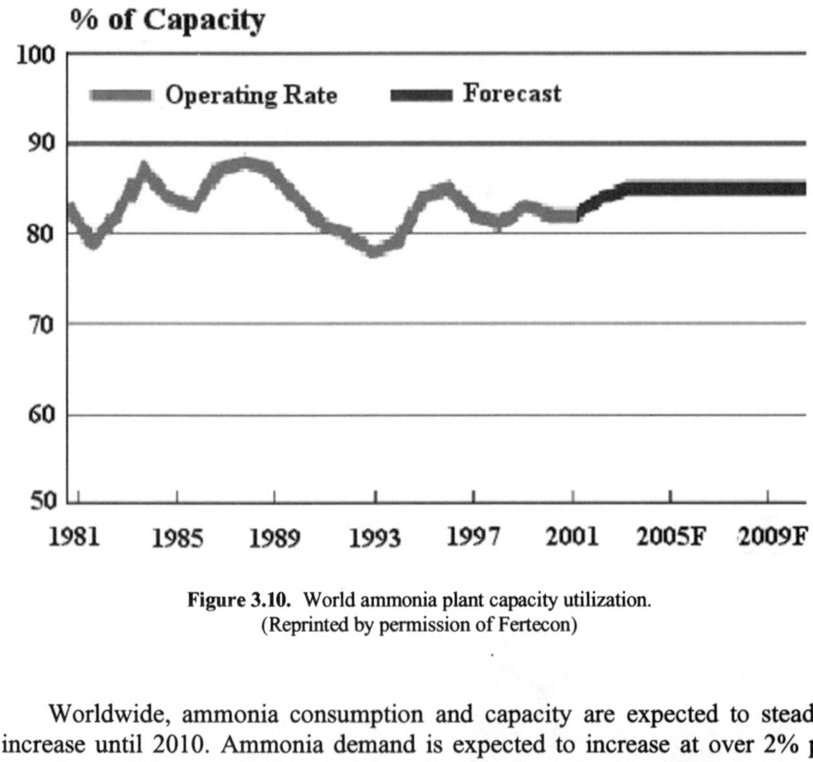

Figure 3.10. World ammonia plant capacity utilization.
(Reprinted by permission of Fertecon)

Worldwide, ammonia consumption and capacity are expected to steadily increase until 2010. Ammonia demand is expected to increase at over 2% per year from 2002 through 2007, but capacity is projected to grow at a slower rate (see Figure 3.12)[153]. According to Figure 3.13, the capacity increases between 1997 and 2003 are expected to be distributed as follows: 47% in Asia, 27% in North and South America, 15% in the Middle East and the balance in other countries[36]. So 89% of the new capacity is forecast to be in only these three regions. The forecast capacity increases between 2002 and 2007 are shown in Figure 3.14[153].

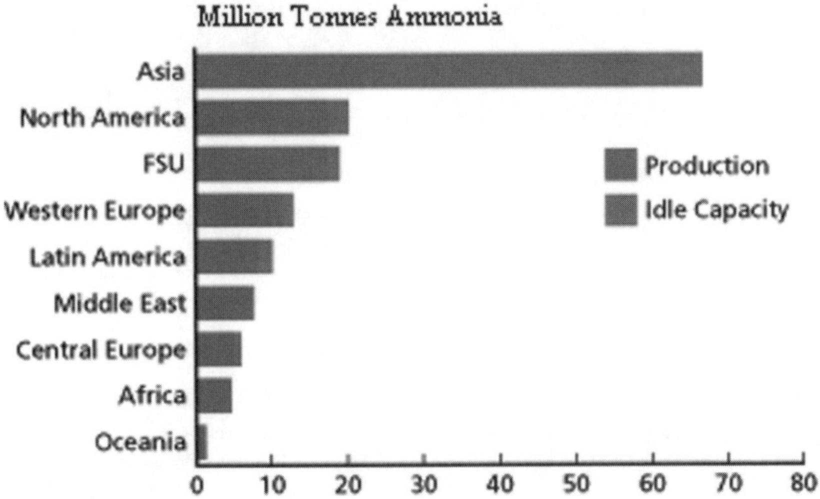

Figure 3.11. 2003 World ammonia production & idle capacity. (Reproduced by permission of Fertecon)

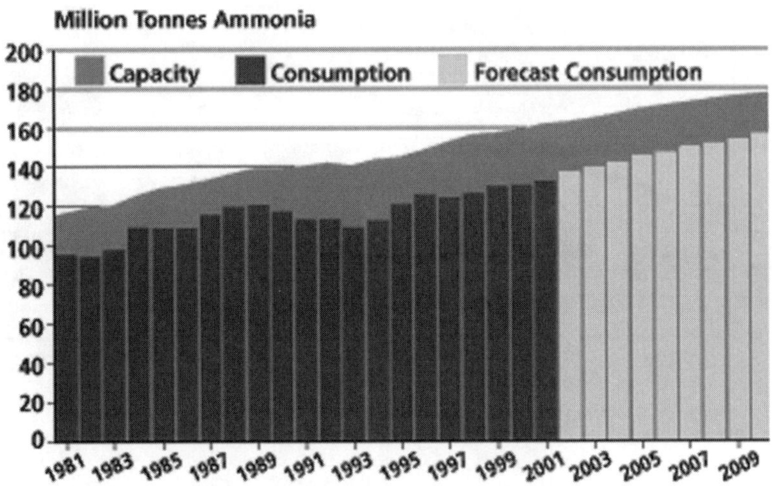

Figure 3.12. World ammonia capacity compared to consumption. (Reproduced by permission of Fertecon)

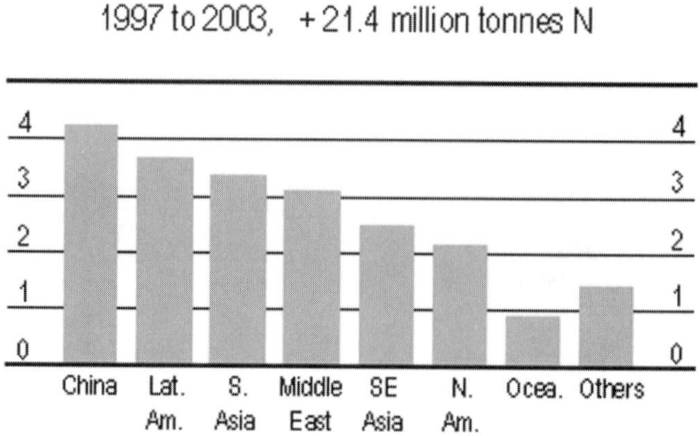

Figure 3.13. Forecast ammonia plant capacity increases: 1997 to 2003. (Reproduced by permission of International Fertilizer Industry Association)

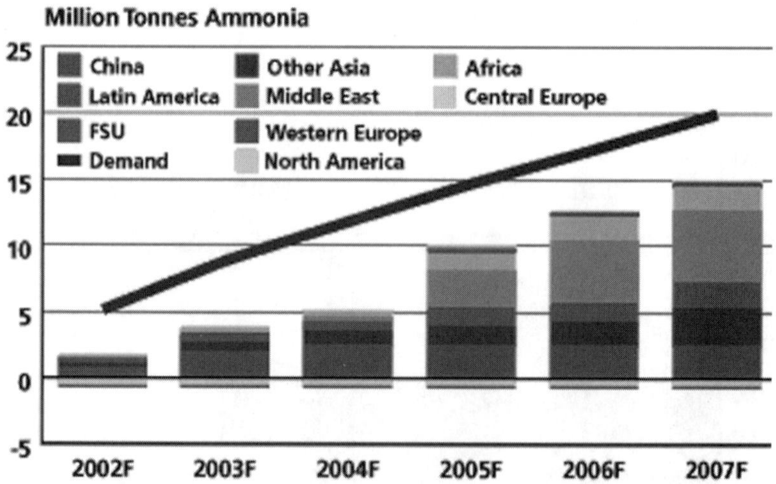

Figure 3.14. Forecast ammonia capacity increases compared to ammonia demand. (Reproduced by permission of Fertecon)

3.4. INDUSTRIAL NITROGEN

Industrial nitrogen is used a several ways. It is used to maintain an inert atmosphere in the chemical industry, in the electronics industry, in production of primary metals and in petroleum refining. It is used for advanced oil recovery and to prevent vacuum formation in the drill hole from which oil and gas are being extracted. In the food industry it is used for refrigeration of foods, and in the medical field it is used to refrigerate medical samples. Industrial nitrogen uses are summarized in Table 3.8[150].

Table 3.8. Uses of Industrial Nitrogen

Application	Market Share (%)
Chemical Industry	33
Oil & Gas Extraction	14
Electronics	13
Primary Metals	11
Petroleum Refining	10
Food Industry	5
Glass	2
Rubber And Plastics	1
Miscellaneous	11

The industrial nitrogen market includes all non-fertilizer applications including animal feed. Worldwide industrial uses in 1996 consumed an estimated 10.7 million tonnes of nitrogen and accounted for about 12% of total world nitrogen consumption[35].

Worldwide industrial nitrogen consumption declined by an estimated 7% between 1984 and 1996 (from about 11,500 tonnes to about 10,750 tonnes of nitrogen). Industrial nitrogen consumption is expected to return to the 1984 level by 2002[35]. Table 3.9 summarizes industrial nitrogen use and values in the United States as reported by the U.S. Census Bureau[151].

In Figure 3.15 the growth in nitrogen use is illustrated while Figure 3.16 shows how the purchase price of nitrogen has varied. With a typical nitrogen price of $US 0.14 per 100 scf in 1997 and a production volume of about 62 billion pounds, the 1997 commercial value of industrial nitrogen was $1.2 Billion[150].

In the United States steel industry (which is part of the primary metals application), nitrogen use in 1990 was 133,250 million scf while consumption had risen to 135,665 million scf by 1995. Nitrogen is used in the heating, stirring, inerting (i.e., using nitrogen to keep oxygen and other potentially reactive gases way from the steel) and injection functions in steel making. Nitrogen purity for most applications is typically 99.999%. Occasionally, nitrogen with a nominal purity of 97% to 99% may be used for inerting to prevent explosive mixtures in confined spaces. High purity nitrogen, that contains 5% to 10% hydrogen, is used to provide a protective reducing atmosphere in batch and continuous-type, bright annealing furnaces for carbon steel grades[152].

Table 3.9. Industrial Nitrogen Use in the United States

Year	Production[a] (Billion scf)	Shipments[a] (Billion scf)	Value of Shipments (Millions of Dollars)
1989	744.1	705.9	766.2
1990	749.5	723.5	791.1
1991	769.9	753.1	953.7
1992	818.5	792.9	981.5
1993	795.5	751.3	1,070.8
1994	870.0	827.2	1,057.6
1995	858.3	819.6	1,134.8
1996	1,030.0	Not Available	Not Available
1997	Not Available	Not Available	1,198.9
1998	Not Available	Not Available	1,293.4
1999	Not Available	Not Available	1,172.8
2000	Not Available	Not Available	1,155.7

[a] Production and Shipments volumes include both High and Low Purity Nitrogen, but the volumes were adjusted to a 100% nitrogen basis.

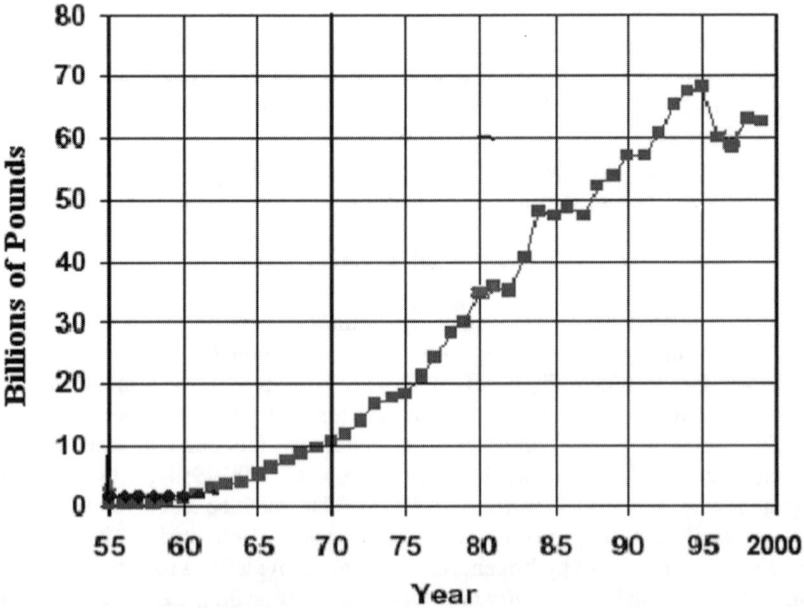

Figure 3.15. Industrial nitrogen production in the United States.

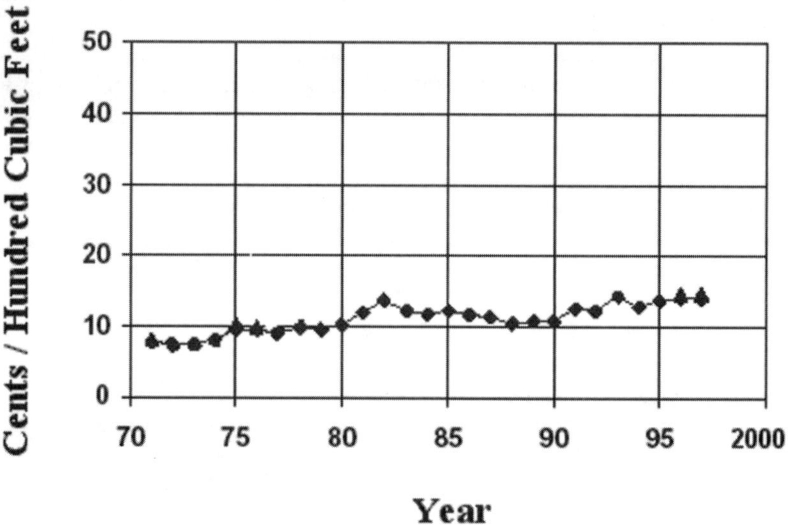

Figure 3.16. Industrial nitrogen value in the United States.

The amount of nitrogen that is needed for steel making depends on the type of plant. In 1995, Integrated US steel plants consumed 130 billion scf of nitrogen to produce 62.5 million tons of raw steel. This gives an average consumption of 2,085 scf per ton. On the other hand, Electric Arc Furnace (EAF) plants consumed only 5.2 billion scf of nitrogen to produce 42.4 million tons of raw steel. This gives an average consumption of only 122 scf per ton. So an Integrated plant needs seventeen (17) times more nitrogen to make steel than is needed in an EAF plant[152].

3.5. WORLD TRADE IN NITROGEN PRODUCTS

3.5.1. Ammonia Trade

Close to 90% of world ammonia production is processed in or used in the countries where it is produced. The remaining 10% to 12% of production enters international trade. The major net exporters in 1997 were Russia and the Ukraine (43% of world exports), Trinidad (22%) and the Middle East (13%). The predominance of the exports from gas-rich regions is evident. The major net importers were the USA (43% of world imports), Western Europe (26%) and certain Asian countries. Most of the ammonia that is traded is used to make fertilizers, but considerable quantities are used to produce industrial chemicals[36].

Between 1986 and 1998 world ammonia trade increased from 10.0 million to 13.6 million tonnes. Canada exports about 1.1 million tonnes of ammonia to the United States each year[36].

Natural gas price has a tremendous impact on the amount of ammonia that is traded. In regions with abundant natural gas reserves and relatively low demand, or where natural gas is produced as a byproduct of oil production, natural gas prices are low. Prices are higher in North America and Europe, where domestic and industrial demand for natural gas is high[153].

The impact of natural gas prices on ammonia prices on the US Gulf Coast is shown in Figure 3.17. This figure also illustrates the high percentage of ammonia cash costs that are due to the natural gas feedstock price.

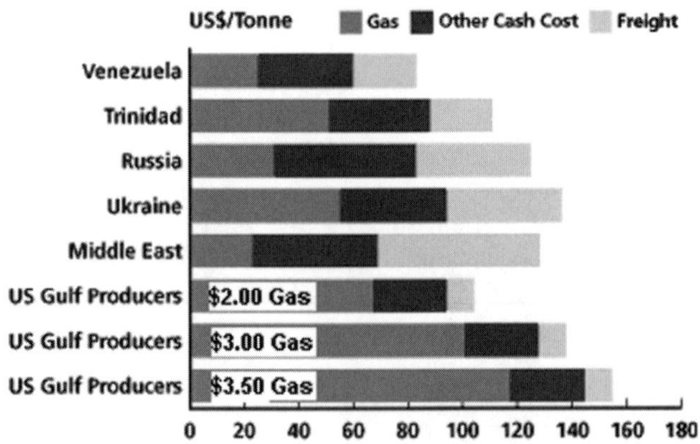

Figure 3.17. Ammonia costs compared to natural gas costs - US Gulf Coast [153].
(Reprinted by permission of Fertecon)

Figure 3.18 illustrates the impact of natural gas prices on imports to the USA. The US nitrogen industry generally experienced high operating rates prior to the escalation in natural gas prices after mid-2000. Ammonia operating rates dropped as natural gas prices rose. The lowest operating rate was reached in January 2001, when it dipped close to 50%. This low operating rate occurred when gas prices were peaking at just below \$10/MM Btu. The ammonia operating rate then rose to meet 2001 spring demand as gas prices declined[153].

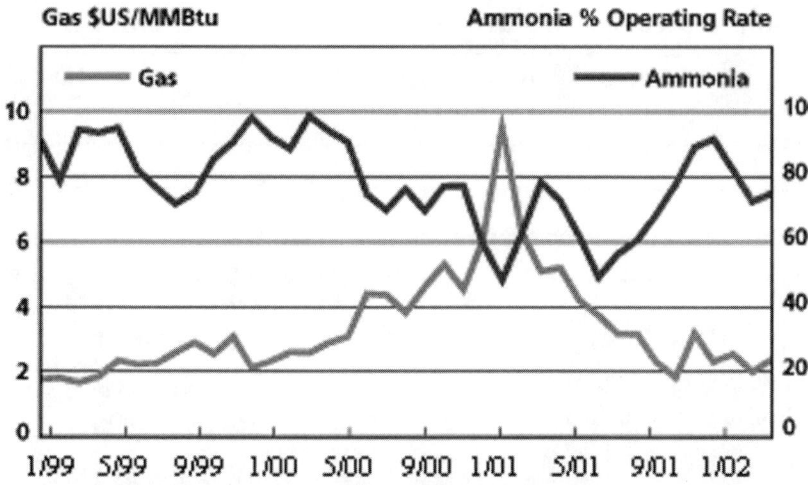

Figure 3.18. USA ammonia operating rate compared to natural gas price. (Reprinted by permission of Fertecon)

Following the spring planting season, when nitrogen products typically experiences a lull in demand, ammonia operating rates tapered off for the summer. As the fall season approached and gas prices moved downwards, ammonia operating rates again picked up. In the first quarter of 2002, gas prices exceeded $3 and ammonia operating rates declined[153].

Tables 3.5 and 3.6 show that US consumption of nitrogen products has been flat to growing slightly since the mid-1990's. It should be noted that ammonia consumption in the United States decreased by 1 percent per year from 1996 to 2001. However US ammonia consumption was forecast to increase by 2 percent per year from 2003 through 2005[223]. To maintain the US ammonia supply at the level that is needed to satisfy consumers, ammonia imports have increased whenever the operating rate declines (as shown in Figure 3.18). Therefore ammonia imports to the USA increase whenever natural gas prices increase (see Figure 3.19). Western and Central European countries experience similar dynamics in their ammonia markets.

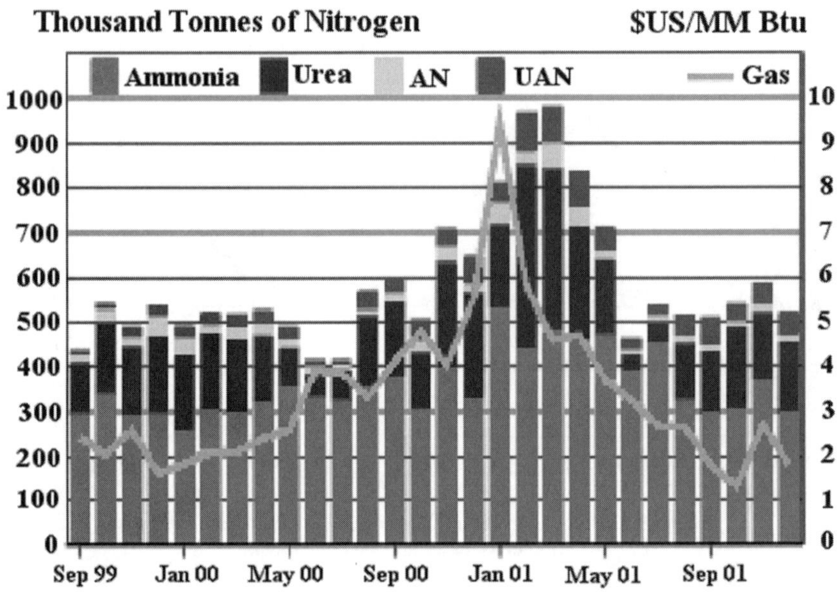

Figure 3.19. Monthly nitrogen fertilizer imports to the United States compared to natural gas prices.
(Reproduced by permission of Fertecon)

3.5.2. Trade in Major Nitrogen Products

A notable feature of international trade in fertilizer materials is the development of global movements of urea and ammonia phosphates. It is no coincidence that these products all have a high nutrient concentration. A simple calculation demonstrates that an increase in nutrient concentration from 30% to 40% results in a 25% reduction in the transport and handling costs. And these costs can account for up to half of the farm-delivered cost of a fertilizer. A summary of the nitrogen content in some nitrogen fertilizers is shown Table 3.10. This illustrates one reason why some nitrogen products have higher trade volumes, i.e., products with a higher nitrogen content deliver more nitrogen per pound of product plus the shipping costs (per unit weight of nitrogen) is lower[36].

Trade is an important component of the world nitrogen industry, and trade in the largest volume nitrogen products increased substantially between 1986 and 1998 (see Table 3.11). The regions that export most of the large volume products are shown in Table 3.12[36].

Table 3.10. Nutrient Content in Some Nitrogen Fertilizers

Nitrogen Product	Nitrogen Content (Wt. %)	P_2O_5 Content (Wt. %)
Urea	45.0 to 46.0	0
Ammonium Sulphate	21	0
Ammonium Nitrate	33.0 to 34.5	0
Calcium Ammonium Nitrate	20.4 to 27.0	0
Di-Ammonium Phosphate	18	46

Table 3.11. World Trade in Nitrogen Products

	(million tonnes of Nitrogen)	
Nitrogen Product	1986 to 1988	1996 to 1998
Ammonia	8.2	11.2
Urea	8.8	10.3
Ammonium Nitrates	-	4.0
Ammonium Phosphates (as P_2O_5)	4.2	8.7

(Reproduced by Permission of International Fertilizer Industry Association)

The regions with the largest trade balances are the EECA (Eastern Europe and Central Asia), the Caribbean Sea and Central America (Primarily Trinidad) and the Middle East (see Figure 3.20). The large trade balances are primarily due to the large resources of low cost natural gas in these regions[36].

Total nitrogen fertilizer exports from Central Europe and EECA increased from 18% of the world total in 1980 to about 33% in 1996. The products available for export increased as the domestic consumption in these countries declined. During the same period West Europe switched from being a net exporter of nitrogen fertilizers to being a net importer of nitrogen fertilizers. In 1995/96, 9.7 tonnes of nitrogen-based products were used in agriculture in the countries of the European Union, and 2.6 tonnes of nitrogen-based products were imported. Of the 2.6 tonnes, the FSU and Central Europe accounted for over 1.7 tonnes or 65% of the total imports.

Table 3.12. Regions with Largest Share of Nitrogen Product Exports: 1996 to 1998

Country / Region	Ammonia	Urea	Ammonium Phosphates
Former Soviet Union	36%	27%	17%
Trinidad / Mexico / Venezuela	19%	7%	-
Near East	12%	24%	-
Central Europe	-	8%	-
Southeast Asia (Indonesia & Malaysia)	-	10%	-
United States	-	-	61%
Morocco & Tunisia	-	-	11%
Others	33%	24%	11%

(Reproduced by permission of International Fertilizer Industry Association)

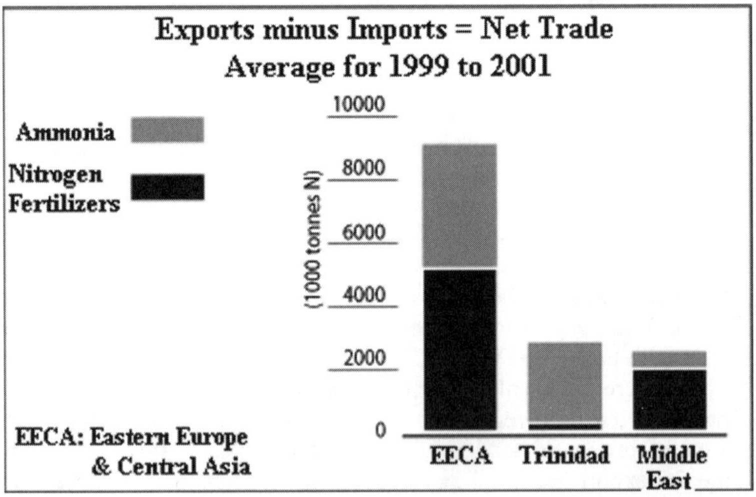

Figure 3.20. Regions that export the largest volumes of nitrogen fertilizers and ammonia: 1999 to 2001.
(Reproduced by permission of International Fertilizer Industry Association)

With Trinidad, Venezuela, Canada and other exporting countries fairly close to the United States, nitrogen fertilizer imports into the United States are growing (see Figure 3.21). Because of the energy markets (i.e., natural gas) in the United States, natural gas prices have a very strong influence on the volume of nitrogen products that are imported into the United States (see Figure 3.19).

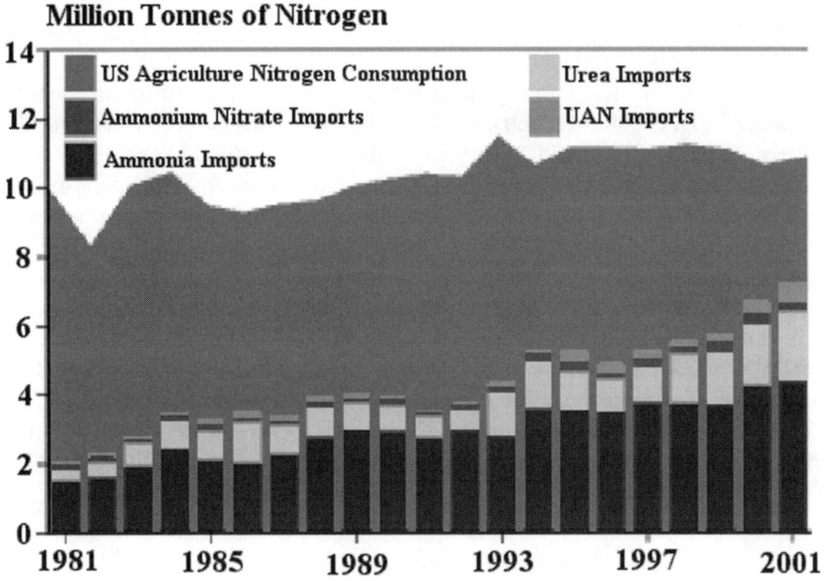

Figure 3.21. Total nitrogen fertilizer imports to the United States. (Reproduced by permission of Fertecon)

As shown by Figure 3.19 and Figure 3.21, an increase in natural gas prices results in curtailment of domestic nitrogen production and an increase in nitrogen fertilizer imports. Because natural gas prices are projected to increase through 2007, the United States can probably expect to see more curtailments and increasing imports. The projected increases in US gas prices are expected to lead more American producers to join those that have permanently shut down nitrogen-based operations in the late 1990's and early 2000's[153].

Shortly after natural gas prices spiked, nitrogen imports into the US reached record levels. When nitrogen prices failed to keep up with gas prices, nitrogen producers began curtailing production and imports became necessary. When prices for natural gas and nitrogen products decreased, imports returned to normal levels[153].

Asia dominates the world import market for urea and ammonium phosphates. China and India are particularly large importers and variations in their import requirements have a major impact on world prices. The agricultural and fertilizer sectors of these two countries could be affected by the World Trade Organization (WTO) agreements with China. The agreements will probably have a beneficial effect globally. However, some individual countries and companies may experience adverse effects[36].

While Chinese farmers have known for many centuries about the value of fertilizer, China's entry into the World Trade Organization (WTO) should finally provide them with the opportunity to fully capitalize on the benefits of crop nutrients. Prior to China's entry into the WTO, central planning distorted China's fertilizer market and interfered with the farmers' desires to achieve low-cost, high-yield production[154].

After years of negotiation, China formally became a member of the WTO on December 11, 2001. Its entry means 1) foreign fertilizer producers will have greater access to China and 2) import levels for urea and DAP are expected to increase. As a member of the WTO, China has agreed to implement a Tariff Rate Quota (TRQ) system. The TRQ will be applied initially to DAP and urea. This should allow increased imports of urea and DAP (see Figure 3.22 and Figure 3.23). The TRQ replaces China's previous import quota system, which banned urea imports from 1997 to December 2001. China has also agreed to uniformly apply its tariffs, to provide the same tax treatment to domestic and foreign products, and they will not enforce any artificial programs that stand in the way of importation or investment. While China will maintain control over distribution within its borders in the short term, it has agreed to allow foreign enterprises to engage in full distribution and warehousing within five years[154].

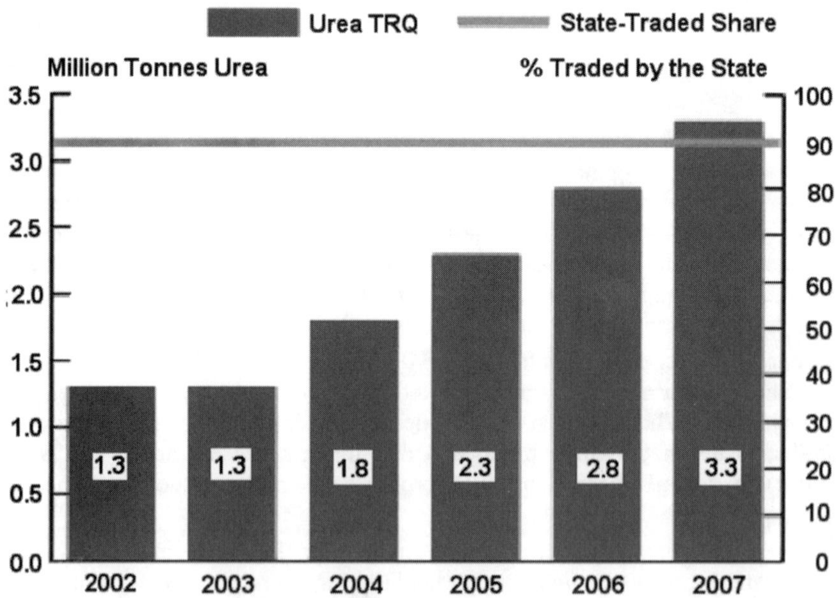

Figure 3.22. China's tariff quota for urea.
(Reproduced by permission of Fertecon)

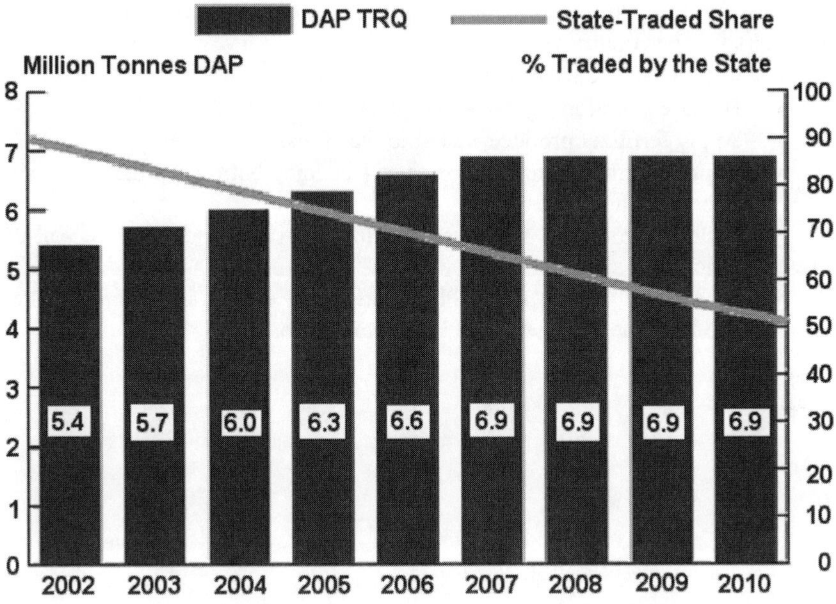

Figure 3.23. China's tariff quota for Di-ammonium phosphate (DAP). (Reproduced by permission of Fertecon)

Product volumes that are imported up to the TRQ are subject to only a 4% import tariff. However imports above this level are subject to an import tariff that can be set as high as 50%. While state agencies will still be responsible for a share of the product that is sold, the TRQ system places no quantitative restriction on imports. Hopefully this should lead to increased sales[154].

The effect that the WTO agreement with China might have on nitrogen products was illustrated in November, 2001 during a speech by a spokesman from China's Ministry of Agriculture. At a Potash & Phosphate Institute (PPI) conference, Guo Shutian outlined his country's strategy for agriculture and fertilizer following their entry into WTO[154].

Guo said China's first priorities are to ensure that enough food is available, and that the country is open to new approaches to ensure its food security. In the process, China would like to increase industrialization and value-added processing in its rural regions[154].

Guo also said China has identified a number of key action points relating to fertilizers[154]. The country will:

- Monitor soil nutrient levels;
- Balance use of inorganic and organic fertilizers;
- Apply fertilizer products suited to the crops;
- Improve fertilizer application rates (reducing Nitrogen while increasing Phosphate and Potassium);
- Educate producers to reduce fertilization costs and improve soil-use efficiency; and
- Close small Nitrogen and Phosphate factories that are severe polluters and increase Nitrogen and Phosphate imports.

4

ENVIRONMENTAL ISSUES

Since 1980 the nitrogen industry has made major improvements in environmental controls to reduce the release of atmospheric emissions and liquid effluents. Some of the environment issues for specific nitrogen products are summarized below. Other environmental issues will be discussed in the chapters devoted to the specific nitrogen products.

The production, distribution and use of ammonia, either for direct application or for conversion into downstream products, pose a number of special problems related to the handling, transportation and storage of the product. For storage at atmospheric pressure at sea level, ammonia must be cooled to –33°C. At higher altitudes, even lower storage temperatures are required. As a result the handling, transportation and storage of ammonia require specialized equipment and well-trained personnel[35].

The primary pollution problem in nitric acid manufacture is the abatement of nitrogen oxides (NO_X) in tail gases. In the United States in 1998, gaseous emissions from newly constructed nitric acid plants must be limited to 1.5 kilograms of NO_X per tonne of nitric acid (100% basis) produced, with a maximum stack opacity of 10%. Modern acid towers, with extended sections, can reduce NO_X emissions to less than 200 parts per million[35]. Production of ammonia also involves the combustion of hydrocarbons so NOx abatement must also be addressed for this product as well.

So what are nitrogen oxides? Where does they come from? And why is there a concern about the amount that enters the atmosphere? Nitrogen dioxide (NO_2) is a brownish, highly reactive gas that is present in all urban atmospheres. NO_2 can irritate the lungs, cause bronchitis and pneumonia, and lower resistance to respiratory infections. Nitrogen oxides are an important precursor both to ozone (O_3) and acid rain, and may affect both terrestrial and aquatic ecosystems. The major mechanism for the formation of NO_2 in the atmosphere is the oxidation of the primary air pollutant, nitric oxide (NO). NOx plays a major role, together with VOCs (Volatile Organic Compounds), in the atmospheric reactions that produce O_3. NOx forms when fuel is burned at high temperatures. The two major emissions sources are transportation and stationary fuel combustion sources such as electric utilities and industrial boilers.

In September, 1998 the United States Environmental Protection Agency (EPA) issued a rule (known as the NOx SIP Call) that required 22 states and the District of Columbia to submit State Implementation Plans (SIP's). These SIP's would reduce emissions of nitrogen oxides (NOx) - compounds that react with other chemicals in the atmosphere to form ozone (or smog). EPA required these reductions after determining that NOx and ozone are transported by the wind to downwind states, which results in unhealthy air quality in the downwind states[156].

The original EPA rule required that NOx reductions be in place by May, 2003. However, several lawsuits have challenged various aspects of this rule. Therefore in April, 2002 the implementation date was delayed until May 31, 2004. The levels of NOx reductions in several states have also been challenged in court. The initial reductions that were proposed in 1998 averaged ~25%. West Virginia was required to reduced NOx emissions by the largest percentage (over 50%), and Ohio was required to make the largest reduction in NOx emissions (almost 133,000 tons per year). The 1998 rule also proposed that over 80% of the NOx reductions would come from sources that generate electricity (i.e., utilities) while less than 20% would come from other industrial sources. Due to ongoing legal activities in 2002, it is not possible to determine the final NOx reductions that each state will have to achieve.

Numerous technologies are available to control NOx emissions from stationary point sources as illustrated in Table 4.1[157].

The technologies in Table 4.1 are listed in approximate order from lowest to highest cost. All technologies (except for SCRs) have relatively low investment requirements. For example, the installation of ULNBs in an existing fired heater is in the \$100,000 to \$1 million per heater range (in 2002 US dollars). There is essentially no operating cost increase to install new burners, including ULNBs, and there is even a potential for energy savings. FGR, SNCR, Catalytic Scrubbing and SCRs have significant operating costs with Catalytic Scrubbing having the highest operating cost. SCRs have a high investment cost (\$1 million to \$4 million per fired unit[157]).

The most important environmental concern about the use of nitrogen fertilizers is the loss of nitrogen to groundwater due to nitrification and leaching. This is having a significant impact on the use of nitrogen fertilizers, particularly in Western Europe where the member states have introduced restrictions on agriculture in those catchment areas where the nitrate content of water exceeds 50 milligrams per liter or is at risk of doing so. There is a lesser but still significant concern about nitrate levels in groundwater in the United States. Other than this runoff aspect, which most seriously affects ammonium nitrate, no serious environmental concerns exist with the use of the major nitrogen products such as fertilizer materials.[35]

Table 4.1. NOx Reduction / Control Technologies

Technology	Approximate Reduction in NOx (%)	Approximate Emissions (lb / MM Btu)	Approximate Emissions (ppmv @ 3% Excess O_2)
Standard Burners	Base Case	0.14	120
Process Modifications	Up to 100%	N/A	N/A
Low-NOx Burners (LNB)	60%	0.06	48
Ultra-Low-NOx Burners (ULNB), First Generation[a]	80%	0.03	24
ULNB, 2002 Guarantee Level	89%	0.015	13
ULNB, Being Developed in 2002	95%	0.007[b]	6
Flue Gas Recirculation (FGR)	55%[c]	0.025	22[d]
Water or Steam Injection at Gas Turbines	50% to 60%	0.06	20
Selective Noncatalytic Reduction (SNCR)	40%	0.033[d] to 0.085	27[d] to 70
Catalytic Scrubbing	70%[e]	0.017[d, e] to 0.044	14 [d, e] to 36
Selective Catalytic Reduction (SCR)	90% to 97%	0.006[d] to 0.015	2.5[d] to 12

[a] Can potentially fit in the opening for the existing burner - which would avoid a shutdown.

[b] Levels achieved during "Shop Tests". When this table was prepared, these NOx levels had not been demonstrated in fired heaters in the field. Emissions of 0.012 lb/MM Btu have been demonstrated in boilers with FGR while they were firing natural gas.

[c] On boilers; Little data is available for fired heaters.

[d] Level that can be achieved assuming LNBs are installed in combination with the technology listed.

[e] The injection of ozone upstream of the scrubber can reportedly achieve a 95% NOx reduction, but low operating temperatures are required.

Environmental issues also play an important role in the ammonium sulfate industry. This is because a significant portion of the world's ammonium sulfate production is the direct result of the necessity to remove sulfur dioxide from stack gases at various metal smelting and refining operations in order to conform to government regulations on sulfur dioxide emissions. A large potential source of additional by-product ammonium sulfate production is sulfur dioxide recovery from coal-fired electrical generating stations. It is estimated that a 500 megawatt plant that is burning coal containing 3.5 wt % sulfur could generate 250,000 tonnes of ammonium sulfate per year[35]. This is equivalent to ~7% of the total worldwide ammonium sulfate production in 2000.

5

HYDROGEN PRODUCTION

Ammonia production is the single largest consumer of hydrogen for use in chemical manufacture. Hydrogen production is typically integrated into the overall plant design. Therefore most ammonia producers regard themselves as consumers of natural gas (or another feedstock) rather than hydrogen.

Worldwide, a few ammonia plants have been located near by-product sources of hydrogen. In the United States, Coastal Refining & Marketing, Inc. brought a new ammonia plant on-stream in late 1997 in Freeport, Texas using purchased pipeline hydrogen. Air Liquide America purifies 45 million standard cubic feet per day of by-product hydrogen from The Dow Chemical Company for pipeline delivery to Coastal's Freeport facility. In 2002 this was the only U.S. ammonia plant using purchased hydrogen[40].

Hydrogen pipelines are used throughout the world. In the United States about 720 kilometers (450 miles) of hydrogen pipelines exist while Europe has about 1500 kilometers (950 miles). Thailand and Brazil also have hydrogen pipelines that are less than 10 miles long[173].

Hydrogen pipelines that were used in 2002 are usually 10-12" (25-30 cm) steel pipe and operate at 10 to 20 bar (150-300 psig). The oldest existing hydrogen pipeline system is in Germany's Ruhr region. It has been used for over 50 years. The longest pipeline is 400 km (250 miles) long and runs between France and Belgium[173].

Figure 5.1 shows the main chemical reactions that occur in the production of ammonia along with some of the other key reactions that are used to make additional products from ammonia[150].

Hydrogen is a non-toxic, colorless, odorless and tasteless gas. It is the lightest and most abundant element (making up over 90% of the atoms in our universe), but it is present at only extremely low levels (0.1 ppm) as a pure element in the earth's atmosphere. More than 50% of the atoms in our environment (the soil beneath our feet, the atmosphere, the oceans, petroleum-based gases and liquids) are hydrogen.

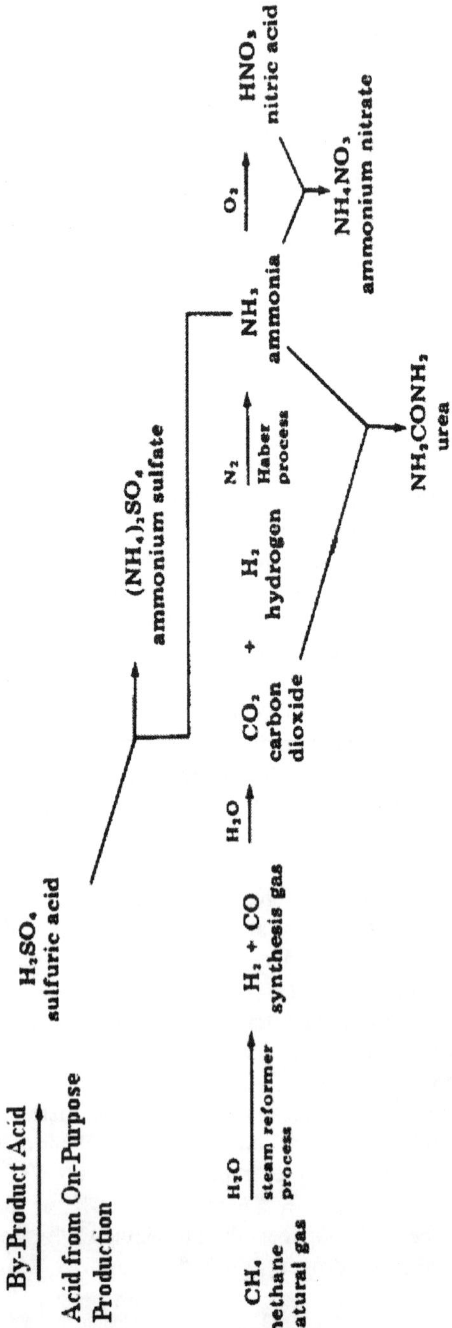

Figure 5.1. Ammonia production: chemical reactions and steps to produce additional products.

Some of the physical properties of hydrogen are shown in Table 5.1

Table 5.1. Physical Properties of Hydrogen[259]

Chemical Symbol	H_2	
Molecular Weight	2.02	
	Metric Units	**English Units**
Specific Gravity, Liquid Phase at Boiling Point (Water = 1.0)	0.071 at 1.013 bar	0.071 at 1 atm
Specific Heat Capacity - Liquid	9.668 kJ/kg °C	2.309 BTU/lb °F
Boiling Point Temperature	-252.8°C at 1.013 bar	-423°F at 1 atm
Latent Heat of Vaporization	446.0 kJ/kg	191.7 BTU/lb
Specific Gravity, Gaseous Phase (Air = 1.0)	0.06998	0.06998
Specific Heat Capacity - Gas	14.34 kJ/ kg °C	3.425 BTU/lb °F
Density, Gas	0.08988 kg/m^3	0.005611 lb / ft^3
Triple Point Temperature	-259.2°C	-434.6°F
Triple Point Pressure	7.205 kPa abs	1.045 psia
Critical Point Temperature	-239.96 °C	-399.93°F
Critical Point Pressure	1315 kPa abs	190.8 psia
Critical Point Density	30.12 kg/ m^3	1.88 lb/ ft^3
Thermal Conductivity	0.1869 W/m °K at 300°K	0.1081 BTU/ft hr °F at 80°F
Flammability Limits (Volume %)		
Lower	4.1	
Upper	74.0	
Stoichiometric Air/Fuel Ratio, Weight Basis	34.3	
High Heating Value (Liquid Fuel - Liquid Water Product)	141,890 kJ/kg	61,002 BTU/lb
Low Heating Value (Liquid Fuel - Water Vapor Product)	119,972 kJ/kg	51,532 BTU/lb
Ignition Temperature	530°C	1085°F

5.1. HYDROGEN PRODUCTION AND USES

Total hydrogen production is difficult to estimate because of the wide variety of sources that produce by-product hydrogen. A few of these by-product sources are crude oil refining processes, formaldehyde production, chlorine-alkaline electrolysis and other chemical processes. In oil refineries, large quantities of by-product hydrogen come from the catalytic reforming of cyclic aliphatic hydrocarbons to produce aromatics such as benzene, toluene and xylene. Refineries generally use all by-product hydrogen output captively and often need to supplement their needs from other sources[160].

However two independent sources estimated worldwide hydrogen production in 2002 to be 500 billion Nm^3 per year (19.0 trillion scf) [161, 162]. Another source estimated worldwide hydrogen production in 2002 to be 454 billion Nm^3 per year (17.3 trillion scf) [173]. By-product hydrogen accounts for 190 billion Nm^3 (7.23 trillion scf) or 38% of worldwide production[161]. Another source estimates that worldwide ammonia and methanol production consumed over 5.475 trillion SCF per year of hydrogen in 2001, and consumption is growing at 2% to 4% per year. This source also estimated that worldwide oil refineries consume 3.650 trillion SCF pèr year of hydrogen in 2001, and their hydrogen use is growing at a rate of 5% to 10% per year[164].

Two sources estimate that 2001 USA hydrogen production is 8 trillion SCF per year (23.1 billion SCF per day) although one source believes this hydrogen volume may be high[160, 163]. Data from some other sources can be used to estimate a hydrogen production figure of 4.3 to 4.7 trillion SCF per year[160]. So a lot of variability also exists in reported hydrogen production in the USA. Table 5.2 summarizes another estimate of USA hydrogen capacity - excluding by-product hydrogen[165].

Table 5.2. Hydrogen capacity in the United States (excluding by-product hydrogen)[165]

	scf / day	% of Production	1000 scf / year	Number of Plants
Locations				
Liquid Hydrogen	109,100	7	39,821,500	11
Gaseous Hydrogen	1,444,470	93	527,231,550	75
Total Merchant Production	1,553,570	100	567,053,050	86
Captive Capacity	2,800,000		1,022,000,000	143
Totals	4,353,570		1,589,053,050	229

Two sources estimate that merchant hydrogen production in the USA constitutes 10% to 12% of total hydrogen production[163, 165]. (Note: Merchant hydrogen refers to hydrogen that is produced by one company and sold to another.) From 1994 to 1999 merchant hydrogen production grew at a rate of

23% per year. And from 1999 through 2003 it is forecast to grow at a rate of 14% per year. Demand for merchant hydrogen in the USA is summarized in Table 5.3[160, 165]:

Table 5.3. Demand for merchant hydrogen in the United States[160, 165]

Year	Billion SCF per Year	Billion Nm^3 per Year
1998	333.6	8.77
1999	381.9	10.04
2001	400 to 450	10.52 to 11.83
2003 Forecast	656.5	17.26

The US Census Bureau reports the hydrogen production in Table 5.4. But these figures do not represent the overall hydrogen production in the USA. These production figures do not include hydrogen produced in refineries for captive use, hydrogen produced and consumed in ammonia manufacture, hydrogen produced by ammonia dissociation, hydrogen used as fuel and hydrogen that is vented[160]. The large changes in Growth Rates suggests that the data may not be very accurate or that the data collection criteria may have changed in some years[151, 160].

Table 5.4. Hydrogen Production in the United States[151, 160]

Year	Production (Billion Standard Cubic Feet per Year)	Growth Rate (% per Year)
1980	106.0	-
1985	121.0	2.8
1989	145.5	5.1
1990	147.8	1.6
1991	152.7	3.3
1992	162.2	6.2
1993	212.9	31.3
1994	331.0	55.5
1995	351.6	6.2
1996	386.0	9.8
1997	526.0	36.3
1998	552.0	4.9
1999	454.0	-17.8
2000	481.0	6.0
2001 Estimate	510.0	6.0
2003 Forecast	625.0	11.3

If you wish to convert the values in the above tables into different units, Table 5.5 provides the conversion factors:

Table 5.5. Data Conversion for Hydrogen[a, b, c, 259]

	Weight		Liquid Volume		Gaseous Volume	
	Pounds	Kilograms	Gallons	Liters	ft^3	m^3
1 Pound Solid	1.0	0.4536	1.6928	6.408	192.0	5.047
1 Kilogram Solid	2.205	1.0	3.733	14.128	423.3	11.126
1 SCF Gas	0.005209	0.002363	0.008820	0.03339	1.0	0.02628
1 Nm^3 Gas	0.19815	0.08988	0.3355	1.2699	38.04	1.0
1 Gallon Liquid	0.5906	0.2679	1.0	3.785	113.41	2.981
1 Liter Liquid	0.15604	0.07078	0.2642	1.0	29.99	0.7881

[a] SCF (standard cubic feet) gas is measured at one atmosphere and 70 °F (21.1°C). Nm^3 (normal cubic meters) gas is measured at one atmosphere and 0 °C. Liquid is measured at 1 atmosphere and the boiling temperature.

[b] The conversion factor for cubic feet to cub ic meters is normally 35.31. However, SCF is measured at 21.1 °C and Nm^3 is measured at 0°C. Therefore the conversion factor must be adjusted for temperature. All values are rounded to the nearest four or five significant figures. Gas values are expressed in the stable hydrogen condition: 75% ortho, 25% para [41].

[c] The two nuclei of the hydrogen atom may spin in the same direction (parallel) and this form is known as ortho. The nuclei may also spin in opposite directions and this form is called para. Para-Hydrogen is the form that is preferred for rocket fuels. Gaseous hydrogen at room temperature is 75% ortho and 25% para [40].

The distribution of hydrogen use in different industries in the United States is summarized in Table 5.6. Two references list similar distributions while another reference shows a different distribution[160, 163, 165]. Worldwide hydrogen use in 2002 is estimated to be as follows: Ammonia - 50%, Refining - 37%, Methanol - 8%, Space / Government - 1% and Other - 4%[173].

The most significant market development that is expected to impact demand for hydrogen between 1998 and 2003 is a change in refinery hydrogen balances. This will occur because environmental regulations in most industrialized countries will result in increased hydrogen requirements at refineries. At the same time, heavy crudes (which are hydrogen-deficient compared to lighter crudes) are making up an increasing proportion of refinery runs. This means the availability of by-product hydrogen at refineries is likely to decline while hydrogen requirements increase[40]. This shortfall in hydrogen at refineries is being made up by an increase in merchant hydrogen capacity - as shown by the large growth rates in the above tables. In 2000 it was estimated that 45% of North American and 35% of European refinery hydrogen was supplied from on-purpose hydrogen facilities and over 95% of this hydrogen came from steam reformers.

Table 5.6. Hydrogen Uses in the United States[160, 163, 165]

	January, 2001	August, 2001	2002	Growth Rate (% per Year)
Refining	66.8	65	37	14
Chemicals	26.2	22	48	
- Ammonia			38	
- Methanol			10	
Total: Refining & Chemicals	93.0	87	85	10
Metals	2.7	4	-	15
Electronics	1.5	3	-	12
Government / NASA	1.2	2	-	15
Hydrogenation of Fats & Oils	0.7	1	-	
Float Glass	0.3	1	-	
Power Generation	0.2	-	-	
Miscellaneous	0.4	2	10 (Merchant)	

5.2. MANUFACTURING PROCESSES

Hydrogen is present in fossil fuels and water in sufficient quantities that it can be produced on a large scale by three different methods: 1) Petrochemical Processes, 2) Coal-based Chemical Processes and 3) Electrochemical Processes (Electrolysis). In Table 5.7, the percentage of hydrogen production is broken down by type of manufacturing process for the years between 1974 and 1988[46]. A similar distribution for 2002 is shown in Table 5.8[162].

Table 5.7. World hydrogen production and type of manufacturing process (1974 to 1988)

<table>
<tr><th>Process</th><th>1974</th><th>1984</th><th>1988</th></tr>
<tr><td>Cracking of Crude Oil</td><td>48</td><td rowspan="2">77</td><td rowspan="2">80</td></tr>
<tr><td>Cracking of Natural Gas</td><td>30</td></tr>
<tr><td>Coal & Coke Gasification</td><td>16</td><td>18</td><td>16</td></tr>
<tr><td>Electrolysis</td><td>3</td><td>4</td><td rowspan="2">4</td></tr>
<tr><td>Miscellaneous Processes</td><td>3</td><td>1</td></tr>
</table>

(Reproduced by permission of Wiley-VCH)

Table 5.8. World hydrogen production and type of manufacturing process (2002)

Source	Hydrogen Production (Billion Nm^3 per Year)	Hydrogen Production (Billion SCF per Year)	Share (%)
Natural Gas	240	9,129.6	48
Oil	150	5,706.0	30
Coal	90	3,423.6	18
Electrolysis	20	760.8	4
Totals	500	19,020.0	100

In the Petrochemical Processes, both natural gas and crude oil fractions can be converted into synthesis gas (a mixture of hydrogen and carbon monoxide) using two basically different methods[46]:

1. With the allothermal steam reforming method, catalytic cracking takes place in the presence of water vapor. The necessary heat is supplied from external sources.
2. With the autothermal cracking process, heat for the thermal cracking is supplied by partial combustion of the feed. Water is used and carbon dioxide (CO_2) may be recycled to attain a desired CO/H_2 ratio.

The processes for the manufacture of synthesis gas were originally based on the gasification of coke from hard coal and low temperature coke from brown coal by means of air and steam. After World War II, the easy-to-handle and valuable liquid and gaseous fossil fuels – oil and natural gas – were also employed as feedstocks. Their value lies in their high hydrogen content that produces higher molar ratios of hydrogen (H_2) to carbon monoxide (CO) in the synthesis gas. By using excess steam with the reforming of methane, the H_2-to-CO molar ratio can be as high as 5.0 (see Table 5.9).

Selection of a process for hydrogen manufacture from hydrocarbons and coal therefore depends on: the raw material and its cost, the scale of operation, the purity of the synthesis gas to be produced, the pressure level of the natural gas feed, and the number and type of processes that will consume the carbon monoxide and hydrogen.

Hydrogen is also manufactured industrially by direct electrolysis of H_2O, HF, and 22% to 25% hydrochloric acid (HCl). However the hydrogen produced by electrolysis only accounts for a small percentage of the total H_2 production (see Table 5.7 and Table 5.8). One advantage of electrolysis (compared to steam reforming of hydrocarbons) is that the hydrogen from the electrolysis is very pure (>99 vol %), which eliminates the costly purification steps needed with reforming.

Table 5.9. H_2/CO molar ratio in synthesis gas

Method of Manufacture	H_2/CO Molar Ratio	Typical Synthesis Gas Composition (% Molar Concentration) H_2	CO	CO_2
Oxygen-Coke-Steam	0.6			
Air-Coke-Steam	0.9			
Oxygen-Coal-Steam	1.0			
Oxygen-Fuel Oil-Steam	1.0			
Propane-Steam	1.33			
Methane-Oxygen (Partial Oxidation)	1.8	62%	35%	3%
Autothermal Reforming[a]	2.1	34%	17%	2%
Oil-Steam	2.1			
Petroleum Ether-Steam	2.4			
Methane-Steam (Steam Reforming)	3.0 to 5.0	75%	15%	10%

[a] Autothermal Reforming is a combination of Steam Reforming and Partial Oxidation. In this example assume 47% nitrogen in the Synthesis Gas.

Hydrogen is manufactured by four principal processes (see Table 5.8 and Figure 5.2)[46]:

1. Steam reforming of natural gas,
2. Partial combustion of natural gas or oil with pure oxygen:
3. Gasification of coal or coke with air (or oxygen) and steam and
4. Recovery of by-product hydrogen from petroleum refinery gases or other cracking operations.

Small amounts of hydrogen also are manufactured by electrolysis. These processes are discussed in more detail in the following sections.

Figure 5.3 is a simplified diagram that shows the basic steps in the different processes that are used to produce synthesis gas. Note that the Sulfur Removal step is just before Purification when Partial Oxidation and Gasification are used. Sulfur Removal is the first step when Steam Reforming is the chosen process. The reasons that the process order is as shown will be explained in the following sections of this chapter[174].

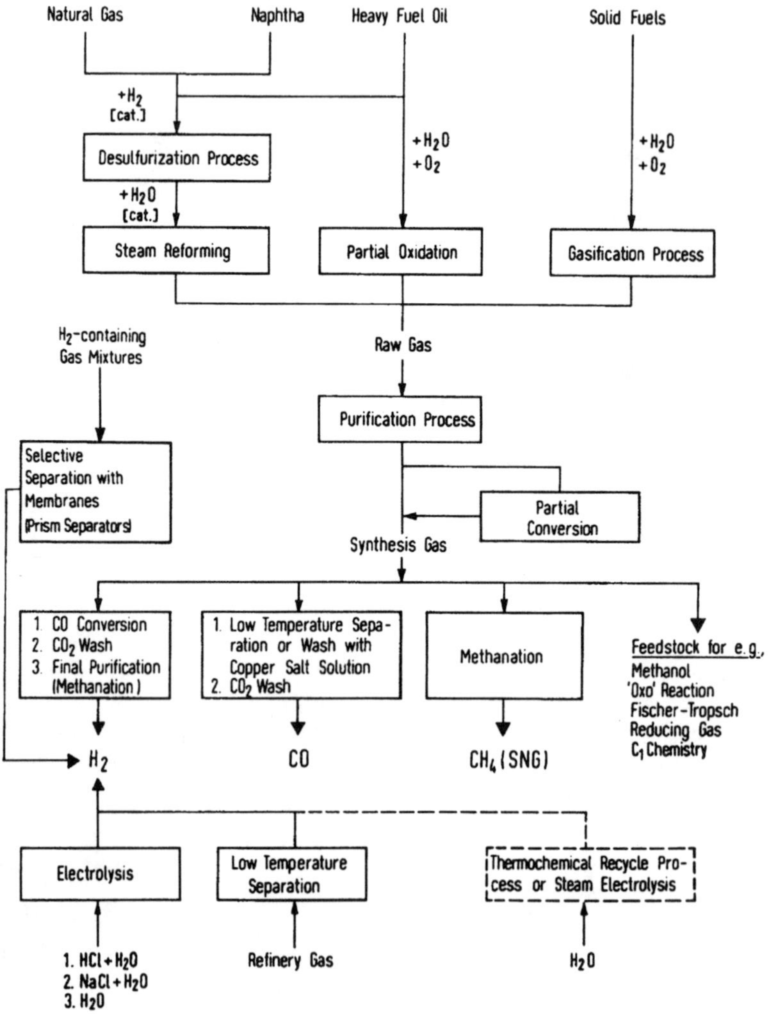

Figure 5.2. Hydrogen manufacturing process steps.
(Reproduced by permission of Wiley-VCH.)

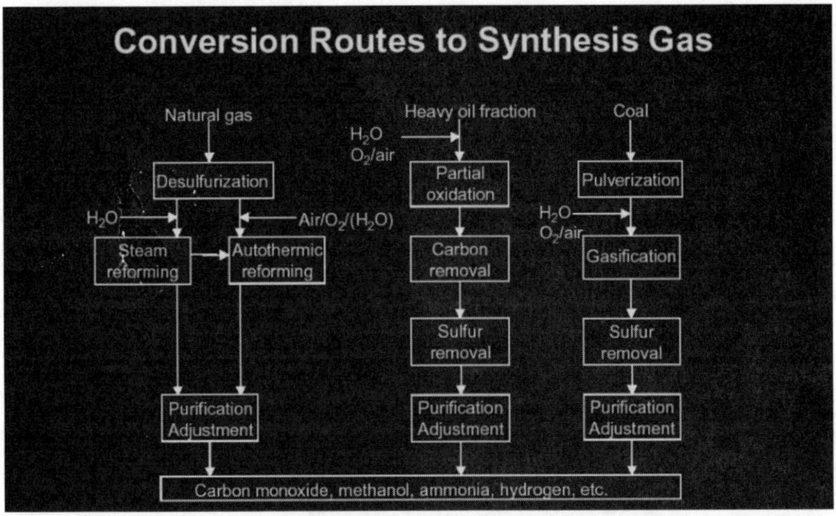

Figure 5.3. Conversion routes to synthesis gas. [174]

5.2.1. Reforming

Reforming is a general name for the reaction of a hydrocarbon, such as methane, with water and/or carbon dioxide, to produce a mixture of carbon monoxide and hydrogen. The different Reforming processes can be divided into the following types[53]:

- Conventional steam reforming with a fired primary reformer and stoichiometric air secondary reforming (stoichiometric H/N ratio).
- Steam reforming with mild conditions in a fired primary reformer and excess air secondary reforming (under-stoichiometric H/N ratio).
- Heat exchange autothermal reforming, with a process gas heated steam reformer (heat exchange reformer) and a separate secondary reformer, or in a combined autothermal reformer, using excess or enriched air (under-stoichiometric or stoichiometric H/N ratio)

The typical steps in the Steam Reforming process are shown in Figure 5.4[53] and Figure 5.5[70]. These steps will be discussed in more detail in the following sections.

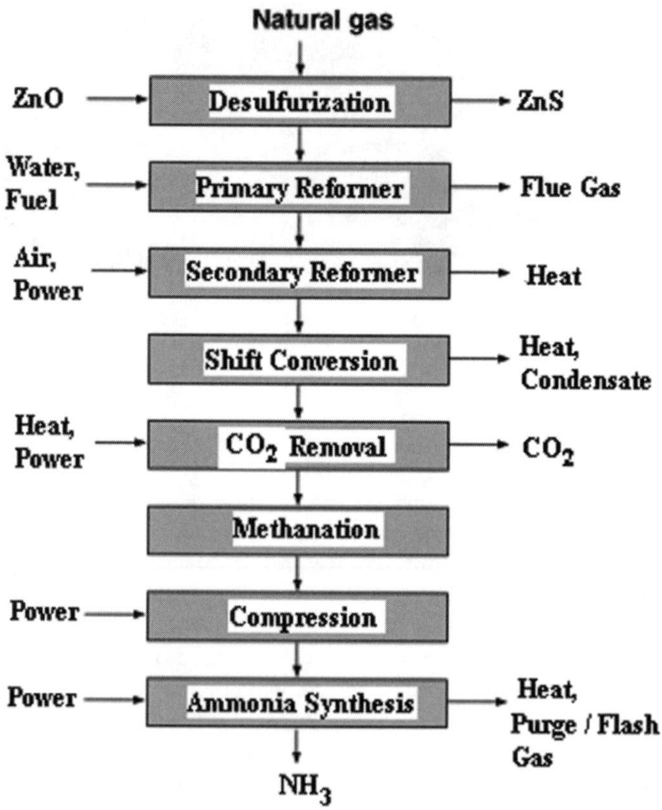

Figure 5.4. Block diagram of the steam/air reforming process [53].
(Reproduced by permission of European Fertilizer Manufactures Association)

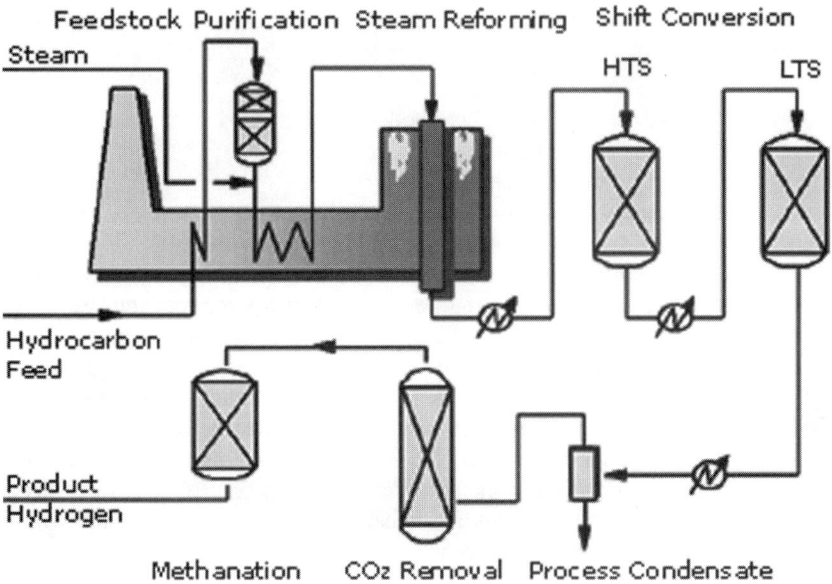

Figure 5.5. Typical steam reforming process[70].
(Reproduced by permission of Johnson Matthey Catalysts. Copyright Johnson Matthey PLC)

If water is used, the process is called steam reforming or steam cracking. The reforming reaction [Eq. (5.1)] is endothermic and requires a catalyst[1].

$$CH_4 + H_2O \leftrightarrow 3\,H_2 + CO \quad \Delta H = +49 \text{ kcal/mol or } 205 \text{ kJ/mol} \qquad (5.1)$$

The steam reforming of hydrocarbons is in principle a reduction of water with the carbon of the organic starting material. In the case of methane, ⅓ of the hydrogen is supplied by water. This share increases with the higher hydrocarbons. Other reactions that proceed at the same time as the reforming reaction are[46]:

$$CO + H_2O \leftrightarrow H_2 + CO_2 \quad \Delta H = -10 \text{ kcal/mol or } 42 \text{ kJ/mol} \qquad (5.2)$$

The Homogenous Water Gas Reaction or Water Gas Shift Reaction

$$CH_4 \leftrightarrow 2\,H_2 + C \quad \Delta H = +17.9 \text{ kcal/mol or } 75 \text{ kJ/mol} \qquad (5.3)$$

$$2\,CO \leftrightarrow CO_2 + C \quad \Delta H = -41.4 \text{ kcal/mol or } 173 \text{ kJ/mol} \qquad (5.4)$$

Boudouard Reaction

The equilibrium composition of the synthesis gas depends on the following factors:

- Steam-to-gas ratio entering the reactor,
- The reaction temperature,
- The reaction pressure and
- The quantity of inerts in the reaction mixture.

To avoid carbon formation [as indicated by Eqs. (5.3) and (5.4)], the steam-to-gas ratio must be kept high enough to favor the reforming reaction [Eq. (5.1)] and the water gas shift reaction [Eq. (5.2)] over the reactions that form carbon.

The thermodynamics of the above reactions are illustrated in Figures 5.6 and 5.7[174]. Both figures assume a steam-to-methane ratio of 1.0. Figure 5.6 illustrates how the feed and product gases interact when the product gas has a hydrogen-to-carbon monoxide ratio of 3.0. Figure 5.7 illustrates the effects of temperature and pressure on the reactions. As pressure increases, lower conversion can be expected and more methane will not be converted and will be found in the reformer discharge stream.

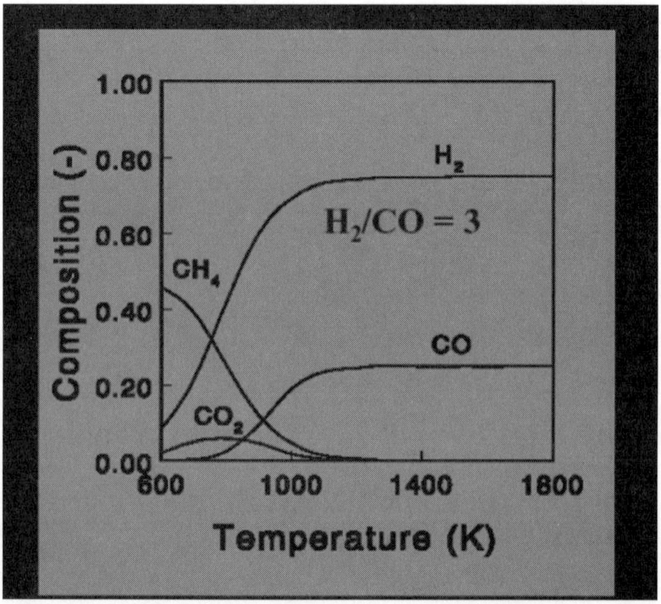

Figure 5.6. Steam Reforming Thermodynamics with Steam-to-Methane Ratio = 1.0 [174].

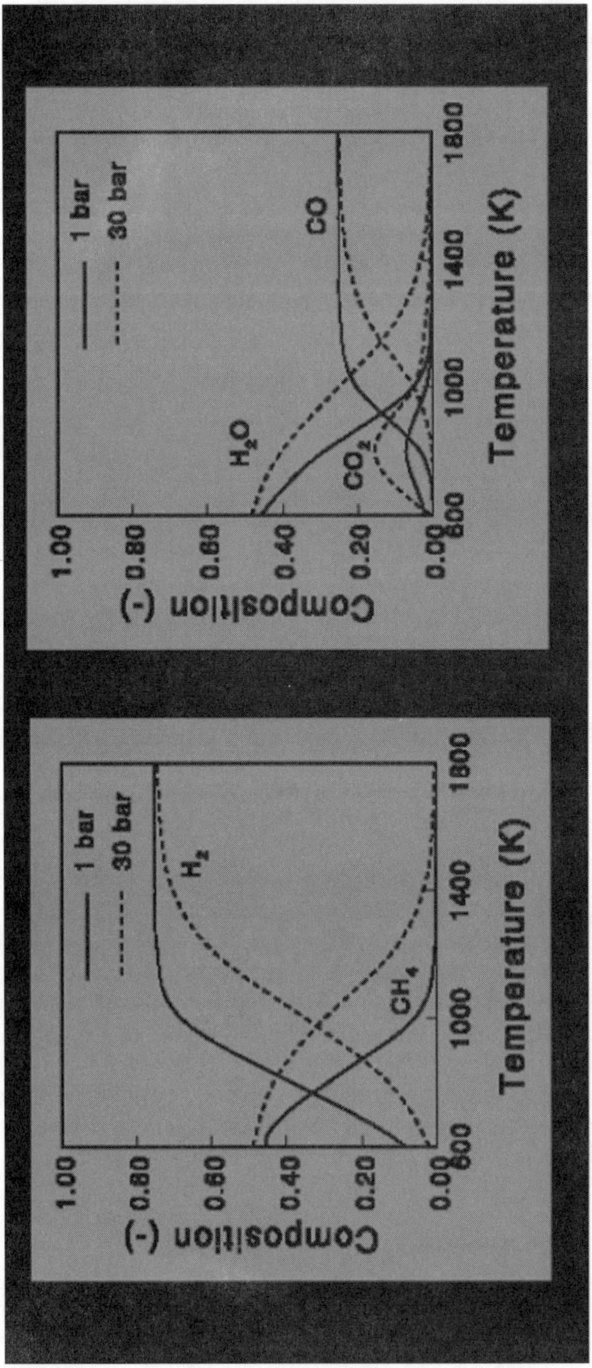

Figure 5.7. Steam Reforming Thermodynamics - Effects of Temperature and Pressure with Steam-to-Methane Ratio = 1.0

The Methane Slip (or % Methane that is not converted) has a big influence on the selection of reformer operating conditions. As shown in Figure 5.8, the best operating conditions will be within the following ranges:

- Steam-to-Methane Ratio: 2.5 to 4.5
- Temperature: 1090 to 1150°K (815 to 875°C)
- Pressure: 7 to 30 bar

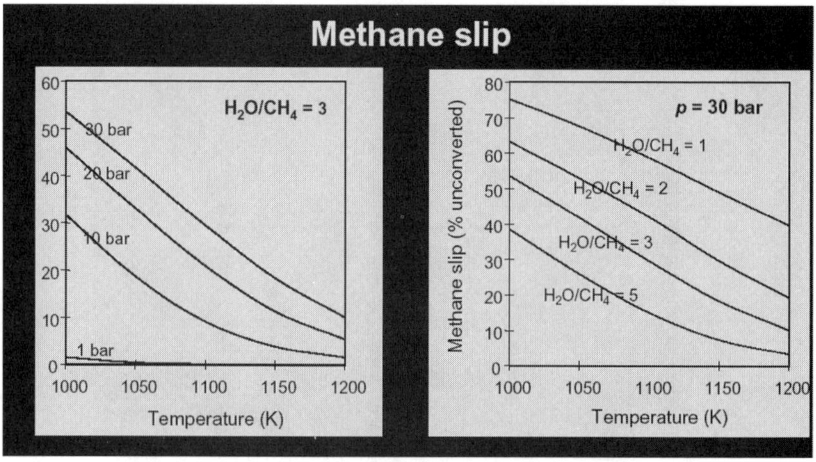

Figure 5.8. Methane Slip at different process conditions [174].

Synetix (a company that in 2003 was owned by Johnson Matthey) licenses one of the large-scale steam reforming processes, which was first operated in 1962. Hydrocarbon feeds with boiling points up to 200°C (i.e., naphtha) can be treated, and the process consists of three steps: Feed Pretreatment, Catalytic Reforming and Reforming of Residual Methane. The advantage of the Synetix process is that no soot forms – even with liquid crude oil fractions as feed. This makes catalyst regeneration unnecessary[46]. Because of these advantages, in 2001 the Synetix technology was use in over 400 reformers in over 30 countries. When North Sea gas and other natural gas reserves around the world were developed, the use of naphtha as a feed stock declined and the need for this technology was reduced.

5.2.1.1. Feed Pre-Treatment

The catalysts used in the steam reforming process are poisoned by trace components in the hydrocarbon feed - particularly sulphur, chlorine and metal compounds. Sulphur is the most common problem. Chlorine compounds are less common and metal compounds are usually found in some heavier LPG and naphtha feeds.

A desulfurization step is necessary in the reforming of natural gas because natural gas is odorized with a sulfur-containing substance - usually methyl mercaptan. This step will prevent damage to (i.e., deactivation of) the nickel or platinum catalysts in downstream processes. Some of these feed streams may also contain chlorides that are not only poisons to the downstream catalysts, but they may also contribute to stress corrosion in the reactors or piping.

The best way to remove sulphur compounds is to convert the organic sulphur species to H_2S over a hydrodesulphurization catalyst. The next step is sulphur removal with an absorbent. The same catalyst can usually convert any organo-chloride species to give HCl and also act as an absorbent for most problematic metal species. A second absorbent is used for chloride removal[70].

As shown in Figure 5.5, the feed stream can be preheated in the reformer before it enters the feed pre-treatment step to enhance the removal process. The design of the removal equipment depends on the impurities that need to be removed, and some potential equipment configurations are shown in Figure 5.9[70].

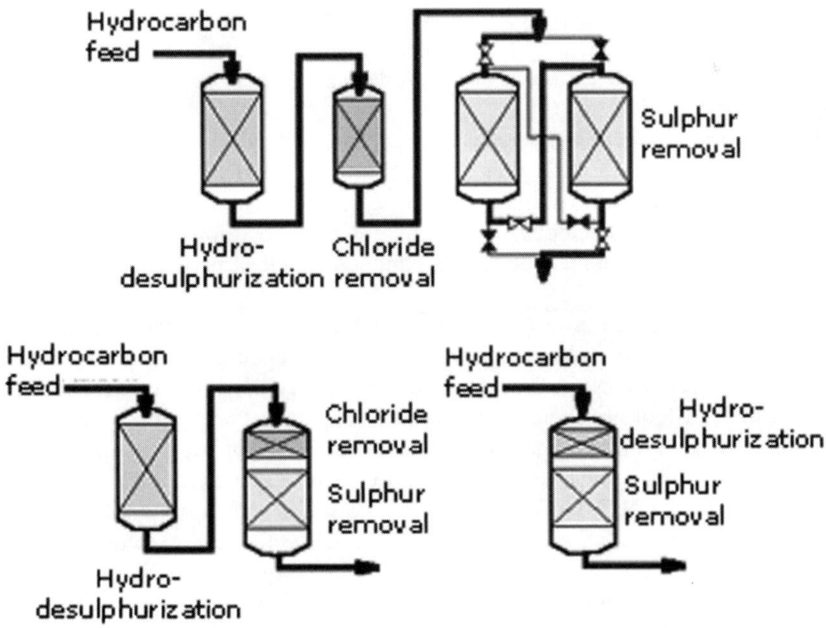

Figure 5.9. Potential configurations of reformer feedstock purification equipment. (Reproduced by permission of Johnson Matthey Catalysts. Copyright Johnson Matthey PLC)

The materials that are used to remove H_2S are based on zinc oxide and are formulated with a very large sulphur capacity for long life. The basic chemical reaction is:

$$ZnO + H_2S \rightarrow ZnS + H_2O \quad (5.5)$$

For single bed arrangements (see Figure 5.9), products with a high sulphur pick-up and a sharp absorption profile are used. This maximizes saturation of the absorbent before sulphur breakthrough. High-density H_2S absorbents are best suited to a lead-lag bed arrangement. Multiple-bed designs use a lead-lag bed arrangement that utilizes high-density products that have high saturation capacity, but a less sharp absorption profile. As shown in Figure 5.9, the lag bed absorbs any sulphur slippage while the lead bed is taken to saturation. The lag bed has sufficient capacity to allow the lead bed to be removed from service while the absorbent is renewed. The renewed bed is then re-commissioned in the lag position[70].

It is important that the sulphur removal absorbent be positioned downstream of the chloride absorbent. This is because HCl reacts with active phases in the zinc oxide absorbent and irreversibly chemically binds the chloride into the absorbent. The resulting structural changes block the porous structure and reduce sulphur absorbent capacity. Since the HCl absorbent has a very sharp absorption profile, frequently a layer of absorbent is placed on top of the sulphur absorbent within the same vessel. This design can help reduce capital costs[70] (see Figure 5.9).

If extra purification is needed to protect and extend the life of nickel-based, pre-forming and steam-reforming catalysts or precious metal-based catalysts, an ultra-purification absorbent may be used. This absorbent can be installed below the ZnO-based, H_2S removal absorbent. It removes more H_2S and organo-sulphur species so that the feed stream contains very low, parts-per-billion levels of sulphur[70].

In the Feed Pretreatment step with a naphtha feed (which may contain up to 5 mg/Nm^3 of sulfur[53] plus chlorine and metal compounds), the feed must be purified because the $Ni\text{-}K_2O/Al_2O_3$ reforming catalyst is very sensitive to these impurities. Chlorine compounds are less common than sulfur while metal compounds are typically found in some heavier LPG and naphtha feeds[70]. The naphtha is treated with H_2 at 350°C to 450°C using a metal oxide catalyst ($CoO\text{-}MoO_3/Al_2O_3$ or a $NiO\text{-}MoO_3/Al_2O_3$) until the sulfur content is less than 1 ppm and in some cases less than 0.1 ppm[46]. The cobalt-molybdenum catalysts are the most common type for sulfur removal. The nickel-molybdenum catalysts are preferred under certain conditions such has high CO content, high content of organic nitrogen compounds and/or high olefin content in the feed[70].

The resulting H_2S is absorbed on ZnO. Simultaneously, any olefins present are hydrogenated[46]. The same catalyst similarly converts any organo-chloride species to give HCl and also acts as an absorbent for most problematic metal species. A second absorbent is used for chloride removal[70]. As shown in Figure

5.10, catalysts can be designed to remove several different chemicals from the feed stream[166].

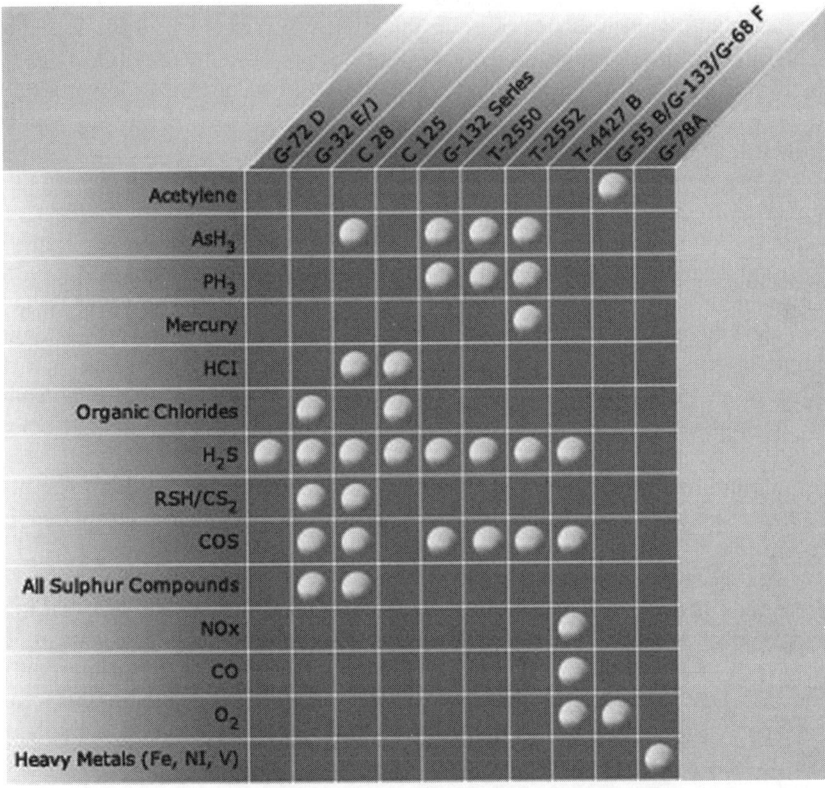

Figure 5.10. Feed purification catalysts for different impurities.
(Reproduced by permission of Süd-Chemie AG)

The petroleum feedstocks that contain sulfur as an impurity are handled under ambient conditions in carbon-steel tanks and pipelines where the corrosion attack by sulfur is less severe. In the desulfurization step, vaporized feedstock is processed at 400°C in the presence of hydrogen sulfide (H_2S) and carbonyl sulfide (COS) – both of which are highly corrosive. Stainless Steels (SS) 304, 316 or 321 (for the fired heater) are used as the material of construction for various pieces of process equipment in this section of the plant. Equipment failures occur because of external corrosion and thinning of fired-heater coils and interior deposition of carbon from coking which leads to overheating. Fuel-gas lines, that contain hydrocarbon vapors and H_2S, should be constructed of SS 304 and heat traced to avoid condensation[88].

5.2.1.2. Pre-Reformer

In some cases a plant may have a pre-reformer. A pre-former is an adiabatic, fixed-bed reactor upstream of the primary reformer. It provides an operation with increased flexibility in the choice of feed stock; it increases the life of the steam reforming catalyst and tubes; it provides the option to increase the overall plant capacity; and it allows the reformer to operate at lower steam-to-carbon ratios[166]. The hot flue gas from the reformer convection section provides the heat required for this endothermic reaction.

A pre-reformer does a small amount of reforming and partially converts the higher hydrocarbons to methane. It accomplishes this upstream of the main steam reformer while 1) operating at a much lower temperature (~900°F) and 2) by using a more active catalyst that contains NiO, MgO, SiO_2 and Al_2O_3. One advantage in new plant designs is that the conventional steam reformer furnace (which is the most costly capital item on the plant) can be smaller[70]. Substantial improvements in thermal efficiency, throughput and operating costs can be achieved in the pre-former. All the ethane, the heavier hydrocarbons and some of the methane are decomposed endothermically and at temperatures below those in the main reformer. The product gas from the pre-reformer is an equilibrium mixture of hydrogen, methane, steam and carbon oxides. The pre-reformer can reduce fuel consumption by up to 10% and increase throughput by up to 15%[71].

Another advantage is that a pre-reformer may be preferred in flowsheets for processing heavier feedstocks (up to naphtha). This is especially true if 1) the end product is CO or town gas, 2) a range of hydrocarbon feeds are to be processed or 3) minimum steam export is a requirement. The pre-reformer also converts heavier hydrocarbons to methane before they are fed to the steam reformer. This allows the steam reformer to be designed for methane service. The pre-reforming catalysts are extremely sensitive to poisons. Therefore a good feed purification section is essential[70]. (Note: In the olden days, town gas, often called manufactured or coal gas, was often made by baking coal in large airtight ovens (gas retorts). The gas bubbles from the hot coals were piped off and used as a household fuel. Natural gas has replaced town gas.)

In addition to the capital cost for a new reactor, the pre-reformer does add complexity to the system - particularly in the requirements for startup and shutdown. It is most attractive in cases 1) where more capacity is needed without increased firing and 2) where the feedstock contains a significant amount of C_3 or higher hydrocarbons, which could crack and form coke in the preheat coil[86].

The Catalytic Rich Gas (CRG) technology from Kvaerner is an example of pre-reformer technology that can be licensed. This technology was developed by British Gas and was first used in 1964. In 2001 over 25 CRG pre-reformers were in operation. If desired, a pre-reduced CRG catalyst is available[205].

5.2.1.3. Reformers

Steam reforming refers to the endothermic, catalytic conversion of light hydrocarbons (methane to gasoline) in the presence of steam [see Eq. (5.1)]. The reforming reaction takes place across a nickel catalyst that is packed in tubes in an externally-fired, tubular furnace (the Primary Reformer). The lined chamber reactor is called the 'secondary reformer', and this is where hot process air is added to introduce nitrogen into the process. Typical reaction conditions in the Primary Reformer are 700°C to 830°C and 15 to 40 bar[46].

The reforming tubes are heated by burning natural gas or ash-free distillates. At a constant temperature, an increase in pressure causes the proportion of methane remaining in the product gas (which is undesirable in synthesis gas) to increase. However, due to construction material constraints, temperatures higher than 830°C cannot be reached in externally heated reforming tubes. For this reason, the product gas from the Primary Reformer is fed into the Secondary Reformer (a lined, chamber reactor filled with a high-temperature-resistant, nickel catalyst). A portion of the product gas is burned in the Secondary Reformer with added air or oxygen so that the gas mixture reaches a temperature that is over 1,200°C. Methane reacts with steam at this temperature until only an insignificant amount remains (0.2 to 0.3 vol %). The methane concentration depends upon the pressure, the temperature, and the quantity of nitrogen and steam that are present. The sensible heat from the resulting synthesis gas is used for steam generation[46].

These operating conditions are necessary 1) to achieve the breakdown of the stable methane molecule and 2) to optimize the size of the reactors by reducing the volume of gases with the higher pressures. Overall, the reactions are exothermic and generate a lot of heat that has to be removed from the system.

One vendor states that they provide a down-flow, down-fired, box-type reformer furnace design for capacities above 10 million SCFD of dry synthesis gas. They provide a cylindrical-type reformer furnace design for capacities below 9 to 10 million SCFD of dry synthesis gas[167].

5.2.1.4. Reformer Catalysts

The catalyst that is used in the secondary reformer does not need to be as active as that in the primary reformer. Hence, the usual nickel concentration is about 15%, compared with 25% in the primary-reformer catalyst.

The optimum steam reforming catalyst selection depends on furnace design, feedstock types and the operating conditions. The selection is usually dictated by the heaviest hydrocarbon feed and may involve two or three catalyst types to obtain optimum performance. The main categories of catalyst are for 1) Light Duty Reforming, 2) Intermediate Duty Reforming and 3) Heavy Duty Reforming. The Light Duty catalysts are for natural gas, refinery off-gas and pre-formed feeds[70]. The optimum catalyst for an air-blown secondary reformer depends on 1) the design of the burner and 2) the distance between the burner tip and the top of the catalyst layer[166].

Intermediate Duty catalysts are for feeds with a significant content of components from ethanes up to liquid petroleum gas (LPG). The heavier feedstock increases the tendency for catalyst deactivation through carbon laydown and requires a special catalyst in the top 30% to 50% of the reformer tubes. This tendency also occurs when light feeds are run at low steam-to-carbon ratios and/or at a high heat flux.

Heavy Duty catalysts are primarily for naphtha feeds that have even more tendency for carbon deposition. These catalysts contain complex stabilized alkali phases and other promoters that maximize carbon gasification activity[70].

Although silica and silica-bearing materials are very useful in making catalyst supports, they cannot be used for the shift catalyst. This is because the silica volatilizes and migrates from the hotter zone to lower temperature zones downstream. Usually it deposits on the waste heat boiler tubes after the secondary reformer.

Normally, the nickel oxide in the catalyst is reduced to nickel and water by the hydrogen that is produced in the operation. In some cases the reduced nickel can be reoxidized to nickel oxide when large amounts of steam and small amounts of H_2 are present[47]:

$$Ni + H_2O \leftrightarrow NiO + H_2 \quad (5.6)$$

The addition of 0.5 to 10 percent by volume of H_2 to the natural gas feed has been shown to keep the nickel in the reduced state, which makes it more active. Also, the hydrogen will retard the formation of nickel sulfide, which will prevent or minimize poisoning of the catalyst:

$$NiS + H_2 \leftrightarrow Ni + H_2O \quad (5.7)$$

Poisons for the nickel catalyst are: sulfur, arsenic, chlorides or other halogens, phosphates, copper and lead. A 15 percent nickel catalyst is poisoned at 775°C if the gas contains as little as 0.005 percent (50 ppm) sulfur.

Carbon can also reduce the effectiveness of the catalyst. When conditions are favorable for the following reactions, carbon will deposit on the catalyst:

$$CH_4 \leftrightarrow 2\,H_2 + C \quad \Delta H = +17.9 \text{ kcal/mol or } 75 \text{ kJ/mol} \quad (5.3)$$

$$2\,CO \leftrightarrow CO_2 + C \quad \Delta H = -41.4 \text{ kcal/mol or } 173 \text{ kJ/mol} \quad (5.4)$$

Thus, if insufficient steam is present, carbon will be deposited on the catalyst and reduce its effectiveness. However, carbon will gasify with steam at 800°C or higher in the reformer:

$$C + H_2O \leftrightarrow H_2 + CO \quad \Delta H = +28 \text{ kcal/mol or } 119 \text{ kJ/mol} \quad (5.8)$$

Reformer operation and changes in operating rate should always be carried out in a way that ensures sufficient steam is over the catalyst. Thus, if the feed rate is to be increased, the steam rate should be increased first. Conversely, if the feed rate is to be decreased, the natural gas rate should be decreased first.

5.2.1.5. Reformer Materials of Construction

The Primary Reformer is a steam-hydrocarbon reforming tubular furnace that is typically externally fired at 25 to 35 bar and 780°C to 820°C on the process side. The reformer tubes function under an external heat flux of 75,000 W/m^2 and are subject to carburization, oxidation, over-heating, stress-corrosion cracking (SCC), sulfidation and thermal cycling. Previously SS 304, SS 310 and SS 347 were used as tube materials. However these materials developed cracks that very frequently led to premature tube failures (see Table 5.10)[88].

Table 5.10. Composition of Various Stainless Steels[88]

Stainless Steel	Composition (%) C	Mn	Si	Cr	Ni	P	S	N	Mo	Others
SS 304	0.08	2	1.0	18-20	8-10.5	0.045	0.03	-	-	-
SS 310	0.25	2	1.5	24-26	19-22	0.045	0.03	-	-	-
SS 316	0.08	2	1.0	16-18	10-14	0.045	0.03	-	2	-
SS 316L	0.03	2	1.0	16-18	10-14	0.045	0.03	0.1-0.3	2-3	-
SS 321	0.08	2	1.0	17-19	9-12	0.045	0.03	-	-	0.4 min Ti
SS 347	0.08	2	1.0	17-19	9-13	0.045	0.03	-	-	0.8 min (Nb+Ta)

(Reproduced by permission of Hydrocarbon Processing)

In the mid-1960's HK 40 alloy (see Table 5.11) was developed and proved to be a good material for vertical reformer tubes. Consequently, plant capacities were extended to 600 tons per day with this tube material. Although this alloy had a design service life of 100,000 hours, overheating considerably reduced tube life. A 55°C excursion above the design temperature could lower tube service life to 1.4 years[88].

In the 1980's HP 25/35 Modified alloys were developed that used metals such as molybdenum (Mo), niobium (Nb) or tungsten (W). These metals increased resistance to creep rupture and offered good ductility and weldability. With stronger alloys, wall thickness of tubes could be reduced. Thinner tube walls offered benefits such as using lighter tubes and tube supports, improved heat transfer, resistance to thermal cycling and capacity increases of up to 30%[88].

HP micro alloys were developed during the 1990's. They contain traces of titanium (Ti), zirconium (Zr) and rare earths that are applied during casting. The micro alloys enhanced carburization resistance and improved high-temperature creep-rupture resistance[88].

Table 5.11. Furnace Alloy Materials by Composition[88]

Alloy Designation	Composition (%) C	Mn	Si	Cr	Ni	Others
HK 40	0.40	1.50	1.5	25	20	-
HP 25/35 (Mod)	0.40	1.50	1.5	25	35	1.5 Nb
HP Micro	0.45	0.70	1.5	25	35	Nb, W present
Alloy 800	0.05	0.75	0.35	20	32	Al + Ti ≤ 0.70
Alloy 800 H	0.10	0.75	0.35	20	32	Al + Ti > 0.85

(Reproduced by permission of Hydrocarbon Processing)

For reformer outlet manifolds the normal metallurgy choice is a wrought type of Alloy 800 H. It has sufficient ductility and thermal-shock resistance during startup and shutdown. The cast version of Alloy 800 H provides a cost-effective, alternate material with a higher creep-rupture strength, low tendency for embrittlement and good ductility. Hot reformed-gas transfer lines are usually refractory-lined with an interior of Alloy 800 sheathing[88].

In the Secondary Reformer air is added to the process stream at operating conditions of 28 to 30 bar and 955°C to 1,025°C. This refractory-lined vessel has an outer shell of a low-alloy steel containing 0.5 Mo. A typical phenomenon (metal dusting) occurs in the secondary reformer outlet sections. With hot gases containing a higher CO content, strong carburizing reactions occur, and carbon will diffuse into the Fe-Cr-Ni alloy. This phenomenon can lead to local mechanical fracturing of surface layers and, subsequent failures, by pitting[88].

Materials such as SS 304 and Alloy 800 are very susceptible to metal dusting in the range of 500°C to 800°C. Besides temperature, carbon activity (ratio of CO/CO_2 in the gas) and gas partial pressure also affect metal dusting. Severe attacks occur when the carbon activity is in the range of 3 to 10. Recirculating CO_2 into the primary reformer along with feedstock can maintain a low CO/CO_2 ratio and avoid the severity of this attack. By maintaining a high oxidizing potential in the gas (i.e., high steam-to-hydrogen ratio), the metal dusting can also be minimized[88].

Hydrogen embrittlement is another important corrosion problem that is encountered in reformed-gas pipelines. Usually a low-alloy steel is used down-stream of the feed-water heaters where the gas temperature is less than 400°C. Internal carburization and fissuring of steel is caused by hydrogen permeating into the steel and forming methane. The methane cannot disuse out and accumulates in voids that are formed at the grain boundaries. This condition contributes to high stresses that ultimately fissure and crack the metal[88].

The Nelson curves in API 941 (published by the American Petroleum Institute) list the operating limits that should be followed to avoid decarburization and fissuring of steel in hydrogen service[88]. However, unsatisfactory service experience with the carbon-$^{1}/_{2}$ moly steel has led to reconsideration of the curves originally provided by API 941. In 2001 API 941 warns against new construction with the alloy and urges inspection and monitoring of existing equipment. The difficulty is that suitable and widely accepted inspection techniques are not available and complete inspection of all components is not practical[168].

To avoid hydrogen attack, steels should contain elements that form stable carbides. For the inspection of existing equipment in hydrogen service, non-destructive methods based on ultrasonics can quantify the hydrogen attack and estimate mechanical properties of hydrogen-damaged steels. The results of such tests can be used to estimate the remaining life of the equipment that has been attacked by hydrogen[168].

At least one company provides ceramic outer coatings 1) for process piping in the radiant section of reformers and 2) for refractories. This coating keeps the tube material from oxidizing and forming layers of scale that provide an effective barrier to the conduction of heat into the tube. As a result, the conductive heat transfer to the process is permanently maximized. The coating on the refractories allows the surfaces of refractory linings to enhance radiant heat transfer by increasing emissivity. Some of the potential benefits are extra capacity (7% to 14%), fuel savings, and the elimination of the effects of over-firing - such as high bridge wall temperatures, high tube skin temperatures and coke formation[169].

5.2.1.6. Waste Heat Recovery

The waste heat recovery system is associated with flue gas from the reformer furnace and process gas from the secondary reformer. It generates high-pressure steam in specially designed boilers. These boilers and steam superheaters operate under very high heat flux on the one side and corrosive water on the other. Consequently, meticulous care is needed to avoid failures. Proper material selections and stringent water quality are two proactive loss prevention methods. Various material grades of SA 213, ASTM A312, ASTM A335 and ASTM A351 provide useful service life in this section[88].

The cooler part of the combustion air preheater – the tail end of the flue-gas-heat-recovery train – is more likely to corrode due to sulfur dioxide (SO_2) condensing from the flue gas. In this area, cast iron or glass will resist the acid attack. Carbon steel preheater tubes, joined with 1.5 to 2 meters of SS 304 tubes at the cold end of the tube sheet, can ensure reasonable service life. Typically, the flue gas temperature to the stack is maintained above the dew point of SO_2 to prevent condensation. During startup and shutdown, condensation of SO_2 will occur[88].

The high-pressure, feedwater heaters are prone to leaks at the tube-to-tube sheet joints. Lining or over-laying the tubesheet with Alloy 600 and using tube materials with 1% Cr / 0.5% Mo can mitigate this risk considerably[88].

5.2.1.7. Hydrogen Production & Capital Costs

The cost of making hydrogen (and ammonia) by steam reforming of natural gas depends primarily on the cost of natural gas. Therefore the 2001 hydrogen cost ranges from \$0.15 to \$0.60 per 100 SCF - depending on whether the hydrogen was produced on-site or delivered as a compressed gas via pipeline[165]. If the hydrogen is sold on the merchant market, the cost of liquefying it and transporting it to the user can make the delivered cost three to five times higher than the cost of gaseous hydrogen[162]. Table 5.12 lists some of the costs associated with hydrogen production, and Table 5.13 shows a way to summarize the costs associated with Steam Methane Reforming.

Table 5.14 shows a way to summarize the costs associated with Steam Methane Reforming in a very large capacity plant.

Two different sources indicate that the cost of a 60 million SCF per day hydrogen plant, based on steam methane reformer technology, will require a capital cost of between \$US 60 million and \$US 90 million in 2001 dollars. Therefore hydrogen plant costs in this capacity range cost \$US 1.00 to \$US 1.50 per SCF of hydrogen per day[164]. Based on current reformer technology, another useful guideline is that the production costs (in \$ per million BTU) - excluding capital charges - are 150% of the natural gas LHV costs (in \$ per million BTU). As illustrated in Table 25, the hydrogen production costs (excluding capital charges) are \$US 5.11 per million BTU (0.73 + 4.38). And 150% of the natural gas cost (\$3.33 per million BTU) is \$US 4.95 per million BTU. So these values are very similar and this guideline appears to be reliable.

Table 5.12. Cost of Steam Reforming of Natural Gas – 2001[164]

	\$ per 100 SCF	\$ per Million BTU (LHV)
Capital Charges[a]	0.06	2.19
Non-Fuel Operating & Maintenance Costs[b]	0.02	0.73
Natural Gas [c]	0.12	4.38
Net Production Cost[c]	0.20	7.30

[a] The Capital Charges are based on these assumptions: i) Investment = \$US 75 Million, ii) Plant Capacity = 60 Million SCF per day of Hydrogen, iii) 85% Uptime and iv) Capital Charges are 15% per year of the invested capital.

[b] The Non-Fuel Operating & Maintenance Costs are based on these assumptions: i) Investment = \$US 75 Million, ii) Plant Capacity = 60 Million SCF per day of Hydrogen, iii) 85% Uptime and iv) Operating & Maintenance Costs are 5% per year of the invested capital.

[c] Natural Gas costs are assumed to be \$US 3.33 per million BTU LHV or \$3.00 per million BTU HHV.

[d] LHV = Low Heating Value, which assumes that water of combustion remains in the vapor state.

[e] HHB = High Heating Value, which assumes that water of combustion is condensed into the liquid state.

Table 5.13. Cost of steam reforming of natural gas [a] – 2000[170]

	$ per 100 SCF	% of Total Costs
Raw Materials		
Natural Gas	0.118	59
Utilities		
Electricity	0.003	2
Water	0.003	2
Steam	-0.007	-4
Variable Cost	0.117	59
Capital Charges (25%)	0.083	41
Net Production Cost	0.200	100

a. Basis for cost estimate:

i. Plant Capacity = 50 million SCFD of Hydrogen
ii. Steam Production = 37,500 lb per hour
iii. Natural Gas Cost = $2.75 per million BTU
iv. Power Cost = $0.045 per kWh
v. Steam Value = $4.00 per 1000 lb

Table 5.14. Cost of steam reforming of natural gas [a] – 2000[175]

	$ per ton H_2 per Year
Raw Materials	
Fuel (Natural Gas)	569.55
Consumables	12.45
Fixed Costs: Operations & Maintenance	9.60
Variable Cost: Operations & Maintenance	1.07
By-Product (Steam) Credit	-61.17
Capital Charges (25%)	145.90
Net Production Cost	677.39
	$5.54 per Million BTU
	$1.80 per Thousand SCF

a Basis for cost estimate:

i. Plant Capacity = 150 million SCFD (417.8 tons /day) of Hydrogen @ 346 psia
ii. Total Plant Cost = $130,998,000
iii. Total Capital Requirement (Plant Cost, Preproduction Costs, Inventory, etc.) = $143,040,000
iv. Steam Production = 220,000 lb per hour
v. Natural Gas Cost = $3.15 per million BTU
vi. Natural Gas Feed Rate = 2,868 million BTU per hr (65.5 million scfd)
vii. Steam Value = $4.84 per 1000 lb
viii. Plant Availability = 90%
ix. H_2 Product Purity = 99.5%
x. Assumes Pressure Swing Absorption to obtain desired H_2 purity.

The theoretical energy requirement per mole of hydrogen produced for the overall steam reforming process is equal to 40.75 kJ/mole of hydrogen. The process does not have any by-product credits, and it does not look environmentally benign due to large CO_2 emissions. The CO_2 emissions (including stack gases) reach up to 0.3 to 0.4 m^3 CO_2 per m^3 of hydrogen produced. The capture and disposal of CO_2 (80-85% of CO_2 is captured from the concentrated streams) adds about 25% to 30% to the cost of hydrogen produced by steam reforming[60].

5.2.1.8. Other Reforming Processes

The RKN process uses steam reforming to make hydrogen from hydrocarbon gases (from natural gas to naphtha). Haldor Topsøe developed this process in the 1960's. By 1974, 24 plants based on this technology were operating[2].

The SMART (**S**team **M**ethane **A**dvanced **R**eformer **T**echnology) makes hydrogen by the steam reforming of methane. It differs from similar systems in that it houses the catalyst in a proprietary heat exchanger. Mannesmann KTI developed the process in 1996, and the first installation was started up in Salisbury, MD, in 1998[2]. KTI and Air Products entered into a technology marketing alliance in 1992 in which both companies contribute to the overall hydrogen plant design. Air Products also offers small-capacity hydrogen plants (> 0.5 million scfd) based on KTI's SMART technology.

The ATR (**Aut**othermal **R**eforming) process makes CO-enriched syngas. It combines partial oxidation with adiabatic steam-reforming and is a cost-effective option when oxygen or enriched air is available. It was developed in the late 1950's for ammonia and methanol synthesis, and then further developed in the 1990's by Haldor Topsøe[2]. The difference between Steam Methane Reforming (SMR) and ATR is in how heat is provided to activate the endothermic steam reforming reaction. In SMR, the catalyst is contained in tubes that are heated by an external burner.

In ATR, a portion of the natural gas is burned to raise the temperature of the process gas before it contacts the catalyst. ATR can be thought of as Partial Oxidation at lower temperatures (900 to 1100°C) or as Catalytic Reforming with oxygen and steam. The overall reaction is:

$$2\,CH_4 + H_2O + \tfrac{1}{2}\,O_2 \leftrightarrow 5\,H_2 + 2\,CO \qquad (5.9)$$

The CAR (**C**ombined **A**utothermal **R**eforming) process is used to make syngas from light hydrocarbons, and the heat is provided by partial oxidation in a section of the reactor. It was developed by Uhde and commercialized at an oil refinery at Strazske, Slovakia, in 1991[2].

The KRES (**K**ellogg **R**eforming **E**xchanger **S**ystem) is a reforming process that provides syngas to the KAAP process. The KAAP (**K**ellogg **A**dvanced

Ammonia Process) process operates at high-pressure process and makes ammonia from hydrogen and nitrogen, but it does not use an iron-containing catalyst. The catalyst was developed by BP and contains ruthenium supported on carbon. It was developed by MW Kellogg Company in 1990 and first installed by the Ocelot Ammonia Company (now Pacific Ammonia) at Kitimat, British Columbia, in 1994[2]. A second KRES unit was scheduled to be started up in late 2002 at an ammonia plant owned by Liaotong in Panjin, Liaoning Province, China[171].

The KRES process can replace primary and secondary reformers with 1) a catalyst-filled shell-and-tube heat exchanger and 2) an autothermal reformer. The heat exchanger uses a unique open-tube design to replace the fired reformer. The catalyst-filled tubes are suspended from a single tube sheet at the cold end of the vertical exchanger. The tubes are typically 2.0 inch OD (50 mm) and contain conventional nickel-on-alumina catalyst. The open-tube design allows each tube to expand freely and also facilitates catalyst loading and unloading[203].

More than 50% of the heat available in the reformer effluent is recovered and used in the reforming reaction. This results in energy consumption as low as 23.4 to 24.8 MM BTU (lower heating value - LHV) per short ton or 5.5 to 5.9 gigacalories (LHV) per metric ton. These values represent a 5% to 10% reduction in energy consumption relative to a conventional ammonia plant[203].

Compared to a fired reformer, catalyst tube damage is less likely. This is because direct flame impingement cannot occur, and flue gas distribution is not a concern. The maximum tube metal temperature is also limited to the temperature of the shell-side gases. Since the reformer burners have been eliminated, process control is less complicated. And the lack of a flue gas stream reduces emissions of NOx and CO_2 by as much as 60% to 75%[203].

The Haldor Topsøe Convection Reformer (HTCR) is a relatively small piece of equipment that combines the radiant and waste heat sections of the conventional reformer. As a result, the heat transfer to the steam reforming reaction is optimized and the process avoids the energy surplus inherent in conventional plants. The gas that leaves the reformer section is cooled by steam production before more hydrogen is formed in a shift reactor. The hydrogen that is produced is purified to 99.9 vol % in a PSA (pressure-swing absorption) unit, and the off-gas from the PSA unit is used in the fired heater/reformer as a supplementary fuel. Typical natural gas consumption is 3.3 to 3.4 Gcal per 1,000 Nm^3 of hydrogen. The HTCR technology is best for small and medium-sized hydrogen plants (500 to 10,000 Nm^3/hr) and can be supplied in a number of skids that offer small plot space requirement, low investment and short erection time[75].

A comparison of the basic operating conditions are shown below for some synthesis gas production technologies[176]:

Table 5.15. Synthesis Gas Production Technologies[176]

Parameter	Units	SMR	ATR	CAR	Oil Gasification
Pressure	bar	<40	<85	<85	<85
Temperature	°C	~860	~1000	~550	~1350
Feedstock		Natural Gas up to Naphtha			Flexible
H_2 to CO Ratio in Product		3 to 6	2.0 to 3.5	1.5 to 3.0	1.3 to 1.8
Catalyst Used		Yes	Yes	Yes	No
Methane Slip	Mol %	2 to 7	0.3 to 1.0	0.3 to 1.0	<0.3
Oxygen to Feed Ratio	Nm^3/Nm^3	N/A	~0.7	~0.6	~1.0
Single Line Capacity (1,000 Nm^3/hr - H_2 + CO)		500	500	70	150

[a] SMR = Steam Methane Reforming
[b] ATR = Autothermal Reforming
[c] CAR = Combined Autothermal Reforming

5.2.1.9. Retrofits of Steam Methane Reforming

Where an increase of 10% to 50% of existing steam reforming capacity is required, revamping is often more cost-effective than a new plant or the purchase of hydrogen over-the-fence. Hydrogen plants generally fall into two groups based on the type of CO_2 removal system. Plants built since the late 1980's tend to use pressure swing adsorption (PSA) for purification while older hydrogen plants use wet scrubbing. However many of the problems encountered in revamps are common to both types of plants[86].

Each plant tends to have areas where the search for extra capacity will pay off as well as areas where the limits to capacity are more fixed. A steam reforming plant contains three functional areas: Reaction, Heat Transfer and Separation. Each of these areas reacts differently to changes in throughput, but changes in one area can affect all the other areas - since they are all related. Within the range of most retrofits, Reaction is only slightly affected by operating rate. Catalyst volume is affected by heat transfer, pressure drop requirements and a requirement for a particular run length. Existing reactors are therefore often able to operate at increased rate[86].

Hydraulic capacity (i.e., pressure drop) shows up in each of the functional areas. For example, reformer pressure drop often becomes limiting at high flow rates. This requires a change in catalyst or provides a reason to install catalyst tubes with a larger diameter[86].

The situation with heat transfer varies. The gas heat transfer coefficient increases with velocity, so existing exchangers are often adequate as the capacity is increased. Whether or not extra heat transfer capacity is needed will depend on the service, the design margins and the operating flexibility[86].

Some reforming processes are designed for reduced primary reforming by moving some of the heat duty to the secondary reformer because of the marginal low efficiency of the primary reformer. Some of the revisions to conventional reforming are[53]: Decreased Firing in the Primary Reformer, Increased Process Air Flow to the Secondary Reformer, and Lower Inert Levels in the Make-Up Syngas.

5.2.1.9a. Decreased Firing in the Primary Reformer. Decreased heat supply in the primary reformer means that the process outlet temperature is lowered to about 700°C; the firing efficiency increases; and the size and cost of the primary reformer are reduced. The milder operating conditions prolong catalyst life, catalyst tube life and outlet header service life. The extent of reforming is reduced according to the lower heat supply and lower temperature. Generally, a slight decrease in steam to carbon ratio (compared to the conventional concept) is acceptable[53].

5.2.1.9b. Increased Process Air Supply to the Secondary Reformer. Decreased heat supply in the primary reformer means that increased internal firing is necessary to achieve approximately the same degree of total reforming. A somewhat higher methane slip (and thus a lower secondary reformer outlet temperature) is acceptable and preferable in this type of process. This is because methane is removed in the final purification.

The process air requirement is about 50% higher than in the conventional process. This means increased compression capacity and energy. The process air compressor is usually driven by a gas turbine with the exhaust gas from the turbine being used as combustion air in the primary reformer. Some excess steam is available for export when using a gas turbine[53].

5.2.1.9c. Heat Exchange Autothermal Reforming. A new development in the late 1980's was Heat Exchange Autothermal Reforming. In this process the heat content of the secondary reformer gas is used in a primary reformer with a new design. This reformer is a gas-heated, heat exchange reformer rather than the conventional fired furnace design. Surplus air or oxygen-enriched air is required in the secondary reformer to meet the heat balance in this autothermal concept[53].

Emissions to the atmosphere are reduced significantly by eliminating the flue-gas from the primary reformer. NOx emissions may be reduced by 50% or more compared to conventional steam reforming. The level of NOx reductions depends on the extent of auxiliary combustion in the plant[53].

Two processes of this kind are in operation, and some others are at the pilot stage. As of 1999, single train capacities have been limited to 1,000 tonnes per day or less[53], and four plants had been installed that used this type of technology from Synetix. This type of reformer, referred to as a Gas Heated Reformer (GHR) is shown in Figure 5.11[72].

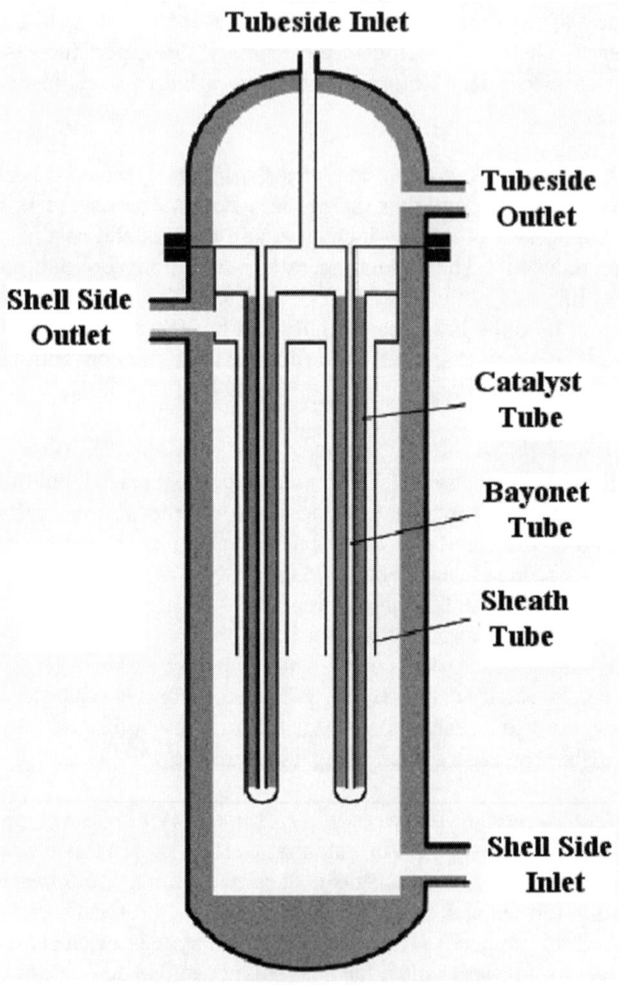

Figure 5.11. Gas Heated Reformer (GHR).
(Reproduced by permission of Johnson Matthey Catalysts. Copyright Johnson Matthey PLC)

Synetix has developed a modified design of the GHR known as the Advanced GHR or AGHR. BHP Petroleum of Australia became the first operator of the AGHR in April, 1998. The design of the AGHR has many features in common with the original design of the GHR, but the key difference is that it uses a single-pass tube arrangement while the GHR uses a bayonet tube reformer. Based on the experience at BHP Petroleum, the AGHR design results in a reformer that is lower in cost, easier to operate, easier to fabricate and allows scale-up to capacities in excess of current world-scale throughputs. (see Figure 5.12)[72].

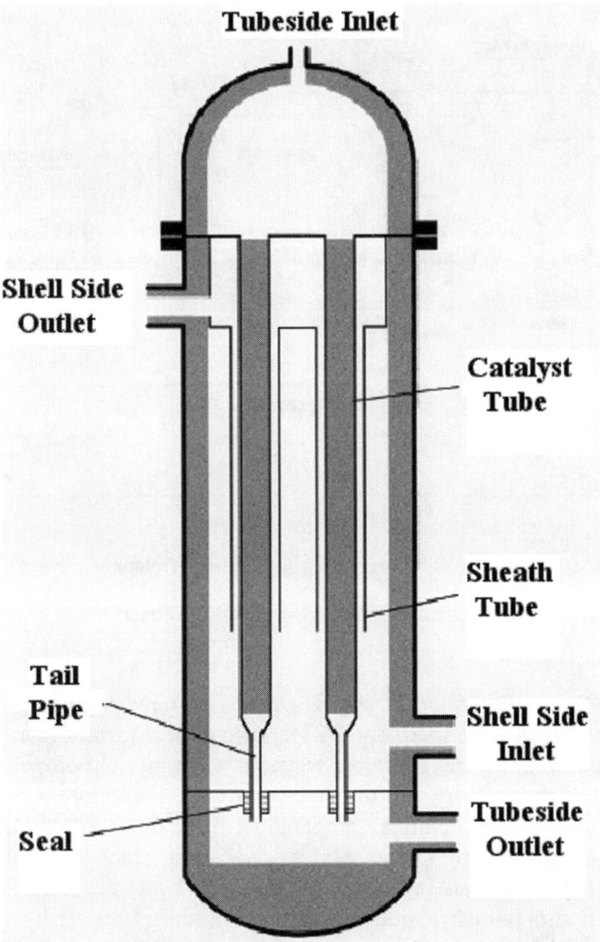

Figure 5.12. Advanced Gas Heated Reformer (AGHR).
(Reproduced by permission of Johnson Matthey Catalysts. Copyright Johnson Matthey PLC)

5.2.1.9d. Steam Reforming with Pressure Swing Absorption (PSA) Purification. A hydrogen plant based on a new design was installed in Venezuela at the Compagnia Hidrogeno de Paraguana (CHP) plant to supply hydrogen to a refinery. The plant is a joint venture between The BOC Group and Foster Wheeler and was started up in September, 1997. This plant has a reduction in total fuel firing that is estimated to be 4%, and an installed cost that is estimated to be 23% below that of previous plants of the same size. The plant uses steam reforming with purification by PSA to produce 50 million SCFD of hydrogen at 99.5% purity and 400 psig. A schematic flow diagram of the process is shown in Figure 5.13[87].

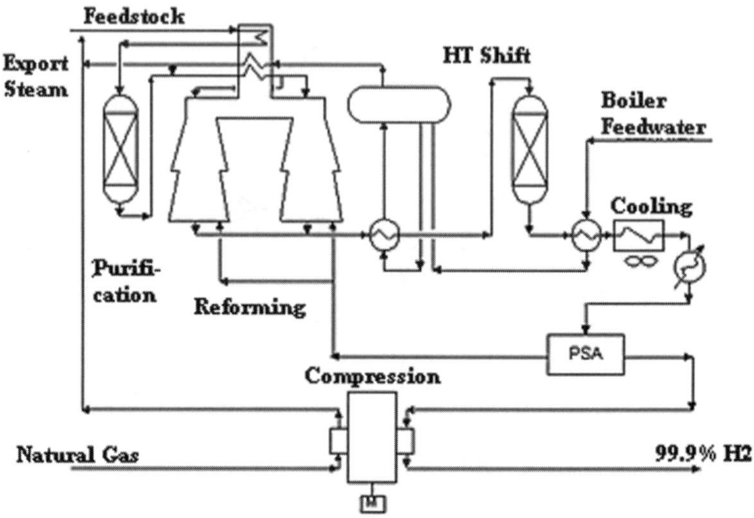

Figure 5.13. Schematic diagram of hydrogen plant based on steam reforming of natural gas with purification by PSA[87].

(Reproduced by permission of Foster Wheeler)

5.2.1.9e. Increase Reformer Firing & Other Reformer Improvements. Plant throughput can usually be increased by increasing the reformer firing rate. This is the first option that many operators consider when more hydrogen production is needed. As flow rate through the tubes increases, firing rate increases. This results in an increase in the metal temperature of the tubes. Since tube life is a function of tube metal temperature at a given wall thickness, tube life will probably be reduced. This problem can be solved by 1) reducing the reformer outlet temperature (which results in lower conversion), or 2) by accepting a shorter tube life. For more information on tube materials, see Section 5.2.1.5 that describes Reformer Materials of Construction or obtain Reference 88. Other equipment limits also need to be examined, such as induced draft (ID) fan capacity, burner capacity, valve sizes and others[172].

With a Terrace Wall or side-fired unit, the limit to capacity will generally not be apparent in the radiant section. The increase in firing is matched by an increase in heat transfer on the inside of the catalyst tube. So the actual tube temperatures show only a minor increase[86].

If the tubes in the radiant section are near the end of their useful life, upgrading this section may be attractive and it can provide a capacity increase of up to 10%. Changes in metallurgy (see Tables 5.10 and 5.11 for tube materials) since the 1970's have allowed the following changes in operating conditions[86]:

- 1970's: HK-40 Tubes — 300 psig & 1450°F (790°C)
- 1980's: HP Tubes — 325 psig & 1550°F (845°C)
- 1990's: Microalloy Tubes — 450 psig & 1575°F (855°C)

If Microalloy tubes are installed, the reformer can operate with increased flux rates and higher reformer outlet temperatures. The reformer can then be operated to provide either 1) more steam and more hydrogen or 2) milder operating conditions. If more steam and more hydrogen are needed, the reformer is operated with a lower steam-to-carbon ratio while keeping the hydrogen purity the same. And more steam is available to be exported. If the process steam flow is reduced, more hydrogen production is possible without increasing the system pressure drop. In cases where flow capacity is available and more steam is not needed, milder operating conditions can be obtained. A higher steam-to-carbon ratio is used to reduce catalyst tube metal temperatures and provide longer life or an increased margin of safety[86].

New tube materials can also provide better operability and reliability. When HK-40 was the best tube material, some furnaces used staggered rows to pack more tubes and more catalyst into a given firebox. However the unbalanced radiation on the tubes caused bowing and increased stress on pigtails. To overcome this problem, larger diameter tubes (made of higher strength materials) can be used. These new tubes provide more equivalent capacity and fewer tubes are needed[86].

To help evaluate the quality of the tubes, Quest Integrated has developed a tube inspection system using laser measurement for more accurate determination of tube diameter. The new system is called LOTIS (Laser Optical Tube Inspection System), and it provides rapid reformer tube diameter measurement with very high accuracy. The data from LOTIS can be used to estimate remaining tube life, determine appropriate operating adjustments or identify tubes that need to be replaced[82].

5.2.1.9f. Convection Section & Waste Heat Recovery Improvements. While hydrogen is made in the radiant section, the convection section 1) recovers waste heat from the flue gases and 2) cools the gases enough so that they can be removed through the stack or the induced draft (ID) fan. Problems in the convection section commonly show up as high coil temperatures or as limited draft capacity. These problems are caused by high fluegas volumes or temperatures[86].

If a rate increase is desired, the greatest impact will generally be in the convection section of the fluegas system. To obtain the rate increase, the heat flux was increased in the radiant section by increasing the firing rate and the firebox temperature. Not only does the gas flow increase, but the temperature of the fluegas leaving the radiant section increases substantially[86].

The effects of this increased gas flow and temperature are multiplied in the convection section. This is because additional load is placed on the convection coils, fans and the steam system. And these areas usually limit the operating rate of the reformer[86].

The tube supports in the convection section must be considered when up-rating a reformer furnace. They are exposed to the hot flue gas without the cooling effect of process fluids. Tube supports in the hotter regions of the convection section are made of high alloy material and may operate in the creep range - like catalyst tubes[86].

If a reformer has extra capacity in the steam system and high temperatures at the outlet of the process preheat coil are a problem, one option is to remove tubes from the coil to reduce the heat input. In this case the extra heat in the flue gas is absorbed by the steam generation coils downstream[86].

One of the most effective reformer modifications is to use heat from the convection section to preheat radiant section feed. This will reduce radiant section duty and firing rate. The effectiveness of this option is limited only by the risk of coking in the preheat coil, the metallurgy of the preheat coil and the metallurgy of the radiant inlet system. This option has been used to increase capacity by 10% without increasing the arch temperature in the radiant section[86].

Convective Heat Transfer Reformers provide additional reforming capacity by using the heat contained in the primary reformer exit gases. Several designs are available, but not all have been commercialized. These units typically replace a portion or the entire duty of the waste heat boiler. So they significantly reduce the steam capability of the reformer. Potential increases in capacity of between 10% and 30% are possible. The modifications are capital intensive but relatively easy to implement[170].

5.2.1.9g. Improvements Using Combustion Air Preheat. Combustion air preheat reduces the fluegas flow through the furnace. As a result the ID fans and the entire convection section handle a lower volume of hot gases[86].

The most common option is to use heat exchange with the fluegas. For new units this can have the advantage of very high efficiency because the heat sink is ambient air. Whether or not this is feasible to add to an existing reformer will depend on 1) the furnace configuration (downfired or Terrace Wall), and 2) the current gas system. Because of the required changes in fans and ductwork, this system may be too expensive for a retrofit[86].

Another option is to use steam to preheat air. This can be more economical to install because only the air side of the reformer is affected. The fluegas system is left untouched. In one case, 600 psig steam was used to preheat combustion air to 400°F. This allowed a 12% increase in capacity at the same fuel firing rate and combustion air flow[86].

A third option is to use circulating boiler water to preheat the air. If the steam already includes circulating pumps, the existing pump system may not need to be expanded. The last two options have the added advantage of reducing the steam make, which unloads the steam drum[86].

5.2.1.9h. Lower Steam-to-Carbon Ratio in PSA-based Plants. Lowering the steam-to-carbon ratio can be incorporated as a stand-alone option or in conjunction with other debottle-necking options. Plants with feedstocks other than naphtha should have steam-to-carbon ratios between 2.5 and 3.5. Plants have a potential to easily increase hydrogen production if they are operating at steam-to-carbon ratios higher than 3.5. However it should be noted that lowering the steam-to-carbon ratio in a conventional (i.e., non-PSA based) plant will lower the hydrogen purity and reduce the energy available for the CO_2 removal unit[172].

5.2.1.9i. Post Reformer (EHTR and O_2 Secondary Reforming). An Enhanced Heat Transfer Reformer™ (EHTR) uses heat from the effluent of the fired reformer to reform additional feed. An EHTR steam methane reformer has been operated by Air Products since the mid-1980's in California. In this reformer a portion of the steam and hydrocarbon feed by-pass the primary reformer and pass through a heat exchanger containing catalyst in the tubes. The primary reformer effluent provides the heat for the EHTR by passing reformed gas in a counter-current flow pattern on the outside of the EHTR tubes that contain the catalyst. The EHTR can increase existing plant capacity by up to 25% and eliminate or reduce excess steam generation from the steam methane reformer[172].

Another post-reformer design in commercial operation is called Oxygen Secondary Reforming. This technology has primarily been used in conjunction with revamping ammonia plant reformers to hydrogen service. An oxygen secondary reformer is installed immediately downstream of the primary reformer. As with a conventional secondary reformer, a portion of the reforming load is shifted from the primary to the secondary. An oxygen secondary reformer provides a low-methane slip and the ability to lower the outlet temperature of the primary. Therefore the primary reformer can process more feed gas without increasing the reformer's firing rate[172]. For older generation steam reformers that use CO_2 Removal and methanation, Oxygen Secondary Reforming can provide capacity expansions between 30% and 50%. This capacity increase can be attained without increased fuel firing because the process steam can be drastically reduced to allow more throughput. NOx and SOx levels remain unchanged as a result. The cost of oxygen is a prime factor in evaluating the economics associated with this option[170].

The secondary reformer vessel is a refractory-lined vessel that has an oxygen burner in its top neck and a fixed catalyst bed. Installation of a secondary reformer usually requires significant changes to the CO_2 removal system. Hydrogen purity can be increased up to 98%. The economics generally depend on a reliable source of low-cost oxygen[172].

5.2.1.9j. Summary of Expansion Options. Some of the expansion and retrofit options are compared in Table 5.16[172].

Table 5.16. Summary of Some Expansion Options[172]

Expansion Option	Increase in Hydrogen Capacity	Steam Rate	Feed Rate	Fuel Feed Rate	Cost	Issues
Increase Reformer Firing	7 to 10%	Proportional, Assuming Adequate Convection Surface	Proportional to Increased Production	Proportional to Increased Production	Low	Tube Life
Reduce Steam to Carbon Ratio	0 to 4%	Increase	Increase	Decrease	Low	Lower H_2 Purity & Steam
Pre-Reformer	8 to 10%	10 to 15% Decrease	Proportional	Same	Medium	Pre-Reformer Catalyst Cost
Post Reformer	20 to 30%	15 to 30% Decrease	Increase Slightly, Higher Proportion than Increased H_2	Same	High	Metal Dusting (EHTR), O_2 Cost (Secon-dary)

5.2.1.10. Small-Scale Reforming

Small-scale reforming systems are being pursued by a number of companies developing fuel cell electrical generation systems for home and automotive fuel cell applications. Long before these systems become affordable and simple enough to be used as home electrical system components, they may be cost-effective for industrial hydrogen supply systems[48].

Small-scale reforming systems are relatively complex because they need fuel and air-feed systems, the reformer, a hydrogen purification system and various cooling and water processing ancillary systems to make it all work. Additionally, the systems have to employ a specific hydrocarbon that is available at a reasonable cost at a customer's location. These systems probably work best for customers with hydrogen consumption rates in the 1,500 scf/hr to 10,000 scf/hr range. They are most cost effective when employed to produce hydrogen of a 99.9% or lower assay. As of 2001, small-scale reforming systems in this capacity range have not been commercialized. But they are in the advanced development stage[48].

However plants with slightly higher capacities of 20,000 scfh or higher are available as skid-mounted units from Air Products and Hydro-Chem. Different designs are available to meet the specific needs of a particular application. These designs are all based on steam reforming combined with PSA and are available as 1) Low Capital Units, 2) High Export Steam Units, and 3) Low Energy Units (Low Feed & Fuel). Hydro-Chem hydrogen plants are based on in-house steam reforming technology utilizing the round, up-fired, up-flow, "can" type reformer design. Depending on client needs, the process is then followed by a purification step using in-house, 4-bed PSA technology, Linde 5-bed plus PSA technology or an amine wash unit. A process sketch and technical data for each of these designs are shown in Figures 5.14 through 5.16 and Tables 5.17 to 5.19[178].

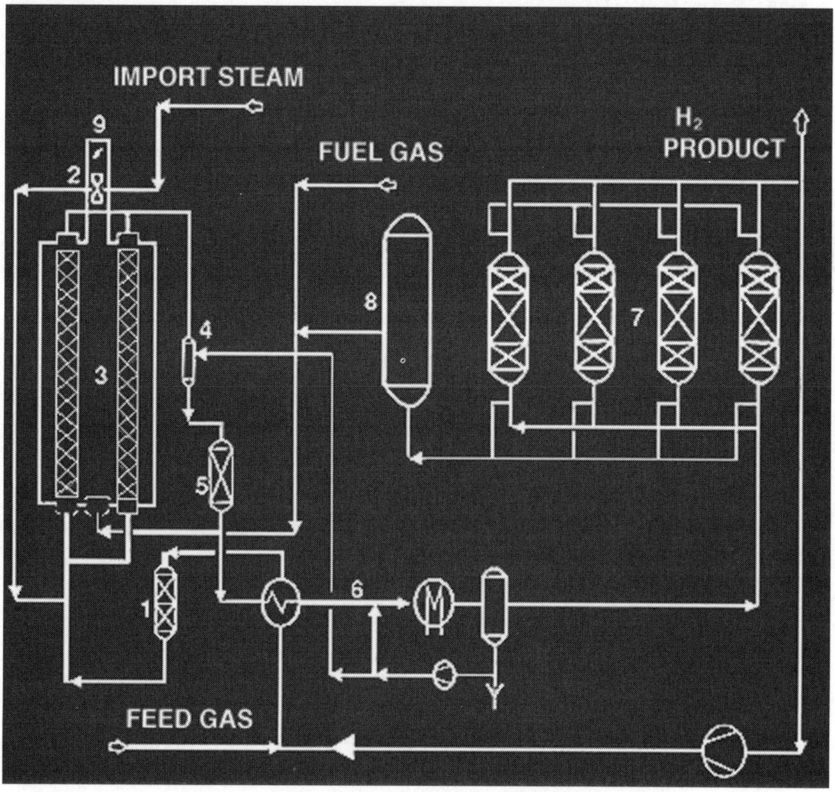

Figure 5.14. Low capital modular hydrogen unit. [178]
(Reproduced by permission of Hydro-Chem, a Division of Pro-Quip Corp.)

Table 5.17. Low capital modular hydrogen unit[178]

Process Features

- No Heat Recovery
- Import Steam Available for the Process
- Low Cost Feed Stock

Technical Data

Feed & Fuel Consumption	Natural Gas	LPG	Naphtha
MM Btu / Hr (LHV)	17.98	18.13	18.48
GJ/Hr (LHV)	18.96	19.13	19.50
Product Hydrogen			
Flow Rate (Nm^3/hr)	1000	1000	1000
MM SCFD	0.896	0.896	0.896
Pressure (bar-g)	14	14	14
Pressure (psig)	203	203	203
Purity (Vol. % H_2)	99.9	99.9	99.9
Utilities			
20 Bar-g Steam Import (mt/hr)	1.0	1.1	1.2
290 psig Steam Import (tons/hr)	1.1	1.2	1.3
Electrical Energy (kW)	6.0	10.0	8.0
Cooling Water			
m^3/hr ($\Delta t = 10^\circ C$)	97	105	106
gpm ($\Delta t = 18^\circ F$)	427	462	467
Dimensions			
Plot Area, m x m	14 x 18	14 x 18	14 x 18
Plot Area, ft x ft	46 x 59	46 x 59	46 x 59

(Reproduced by permission of Hydro-Chem, a Division of Pro-Quip Corp.)

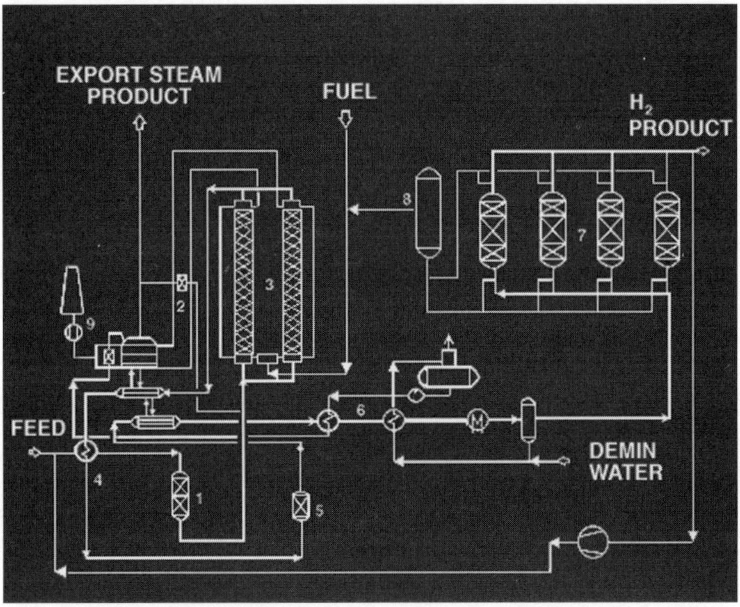

Figure 5.15. High export steam - hydrogen modular unit [178].
(Reproduced by permission of Hydro-Chem, a Division of Pro-Quip Corp.)

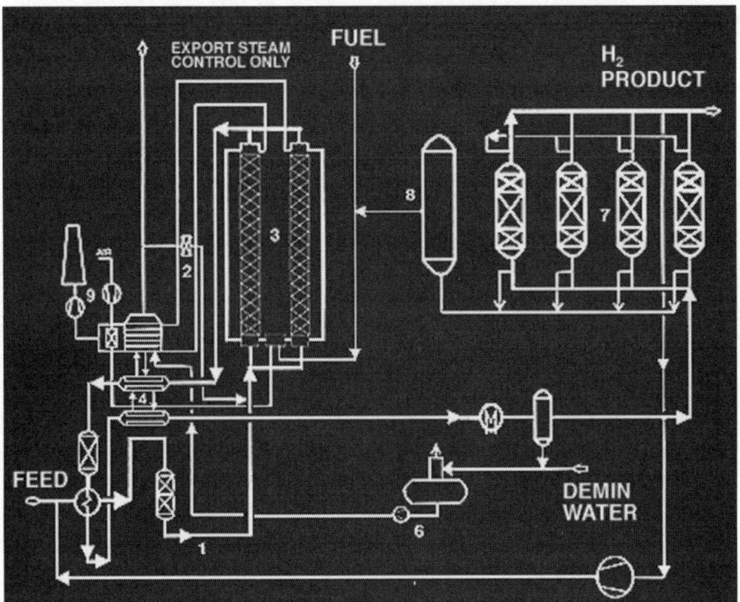

Figure 5.16. Low energy hydrogen modular unit [178].
(Reproduced by permission of Hydro-Chem, a Division of Pro-Quip Corp.)

Table 5.18. High Export Steam - Hydrogen Modular Unit[178]

Process Features

- Maximum heat recovery from both flue gas and process gas streams for steam production
- No combustion air preheat
- Export steam: 100 lbs / 1000 scf of H_2 produced

Technical Data

Feed & Fuel Consumption	Natural Gas	LPG	Naphtha
MM Btu / Hr (LHV)	18.09	18.29	18.65
GJ/Hr (LHV)	19.09	19.30	19.67
Product Hydrogen			
Flow Rate (Nm^3/hr)	1000	1000	1000
MM SCFD	0.896	0.896	0.896
Pressure (bar-g)	14	14	14
Pressure (psig)	203	203	203
Purity (Vol. % H_2)	99.9	99.9	99.9
Utilities			
20 Bar-g Steam Import (mt/hr)	1.8	1.7	1.7
290 psig Steam Import (tons/hr)	2.0	1.9	1.9
Electrical Energy (kW)	32.0	36.0	34.0
Cooling Water			
m^3/hr ($\Delta t = 10°C$)	12	16	19
gpm ($\Delta t = 18°F$)	53	70	84
Demineralized Water, mt/hr	2.4	2.4	2.5
Demineralized Water, tons/hr	2.6	2.6	2.8
Dimensions			
Plot Area, m x m	14 x 20	14 x 20	14 x 20
Plot Area, ft x ft	46 x 66	46 x 66	46 x 66

(Reproduced by permission of Hydro-Chem, a Division of Pro-Quip Corp.)

Table 5.19. Low Energy Hydrogen Modular Unit[178]

Process Features

- Combustion air and reformer feed temperatures are optimized to minimize energy consumption.
- Energy consumption down to 0.4 MM Btu (LHV) per 1000 SCF of hydrogen produced.

Technical Data

Feed & Fuel Consumption	Natural Gas	LPG	Naphtha
MM Btu / Hr (LHV)	14.97	15.16	15.72
GJ/Hr (LHV)	15.80	15.99	16.59
Product Hydrogen			
Flow Rate (Nm^3/hr)	1000	1000	1000
MM SCFD	0.896	0.896	0.896
Pressure (bar-g)	14	14	14
Pressure (psig)	203	203	203
Purity (Vol. % H_2)	99.9	99.9	99.9
Utilities			
20 Bar-g Steam Import (mt/hr)	0.3	0.1	0.2
Electrical Energy (kW)	28.0	32.0	33.0
Cooling Water			
m^3/hr ($\Delta t = 10^\circ C$)	45	48	51
gpm ($\Delta t = 18^\circ F$)	198	211	225
Dimensions			
Plot Area, m x m	14 x 25	14 x 25	14 x 25
Plot Area, ft x ft	46 x 82	46 x 82	46 x 82

(Reproduced by permission of Hydro-Chem, a Division of Pro-Quip Corp.)

In 1998 Praxair installed a small, on-site, hydrogen generating system at Kobelco Metal Powder of America in Seymour, Indiana. This generating system can produce 30,000 cubic feet per hour (800 cubic meters per hour) of 99.99% hydrogen[189].

The hydrogen generating system combines Praxair's non-cryogenic gas purification technology with the natural gas processing technology of International Fuel Cells (IFC). The system can use either natural gas or propane feedstock. Hydrogen production costs are about 20% lower than steam methane reformers and about 15% lower than liquid hydrogen deliveries. System capacities range from 6,000 up to 30,000 scf per hour of hydrogen with purities up to 99.999%[189].

5.2.1.11. Technology Suppliers

Some of the companies that provide technology and contracting services to produce hydrogen via Steam Reforming are listed in Table 5.20:

Table 5.20. World Suppliers of Technology to Produce Hydrogen via Steam Reforming - 2002

Company	Headquarters Location
Air Liquide	Paris, France
Air Products and Chemicals	Allentown, Pennsylvania, USA
Kellogg Brown & Root, Inc. (A Halliburton Company)	Dallas, Texas, USA
Kvaerner Process	Houston, Texas, USA
Foster Wheeler Corporation	Clinton, New Jersey, USA
Haldor Topsøe A/S	Copenhagen, Denmark
Howe-Baker Engineers	Tyler, Texas, USA
KTI Group BV (a Mannesmann Company)	Zoetermeer, Netherlands
Linde AG	
- Process Engineering and Contracting Div.	Munich, Germany
- Lotepro Corp.	Valhalla, New York
- Selas Fluid Processing	Blue Bell, Pennsylvania
- Pro-Quip Corp.	Holly Springs, Georgia
Linde BOC Process Plants LLC	Tulsa, Oklahoma, USA
Praxair	Danbury, Connecticut, USA
Süd-Chemie Ag	Munich, Germany
Synetix	London, England
Technip-Coflexip	Paris, France
Uhde GmbH	Dortmund, Germany

5.2.2. Partial Oxidation

Noncatalytic, partial oxidation processes react hydrocarbons with gaseous oxygen and steam at very high temperatures in a refractory-lined, pressurized reactor to produce synthesis gas. The quantities of oxygen and steam are controlled such that gasification continues without the need for external energy input (hence the autothermal description)[161]. While the process can handle a wide range of feedstocks, including natural gas, it was developed mainly to utilize the low-value "bottom of the barrel" and waste feedstocks that typically have high sulfur and high metal contents[40].

The low fluidity and normally high sulphur content in heavy hydrocarbons do not allow for the practical application of steam reforming to these feedstocks. They are normally processed by 1) partial oxidation or 2) an autothermal, flame reaction process that occurs at 1300 to 1500°C in the presence of steam and oxygen (the Texaco Process). The ratio of oxygen-to-steam is controlled so that the gasification process requires no external energy[173].

One advantage of partial oxidation is that significant reductions in NOx and CO_2 emission rates can be achieved compared to steam methane reforming. In one case a reduction of 20% was achieved when partial oxidation replaced steam methane reforming. This is because the point source emissions related to the furnace flue gas exhaust are not a part of this process. Emissions from partial oxidation processes are relatively small by comparison and primarily come from fired heaters that may be used to preheat the process feed[177].

Synthesis gas manufacture by partial oxidation or autothermal cracking of crude oil fractions was developed by BASF/Lurgi, Texaco and Hydrocarbon Research. Heat for the thermal cracking is supplied by partial combustion of the feed in the presence of water. Recycled CO_2 may also be added to the combustion to attain a desired CO/H_2 ratio[46].

Shell developed a modified version that is referred to as the gasification process. The first commercial plant was started up in 1956. By 2002 a total of 85 Shell gasifiers were in operation world-wide. All hydrocarbons from methane to crude oil residues (heavy fuel oil) can be used as feedstock[46].

In the Shell gasification process, no catalyst is used. The preheated feeds react with H_2O and less than stoichiometric amounts of O_2 in the reactor's combustion section at 30 to 80 bar and 1200 to 1500°C. The heat generated is used to steam reform the oil. Soot, which is formed from a small portion of the oil, is removed from the synthesis gas by washing with water or oil and is made into pellets[46]. The process can also produce elemental sulfur.

As illustrated in Table 5.9, the H_2 to CO molar ratio in the synthesis gas product stream depends on the raw materials and the operating conditions. By burning natural gas with a limited quantity of oxygen (that is less than the stoichiometric requirement), a synthesis gas that has a H_2 to CO molar ratio of approximately 2 to 1 can be produced:

$$CH_4 + \tfrac{1}{2}\,O_2 \rightarrow 2\,H_2 + CO \quad \Delta H = -\,36\text{ kcal/mol or }150.6\text{ kJ/mol} \qquad (5.10)$$

If steam is introduced into the reaction, the H_2 to CO molar ratio can be increased to over 2.0. This reaction is exothermic and can reach temperatures of 1200°C to 1500°C. Therefore the reaction vessels are either refractory-lined or water-cooled. High temperatures favor the formation of carbon monoxide and hydrogen[49].

A typical Partial Oxidation process is shown in Figure 5.17.

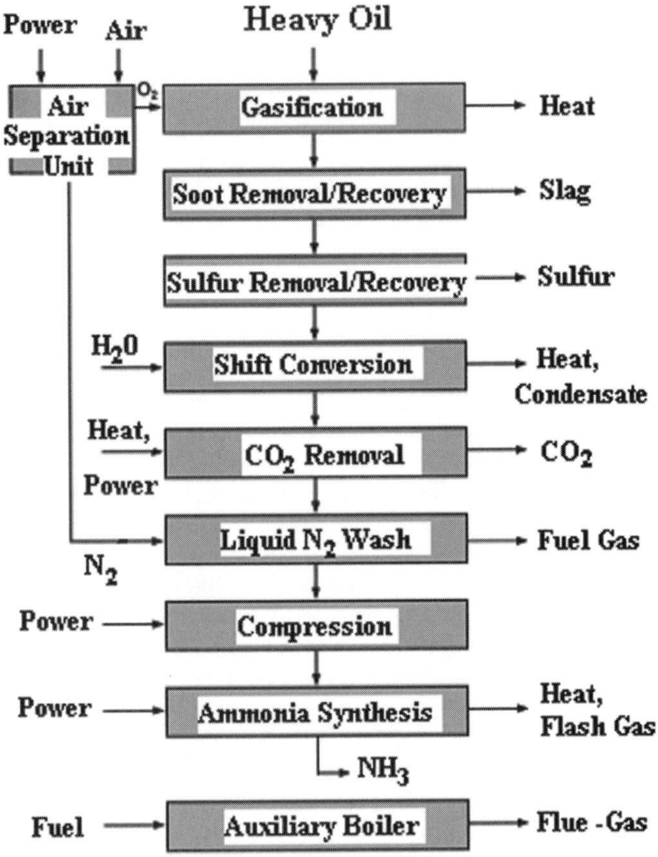

Figure 5.17. Block diagram of partial oxidation process [53]
(Reproduced by permission of European Fertilizer Manufacturers Association)

If fuel oil or crude oil are used as the raw material in partial combustion, care must be taken to provide for the removal of sulfur compounds and ash-containing materials in the crude oil (see Figure 5.17). These substances introduce sulfur into the product stream and have been known to damage the refractory. With crude-oil or fuel-oil partial combustion, the carbon content is much greater than in the case of natural gas, and special design considerations are necessary to produce a satisfactory gas.

Typical gas compositions from the partial combustion of various fuels are given in Table 5.21. The nitrogen that is needed to produce the desired H_2/N_2 ratio for ammonia production is usually introduced later in the processing sequence.

Table 5.21. Product Gas Compositions from Various Feeds by Partial Oxidation (percent by volume - dry)

Feedstocks & Gas Components	Natural Gas	Light Naphtha	Heavy Fuel Oil	Vacuum Residual	Propane Asphalt	Coal
Hydrogen	61.80	52.09	46.74	44.80	43.69	34.39
Carbon Monoxide	33.75	42.59	48.14	49.52	50.09	44.22
Carbon Dioxide	3.10	4.88	3.80	4.16	4.45	18.64
Methane	1.00	0.30	0.30	0.30	0.30	0.38
N_2 + A	0.27	0.13	0.23	0.18	0.30	0.68
H_2S	-	0.01	0.76	1.00	1.12	1.46
COS	-	-	0.03	0.04	0.05	0.10
NH_3	-	-	-	-	0.13	-
H_2/CO Molar Ratio	1.83	1.22	0.97	0.90	0.87	0.78

The thermodynamics of Steam Reforming and Partial Oxidation are compared in Figure 5.18[174]. These graphs illustrate how the equilibrium compositions in the products streams vary (compare Table 5.9 and Table 5.21).

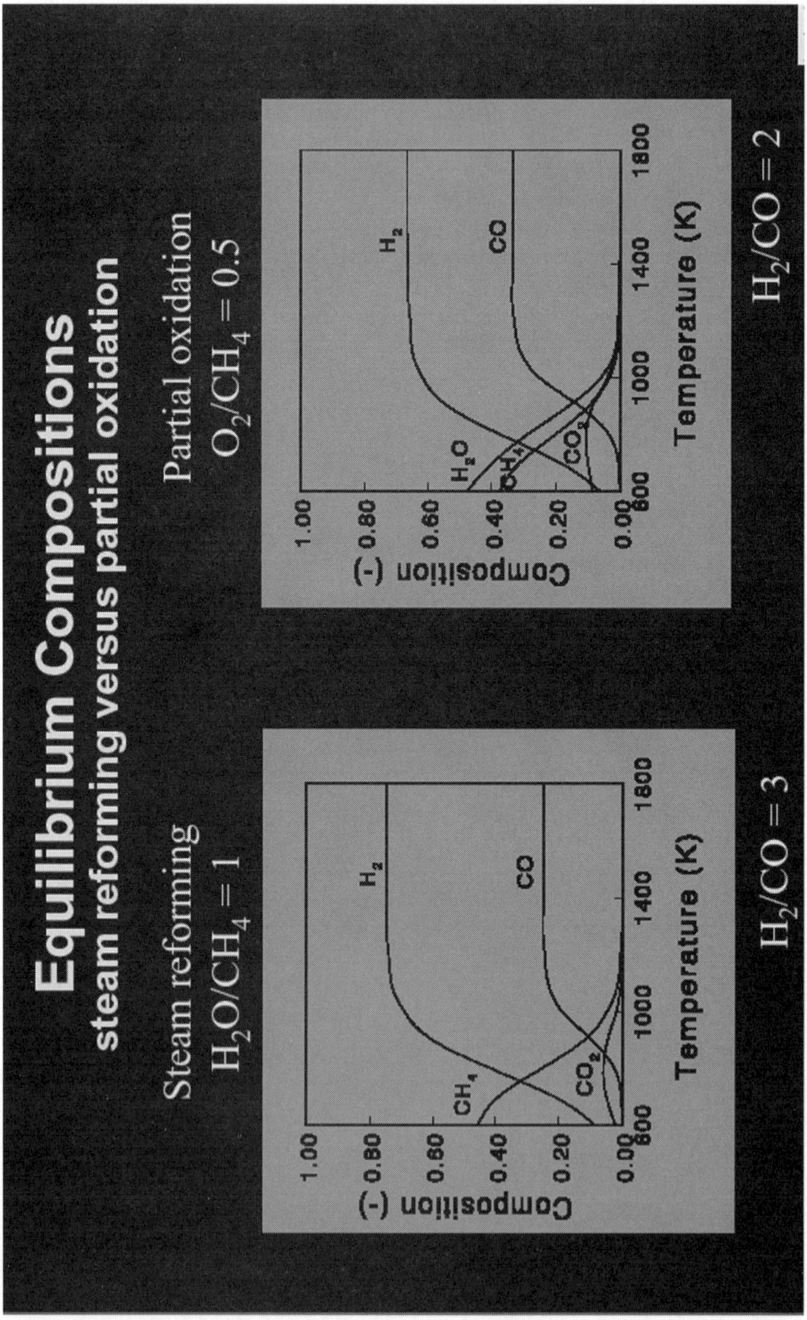

Figure 5.18. Compare thermodynamics of steam reforming and partial oxidation.

Some of the partial oxidation processes that have been developed are summarized below.

5.2.2.1. Partial Oxidation Processes of Academic Interest

The Electropox (**Electro**chemical **P**artial **Ox**idation) is an electrochemical process for oxidizing methane to syngas. It combines the partial oxidation and steam reforming of methane with oxygen separation in a single stage. It was invented in 1988 by T.J. Mazanec at BP Chemicals. An industrial-academic consortium to develop the process further was formed in 1997[1].

The Koppers Hasche process is a cyclic process for converting methane to syngas by partial oxidation over an alumina catalyst:

$$CH_4 + \frac{1}{2} O_2 \rightarrow 2 H_2 + CO \quad (5.10)$$

The process operates in a pair of horizontal catalyst chambers that alternate their functions at one-minute intervals[1].

In the 1990's, the University of Orleans (France) developed a non-catalyzed, plasma-assisted partial oxidation (or POX) process for making syngas[1,50].

5.2.2.2. Texaco Hydrogen Generation Process

The THGP (**T**exaco **H**ydrogen **G**eneration **P**rocess) process can be used to make pure, high-pressure hydrogen from various gaseous and light hydrocarbons. Partial combustion of the hydrocarbons yields syngas. The carbon monoxide in the syngas is converted to carbon dioxide by the shift reaction and is removed by PSA. This process can be licensed from the Texaco Development Corporation[1,51].

In this application of the Texaco Process, natural gas feed is heated in the Process Heater and fed to the Hydrotreater/Desulfurizers for sulfur removal. The purified feed gas is then heated further by passing through the Process Heater again.

Recycle CO_2 from the CO_2 Removal unit (i.e., the amine unit) is mixed with additional CO_2 (that is imported from another process at the Longview Operation) and compressed. Part of the compressed CO_2 is exported to another process, and the remainder is heated in the Process Heater.

The gas feed stream is split, recycle CO_2 is added to each stream and the combined streams are fed to each of the two gasification trains. Oxygen from the Air Separation Unit (ASU) is preheated by steam, split and fed to the two gasification trains. Partial oxidation and reforming reactions take place in the Gasifiers.

The syngas product leaves the Gasifiers at approximately 2500°F (1370°C), is mixed and then cooled in the Syngas Coolers by producing high pressure steam. The syngas is cooled further in the Syngas Coolers by heating a portion of the boiler feed water (BFW). The cooled syngas is then scrubbed with water to remove any by-product soot.

Overall energy recovery and utilization is dependent on the method that is used to control the hydrogen-to-CO ratio in the syngas product. Excess hydrogen, if present, must be removed from the product. This hydrogen may be valued as an additional product stream or as fuel. The most economical means of controlling the hydrogen-to-CO ratio in syngas is to adjust the quantity of CO_2 fed to the gasification reactor. An increase in carbon dioxide feed forces equilibrium towards carbon monoxide production [see Eqs. (5.2) and (5.4)]. However, syngas product availability is impacted by CO_2 availability and by the responsiveness of the process controls.

A simplified diagram of the Texaco Gasification Technology is shown in Figure 5.19. This particular process was installed at Eastman Chemical Company's Texas Operations near Longview, TX. Start-up was scheduled to occur in 2002[177].

Texaco gasification technology offers several options for heat recovery from the syngas. A quench reactor design cools the syngas by direct contact with water. Although this reactor design has the lowest capital cost, the temperature of the quenched syngas is too low to generate high pressure steam. Texaco also provides a reactor design with an integrated syngas cooler. The capital cost for this design is higher than the quench reactor, but it is capable of producing high pressure steam. This design was used for this application.

The cooled process gas that leaves the Syngas Scrubber is fed to the Amine unit. The amine unit consists of an absorption-stripping system plus associated equipment. In this system a circulating amine stream (activated MDEA - or Methyl Diethanolamine) absorbs the CO_2 in the CO_2 Absorber. The amine is regenerated in the CO_2 Stripper with the CO_2 being recycled to the compressor.

The syngas in the CO_2 Absorber overhead stream is water-washed and fed to a MEDAL™ membrane unit. The membrane feed gas is sent to a coalescing filter to remove liquids and is preheated before is enters the permeator. In the permeator, syngas is separated into a hydrogen-rich permeate and the syngas product. The operation of the membrane unit is very simple. The driving force for separation is the difference in partial pressure between the hydrogen in the feed gas and that of the permeate.

The use of a membrane unit to strip off excess hydrogen significantly reduces the influence of CO_2 supply. It is also effective at controlling the hydrogen-to-CO ratio in syngas by selectively removing hydrogen. The heart of the equipment consists of millions of polymeric hollow fibers that are as thin as hair. "Fast" gases, with a high permeation rate, diffuse through the membrane, flow out through the hollow fiber interior and are channeled into the permeate stream. "Slow" gases flow around the walls of the fibers. In this way, a fast gas like hydrogen is separated from slower gases like CO and methane[177].

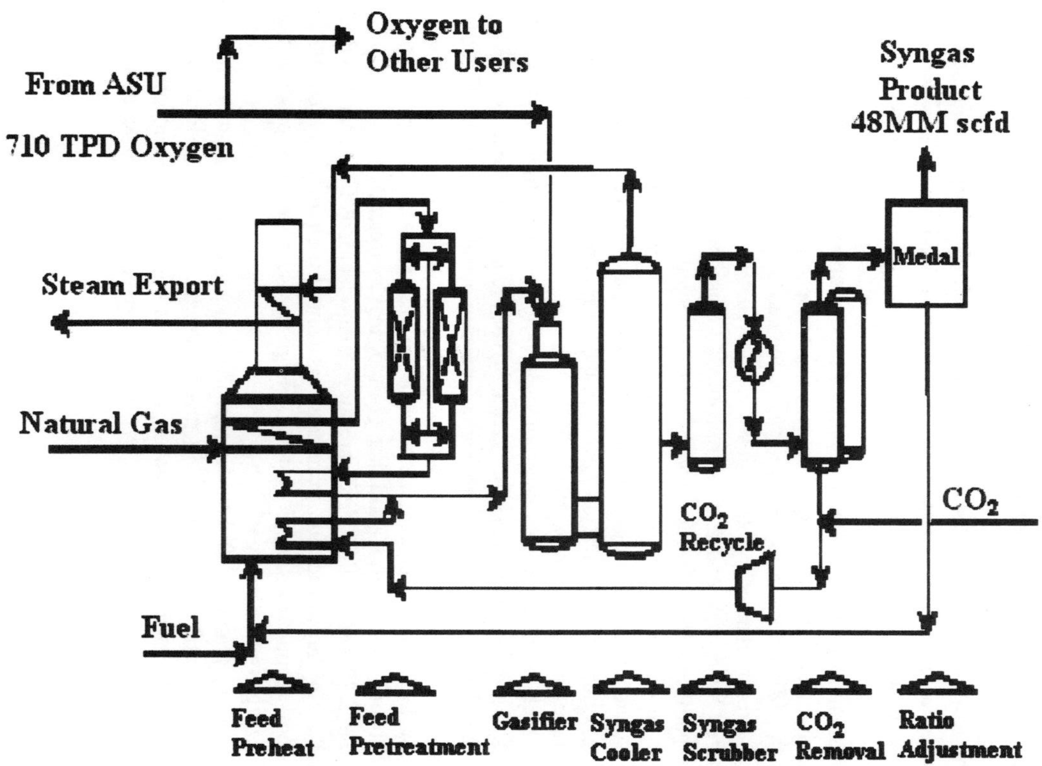

Figure 5.19. Texaco Hydrogen Generation Process (THGP).
(Provided by: Gasification Technologies Council, 2001 Gasification Technologies Conference)

Since Texaco syngas process was not being used to make ammonia, some of the steps would have to be modified to apply it to ammonia production. A CO Shift step might be needed upstream of the CO_2 Removal unit; the membrane unit might be replaced by a Pressure-Swing Absorption (PSA) unit; and a methanation unit might be needed to remove residual CO and CO_2.

5.2.2.3. Lurgi Multi-Purpose Gasification (MPG) Process

The Multi-Purpose Gasification (MPG) process uses partial oxidation with oxygen to generate synthesis gas from different hydrocarbon feeds. These feeds include natural gas, tars and other coal gasification residues, refinery residues, asphalts, coal and coke slurries, and chemical wastes. MPG is an updated development based on technology that Lurgi acquired in 1997 from SVZ (Sekundär-rohstoff Verwertungs Zentrum Schwarze Pumpe)[179].

In the MPG a gaseous feedstock is preheated by the hot, raw gas and an optional fired preheater. Preheat level and installation of the fired heater are determined by energy and cost optimization. This optimization considers the relative values of feed, fuel gas and oxygen. Oxygen also is preheated using a small amount of high-pressure steam from the heat recovery boiler. Feed gas and oxygen enter the reactor together with a minor amount of "shielding" steam via the gas burner. Gasification occurs in the empty, refractory-lined reactor at temperatures between 1200°C and 1400°C[179].

The high-level heat is recovered in a specially designed, fire-tube boiler that produces saturated high-pressure steam. The remaining heat is used for preheating feedstock, boiler feed water and demineralized water. Finally, a water-scrubbing tower removes traces of soot, HCN and NH_3. Soot formation is extremely low, and no special filtration is necessary. The wastewater passes through a sour-water stripper and then goes to final treatment[179].

Table 5.22 shows the MPG product stream composition that would be expected from the gasification of natural gas using high-purity oxygen (99.5%) and a small amount of shielding steam. High-pressure steam is produced at a rate of 2.1 kg/Nm^3 of feed gas[179].

The main features of the MPG process are shown in Figure 5.20. A desulfurization unit is not shown in this flowsheet. It may be located either upstream or downstream of the gasification unit. Its location depends on the heat utilization and on material selection for the equipment in areas that are prone to metal dusting[179].

Table 5.22. Composition of product stream leaving Lurgi MPG process[179]

Component	Composition (Volume %) Feed	Product
CO_2	1.00	4.40
CO		31.09
H_2		60.16
N_2	1.93	0.73
CH_4	88.01	0.38
C_2H_6	6.85	
C_3+	2.21	
Total Sulfur	5.2 ppmv	
Low Heating Value (MJ/Nm^3)	38.37	10.55
Temperature (°C)	15	60
Pressure (bar)	30	68

5.2.2.4. Shell Gasification Process (SGP)

The non-catalytic partial oxidation of hydrocarbons by the Shell Gasification Process or SGP (see Figures 5.21 and 5.22) takes place in a refractory-lined reactor that uses a specially designed burner. The oxidant is preheated (to minimize oxygen consumption) and then mixed with steam (as a moderator) before it is fed to the burner. The burner geometry and reactor geometry are designed so that the oxidant and steam are intimately mixed with the preheated feedstock[180].

As shown in Figures 5.21 and 5.22, the feedstock is gasified in the SGP reactor with oxygen to produce raw synthesis gas at a temperature of 1200 to 1500°C and a pressure of 50 to 70 bar. The operating conditions for natural gas processing are ~1400°C at ~50 bar. Vacuum Residue, vacuum flashed cracked residue and asphalts are processed at ~1320°C and 35 to 65 bar. The synthesis gas is primarily a mixture of hydrogen and carbon monoxide along with lower concentrations of CO_2, H_2S and COS. A small amount of free carbon and ash are also present[180]. Between 0.5% and 1.0% of the carbon in the feed is converted to soot.

The raw synthesis gas leaves the gasifier and is cooled in the Syngas Cooler by generating high-pressure, saturated steam (~100 bar). The raw gas is cooled to about 340°C. Part of the steam is used to preheat the feedstock and the oxygen. The remainder is superheated for use in the CO Shift step and in the steam turbine drives. The raw gas is cooled below 340°C after it leaves the Syngas Cooler when it is used to preheat the Boiler Feed Water that goes into the Syngas Cooler[180].

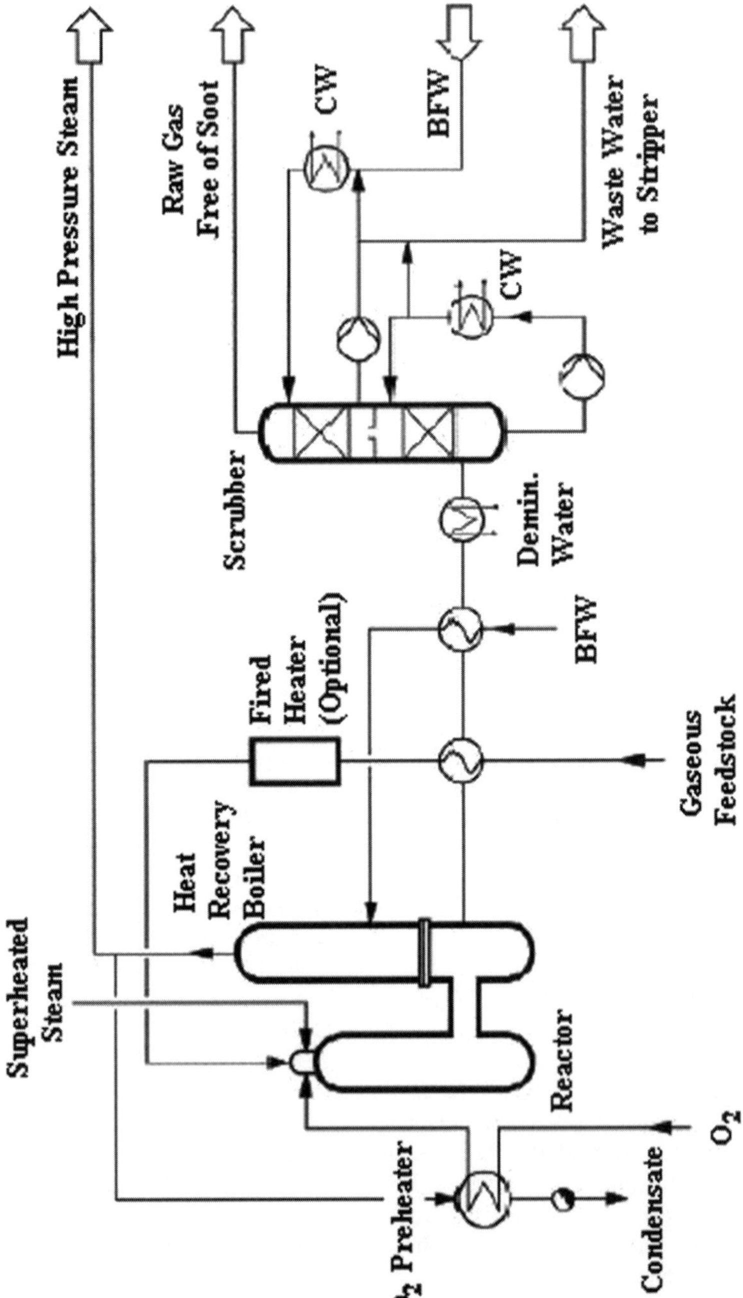

Figure 5.20. Lurgi Multi-Purpose Gasification (MPG) process.
(Reproduced by permission of Lurgi)

The carbon particles in the gas stream are removed together with the ash in a two-stage water wash that is downstream of the Syngas Cooler (see Figure 5.22). This step includes a quench pipe and a soot separator followed by a packed column. In the quench pipe about 95% of the soot is removed by a direct water spray. The remaining soot is removed by the counter-current flow in the packed beds. A carbon slurry is formed together with the ash and the process condensate. This slurry is processed in the Ash Removal Unit. The syngas that leaves the scrubber has a soot content of less than 1 mg per Nm^3, and its temperature is about 40°C[176, 180].

The Soot Ash Removal Unit (SARU) is shown in Figure 5.23. In this step the carbon slurry is flashed to atmospheric pressure in the slurry tank. The slurry is then filtered to produce 1) a filter cake (with about 80% residual moisture) and 2) a clear water filtrate. The filter cake then goes through a controlled oxidation process in a multiple hearth furnace. The furnace allows the combustion of the carbon to take place under conditions where vanadium oxides neither melt nor corrode the furnace. The product is a vanadium concentrate that contains about 75% V_2O_5 and can be sold to metals reclaiming companies. The process is autothermal because the heat of combustion of the carbon is sufficient to evaporate the moisture in the filter cake[180].

Most of the filtrate is recycled as wash water. However the overall system generates a net water surplus, which must be treated before final discharge. First the surplus water passes through a Sour Water Stripper to remove dissolved gases that are then incinerated. The water is then treated in a biotreatment facility[180].

The SARU has several advantages over the Naphtha Soot Recycling Unit (NSRU) that has been used in some Shell Gasification plants. It is fundamentally simple; it has a lower capital cost; its operating costs are lower because of lower energy costs; and feedstock flexibility is increased. The feedstock flexibility increases because the metal-containing soot and ash are no longer recycled to the gasifier. Therefore feedstocks with higher ash contents and higher metal contents can be used[180].

The soot-free gas that leaves the carbon removal scrubber then goes through the following steps: Desulphurization (the Rectisol process is often used), CO Shift, CO_2 Removal and Methanation (see Figure 5.21 and Figure 5.22). These purification steps will be discussed in more detail in later sections.

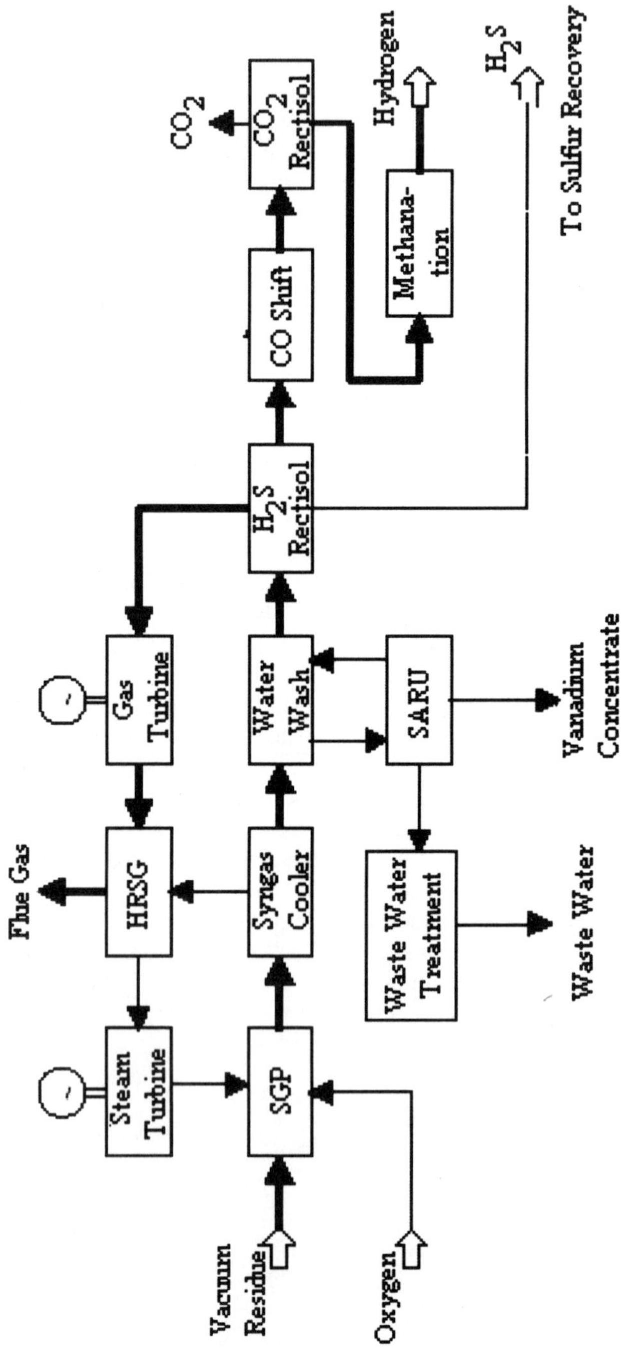

Figure 5.21. Shell Gasification Plant (SGP) - Hydrogen Production from Vacuum Residue. (Reproduced by permission of Chris Higman)

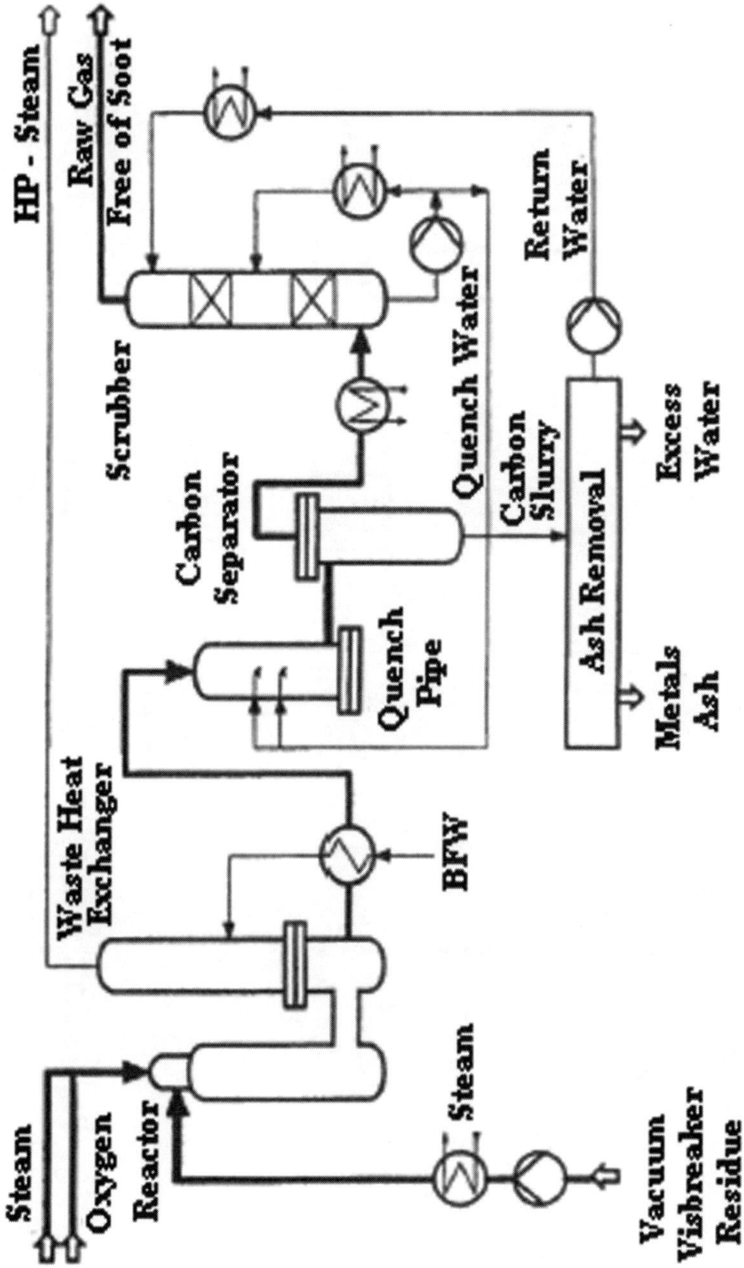

Figure 5.22. Shell Gasification Plant.
(Reproduced by permission of Chris Higman)

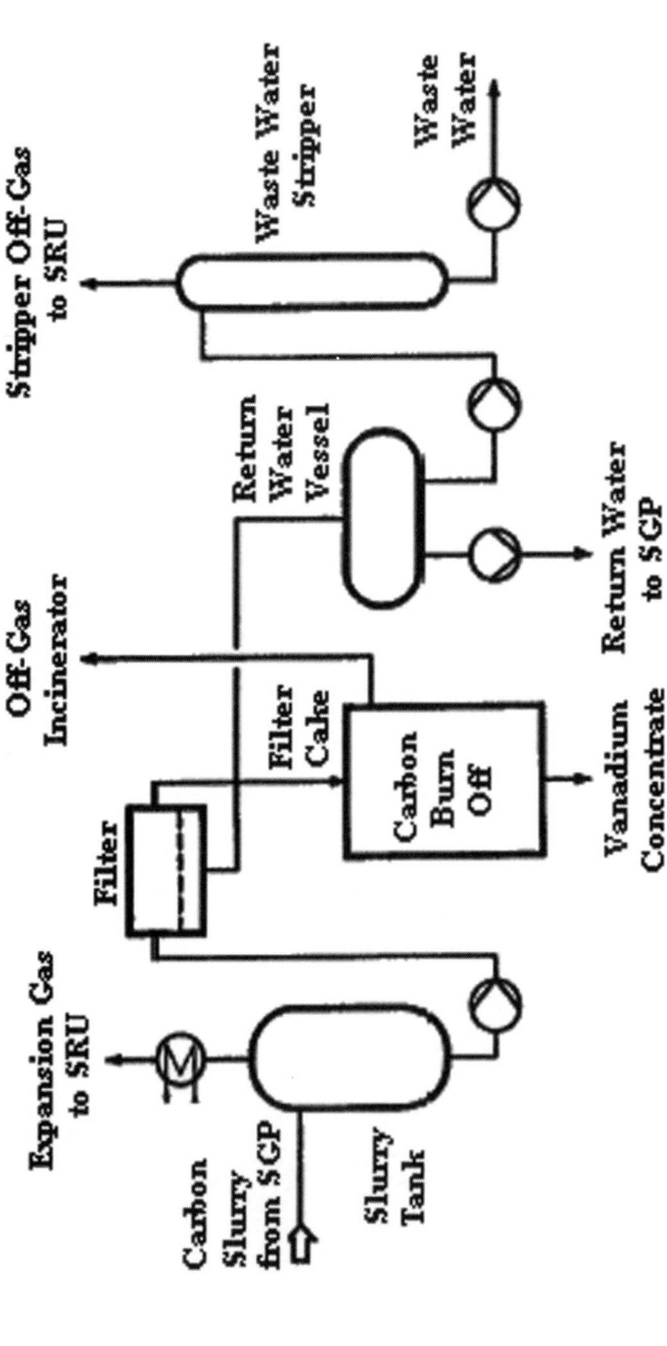

Figure 5.23. Shell Gasification Plant - Soot Ash Removal Unit (SARU). (Reproduced by permission of Chris Higman)

5.2.2.5. Materials of Construction in Partial Oxidation

Operating conditions in partial oxidation reactors and steam reformers vary considerably - depending on the feedstock that is being used. An operating temperature of 1400°C (i.e., the temperature of the inner surface of the lining) is not critical for many refractory materials. However factors other than temperature determine the refractory selection and the refractory design.

The partial oxidation process CO and hydrogen - strongly reducing gasses that are able to reduce free silica if it is present in the hot face bricks. This results in a more porous structure of the brick and an increased surface for the reaction with slags. In addition, the volatile silica leaves the reactor with the gas stream and can be deposited inside the boiler tubes where it increases the potential to plug the tubes. Therefore, a low percentage of silica in bricks is required (i.e., a maximum of 0.25%). The presence of vanadium pentoxide (V_2O_5) in the slags also causes severe chemical attack on the refractory[181].

The refractory lining in partial oxidation reactors is usually composed of dense brick at the hot face and various refractory insulating materials (insulating bricks, castables and other products) between the inner layer of dense bricks and the steel shell. The design of the refractory lining should provide close dimensional tolerances between the bricks. Open joints in the brickwork may cause by-passing and/or slag penetration. Because the reactors are heated up and cooled down fairly frequently during commissioning of the plant, the refractory materials should have sufficient volume stability and thermal shock resistance to accommodate these extreme temperature fluctuations[181].

The production of high-pressure steam by cooling syngas creates the potential for metal dusting. The term "metal dusting" refers to the situation in which a metal surface in contact with carbon monoxide turns to powder. This takes place under certain process conditions in which carbon diffuses into a metal matrix and forms carbides. These carbide formations result in a buildup of stress that causes metal dusting[177].

The potential for carbon formation is based on the gas composition and metal temperature. Industry experience shows that if the gas temperature is less than ~1100°F, the kinetics of the reaction are too low and carbon does not form. The equilibrium Kp for typical gasifier reactor effluent compositions is about 1800°F. If the calculated Kp, which is based on gas composition, is greater than the equilibrium Kp, carbon cannot form. Therefore, gasifier reactor effluent metal dusting potential typically occurs between the temperatures of 1100°F and 1800°F.

For metal dusting to occur in the gasifier reactor effluent, two essential elements are necessary: 1) a metal surface and 2) a metal surface temperature between 1100°F and 1800°F (595 to 895°C). The temperature of syngas leaving the gasification reactor passes through this temperature range. With the piping and equipment providing a metal surface, the remaining element necessary for metal dusting is the metal surface temperature[177].

Fortunately the Syngas Cooler cools the syngas by generating steam. The metal temperature of the steam generating tubes is practically the same as the steam/water temperature, which is substantially less than 1100°F. The cooler metal temperature is maintained by an internal water circulation system[177].

Even though the exact mechanism of metal dusting is not completely understood, the necessary elements are. Using this information, preventative measures can be taken to avoid operating in regions prone to metal dusting. The water circulation design is one way to accomplish this[177].

5.2.2.6. Suppliers of Partial Oxidation Technology

Some of the companies that license partial oxidation technology are listed in Table 5.23.

Table 5.23. World Suppliers of Technology to Produce Hydrogen via Partial Oxidation - 2002

Company	Headquarters	Remarks
British Gas / Lurgi	Berkshire, England	Fixed Bed Gasifier
Carbona	Tampere, Finland	U-Gas Process/ Tampella
Global Energy	Houston, TX, USA	E-Gas (Destec)
Foster Wheeler Corporation	Clinton, NJ, USA	Air Partial Oxidation
Haldor Topsøe A/S	Copenhagen, Denmark	Catalytic Partial Oxidation
KTI Group BV (a Mannesmann Company)	Zoetermeer, Netherlands	Air Partial Oxidation
Linde AG		
- Process Engineering and Contracting	Munich, Germany	Partial Oxidation
- Lotepro Corp:	Valhalla, NY	
- Selas Fluid Processing:	Blue Bell, PA	
- Pro-Quip Corp:	Holly Springs, GA	
Lurgi	Frankfurt, Germany	Partial Oxidation
MW Kellogg Company	Houston, TX, USA	Air Partial Oxidation
Noell KRC	Würzburg, Germany	Entrained Flow Gasifier
Krupp Uhde	Dortmund, Germany	Prenflow Gasifier
Shell Oil Company	Houston, TX, USA	Partial Oxidation
Texaco Inc.	White Plains, NY, USA	Partial Oxidation

5.2.3. Coal and Coke Gasification

Coal gasification to produce syngas can be regarded physically as a gas/solid reaction. From a chemical point of view, it can be considered as the partial oxidation of carbon or as the reduction of water with carbon. However the total process is much more complex and only describable using numerous parallel and secondary reactions. The exothermic partial combustion of carbon and the endothermic water gas formation represent the actual gasification reactions[46]:

$C + O_2 \leftrightarrow 2\,CO \quad \Delta H = -60$ kcal/mol or 246 kJ/mol (5.11)
Partial Combustion

$C + H_2O \leftrightarrow H_2 + CO \quad \Delta H = +28$ kcal/mol or 119 kJ/mol (5.8)
Heterogeneous Water Gas Reaction

Some other important reactions that take place are:

$C + O_2 \leftrightarrow 2\,CO \quad \Delta H = +41.4$ kcal/mol or 173 kJ/mol (5.4)
Boudouard Reaction

$CO + H_2O \leftrightarrow H_2 + CO_2 \quad \Delta H = -10$ kcal/mol or 42 kJ/mol (5.2)
Homogenous Water Gas Reaction (Water Gas Shift)

$C + 2\,H_2 \leftrightarrow CH_4 \quad \Delta H = -21$ kcal/mol or 87 kJ/mol (5.12)
Hydrogenative Gasification

$CO + 3\,H_2 \leftrightarrow H_2O + CH_4 \quad \Delta H = -49$ kcal/mol or206 kJ/mol (5.13)
Methanation

The different gasification processes can be characterized 1) by the type of coal used and 2) by the coal's physical and chemical properties. On the other hand, the processes differ in the technology involved. For example the heat supply can be allothermal (i.e., external heating) or autothermal (i.e., self-heating) and the reactor can be fixed-bed, fluidized-bed or entrained-bed. Furthermore, the actual gasification reaction and the gas composition are determined by[46]:

- The Type of Feedstock,
- The Gasification Agent (H_2O, O_2 or Air, CO_2, H_2O),
- The Process Conditions (pressure, temperature, coal conversion) and
- The Reaction System (parallel or counter flow).

Table 5.24 compares different gasification systems and lists typical syngas compositions upstream of the purification part of the process[182]. Table 5.25[183] and Figure 5.24[184] provide additional ways to compare gasification systems. The Shell gasification process uses a dry coal feed while other processes feed the coal as a wet slurry[76].

Table 5.24. Summary of typical gasification systems.
(Reproduced by permission of AP Energy Business Publications)

	Fixed bed		Fluidised Bed		Entrained flow
Item	Lurgi dry bottom gasifier	BGL gasifier	Lurgi CFB gasifier	HTW gasifier	Shell gasifier (SCGP)
Preferred feedstocks	lignite, reactive bituminous coals, wastes	bituminous coals, petcoke, wastes	lignite, bituminous coals, cokes, biomass, wastes	lignite, reactive bituminous coals, wastes	lignite, reactive bituminous coals, petcokes
Ash content	no limitation	<25% referred	no limitation	no limitation	<25% referred
Preferred ash melting temperature	> 1200°C	< 1300°C	> 1000°C	> 1100°C	< 1300°C
Caking / swelling	non-caking to highly caking acceptable	non-caking to highly caking acceptable	non-caking to highly caking acceptable	non-caking	any
Ash removed as	ash	slag	ash	ash	slag
Fuel size requirements	6-50 mm	6-50 mm	<6 mm	<6 mm	< 0.1 mm
Gasification agent(s)	air + steam; oxygen + steam	oxygen + steam	air + steam; oxygen + steam; CO_2 + oxygen	air + steam; oxygen + steam	oxygen + steam
Gasification pressure (Mpa)	3 (+)	3 (+)	0.15 (+)	up to 3	up to 5
Feedstock	Lignite	Bituminous	High Ash Bituminous	Lignite	Bituminous
Typical gas composition (oxygen blown, main components) -% b.v.					
CO_2	33.0	3.2	16.9	20.0	0.8
CO	14.6	53.5	41.6	36.3	62.1
H_2	40.0	27.4	37.9	33.3	31.2
Hydrocarbons	11.7	7.6	2.9	6.5	Traces
H_2S / COS	0.4	1.0	0.2	0.2	0.3
N_2	0.3	7.3	0.5	3.7	5.6
Plants in operation/ under construction (products)	PRC, NH_3, town gas USA, Sng RSA, Liquid fuels	Great Britain Fuel gas	Germany, Austria Fuel gas	Germany Syngas	Netherlands IGCC (Fuel gas)

Table 5.25. Comparison of gasifier technologies[183]

Type	Moving Bed	Fixed Bed	Fluidized Bed	Entrained Flow
Gasification Temperature	700-900°C	800-1000°C	800-1000°C	1500-1900°C
Gasifier Pressure	~25 bar	10-100 bar	10-25 bar	25-80 bar
Preferred Coal Size	40-5 mm	30-3 mm	3-0.5 mm	<0.1
Coal types	(Lignite) Sub-bituminous, Bituminous		Lignite, Sub-bituminous	All types
Other Solid Fuels	Briquetted Fuels are Possible	Briquetted Fuels are Possible	Small Particulate Fuels	Any Slurried Fuel
Air-Blown Processes	Siemens Mond	Lurgi (dry)	High Temperature Winkler	VEW Energie AG
Oxygen-Blown Processes	British Gas / Lurgi	Lurgi (dry), Westfield	KRW, U-Gas, High Temperature Winkler, GSP	Texaco*, Dow, NEDO, Prenflow, Shell*, GSP*, Koppers-Totzek
Liquid Fuels	None Possible	None Possible	None Possible	Only possible for liquid-entrained processes above marked with *, but Shell cannot take coal slurry.
Advantages	Low O_2 Consumption, High Thermal Efficiency, Good Load Flexibility	Low O_2 Consumption, Low Auxiliary Power Consumption, Good Load Flexibility	Low O_2 Consumption, Good Load Flexibility, HTW-dry ash	High Carbon Conversion, High Coal throughput, No Tars, Oils or Phenolic Liquors
Disadvantages	Volatile Hydrocarbons In Exit Gas, Requires Mechanical Moving Parts, Large in Size	Low Carbon Conversion, High CO_2 Production	Low Carbon Conversion	High O_2 Consumption, Low Thermal Efficiency, Critical Control Characteristics

(Reproduced by permission of Institution of Diesel and Gas Turbine Engineers)
(The Institution of Diesel and Gas Turbine Engineers exists to serve its members in promoting advancement of reciprocating engines, gas turbines and associated equipment)

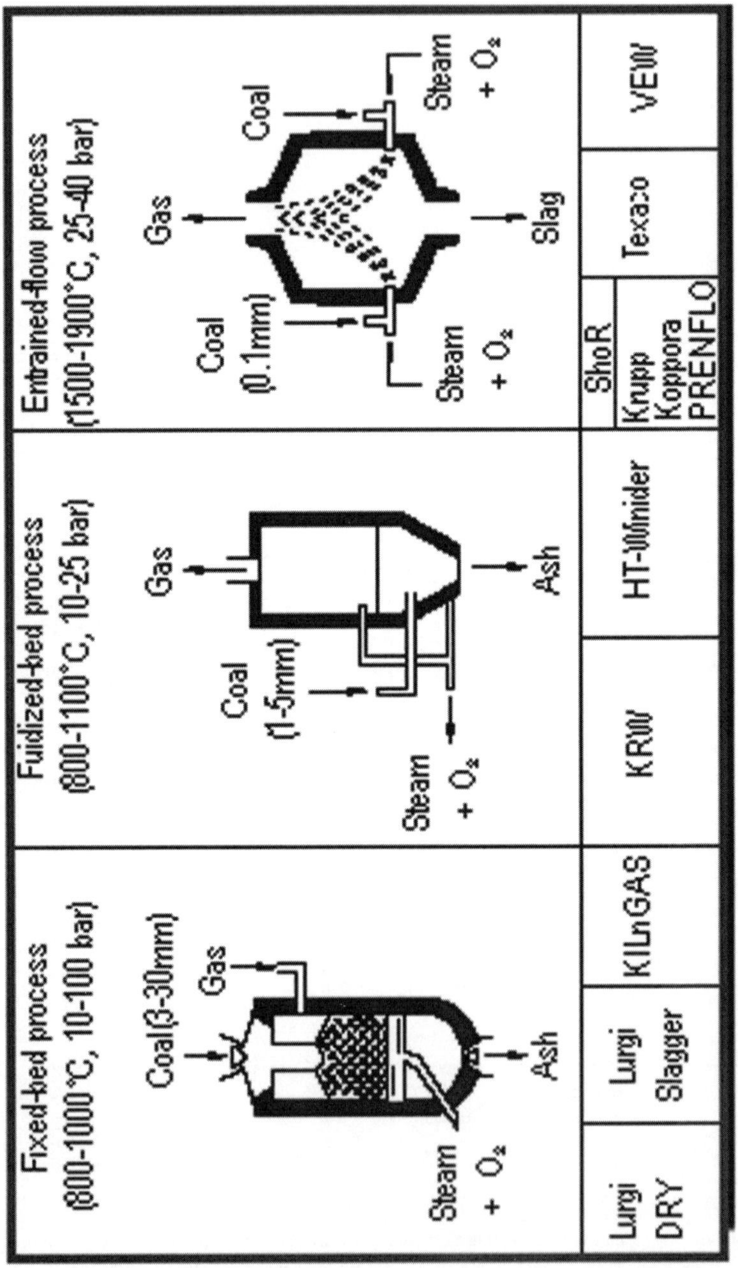

Figure 5.24. Comparison of Coal Gasification Reactors.

Between 1996 and 2002 Texaco captured about 75% of the technology market for new gasification installations. The reasons for their success were[164]:

- They have a simple, direct water quench design,
- Their aggressive marketing has included equity participation in 6 projects,
- They have extensive successful operating experience on various feedstocks,
- They have been very aggressive in reducing capital costs by working with General Electric and Praxair on a standardized gasification technology that is very integrated,
- By working with General Electric and Jacobs Consultancy, they have developed an innovative H_2-Coal Gasification Combined Cycle design that includes CO_2 capture.

The challengers to Texaco's technology are mostly entrained slagging processes. These processes are simple, flexible and clean. Some of the companies that offer this technology are[164]:

- Shell, who is now taking equity positions in projects such as the Sinopec coal-to-ammonia project in China,
- E-Gas (Destec) whose technology uses petroleum coke as a feedstock and produces 'over-the-fence' syngas, and
- Lurgi MPG and Noell KRC technologies can process waste fuels.

The Sinopec plant in Dongting, China will process 2,000 tonnes per day of coal and produce 3.4 million Nm^3 per day of syngas. The investment is estimated to be $US 150 million with startup scheduled for the second half of 2004. The syngas will be used to produce ammonia and urea[187].

Sasol (**S**uid-**A**frikaans **S**intetiese **Ol**ie) in South Africa has one of the largest coal gasification operations in the world. The three plants (Sasol One, Sasol Two, and Sasol Three) use Lurgi fixed-bed gasification technology[1]. The Sasol One plant produces only chemicals while the other plants produce both liquid fuels and chemical feedstocks from coal. Sasol One recovers ammonia from the raw gas that leaves the gasifiers, and ammonia is also produced from the gas that leaves the Low-Temperature Slurry Phase Distillate reactor (see Figure 5.25). In 1994 a new 240,000 ton per year ammonia plant was commissioned to bring their total capacity to over 500,000 tons per year[54].

One of the first commercial-size, coal gasification plants in the United States was the Great Plains Synfuels plant that is operated by Dakota Gasification in Beulah, ND. This plant consumes over 5.5 million tonnes of coal per year and produces over 54 billion standard cubic feet (1.4 billion Nm^3)of natural gas, 365,000 tonnes per year of ammonia, 24 million gallons of

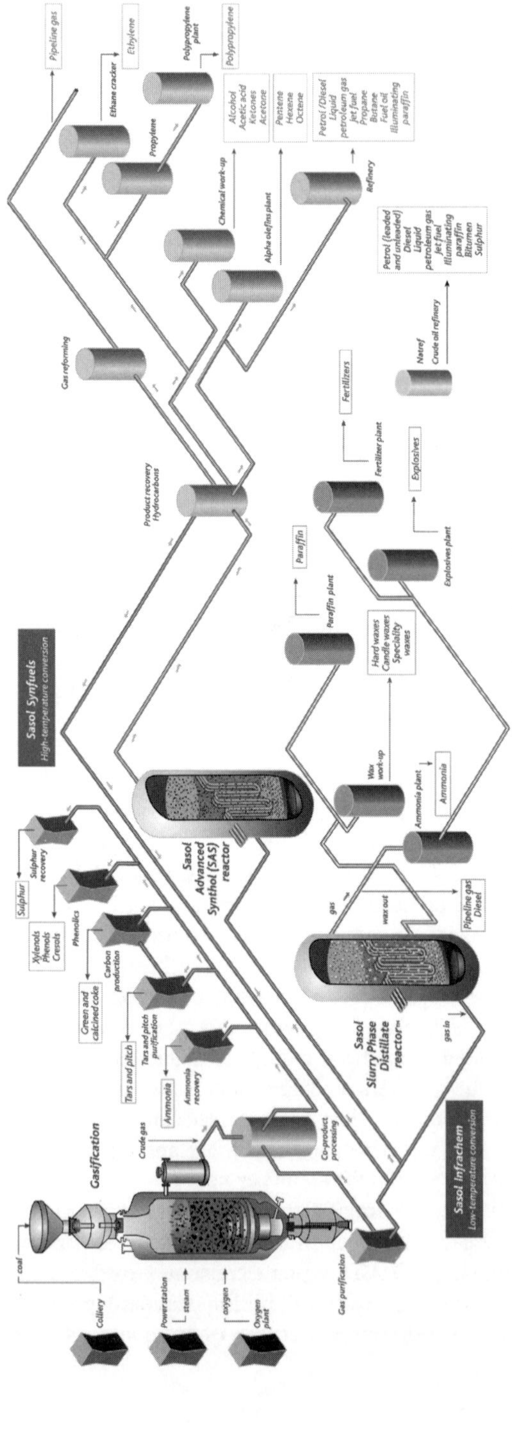

Figure 5.25. Sasol process diagram. (Reproduced by permission of Sasol)

liquid nitrogen per year (68 million kg per year or 150 million lb per year) plus several other chemicals[89]. (See www.dakota.com for more details)

In July, 2000 Farmland Industries started up a petroleum coke gasification unit (see Figure 5.26 and Figure 5.27) adjacent to their existing refinery in Coffeyville, Kansas. The new gasification unit will produce a synthesis gas for the production of ammonia. Approximately 1,000 tonnes per day of petroleum coke from the Coffeyville refinery will be gasified to produce 1,000 tonnes per day of ammonia. A portion of the ammonia (~550 tonnes per day) will subsequently be upgraded to Urea-Ammonium-Nitrate (UAN) solution in a facility with a capacity of 1,360 tonnes per day [185, 186].

The Texaco Hydrogen Generation Process technology is being use to produce hydrogen from coke. The petroleum coke comes from Farmland's refinery. After the coke reaches the gasification unit, it is crushed and slurried before it is fed to the gasifier. The gasifiers (one operating and one spare) were relocated from Texaco's Cool Water Demonstration facility in Daggett, California[185].

As shown in Figure 5.27, coke slurry and oxygen are fed to the gasifier. The raw syngas that leaves the gasifier is scrubbed to remove carbon fines before it enters the CO shift conversion step. The quench section of the gasifier provides sufficient water for high conversion in the two-stage shift reactor. The hot gas from the shift reactor is cooled in several steps. First it is used to generate high-pressure steam and medium-pressure steam. Then it is used to preheat boiler feed water and finally it passes through air and water cooling systems. The cooled syngas is fed to the SELEXOL® unit for acid gas removal.

The SELEXOL® unit removes H_2S and most of the CO_2 in the syngas stream. The final hydrogen purification is accomplished in a downstream PSA unit. The purified hydrogen stream is combined with nitrogen from the air separation unit (ASU) and converted to ammonia. A unique CO_2 purification scheme was developed to remove traces of H_2S and COS from the CO_2 that is fed to the UAN plant[185].

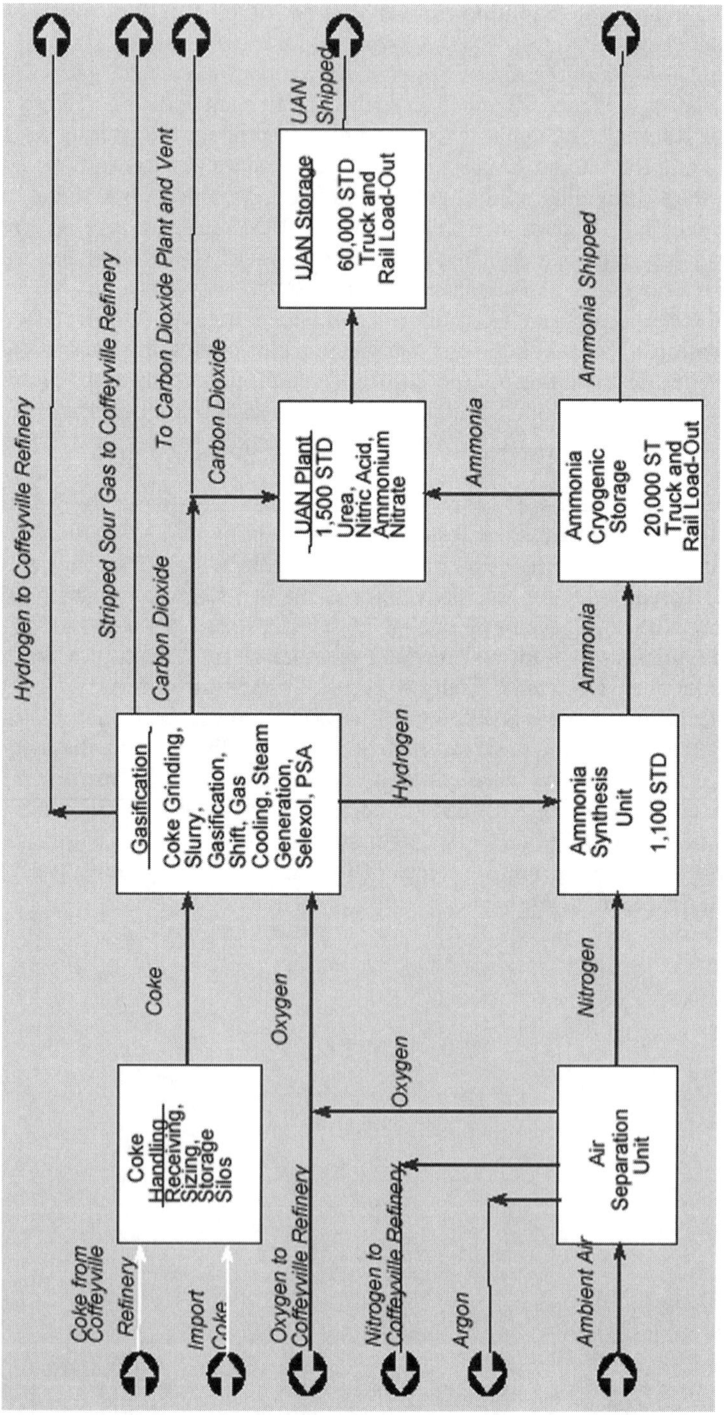

Figure 5.26. Overview of Farmland's petroleum coke-to-ammonia plant in Coffeyville, Kansas. (Reproduced by permission of UOP)

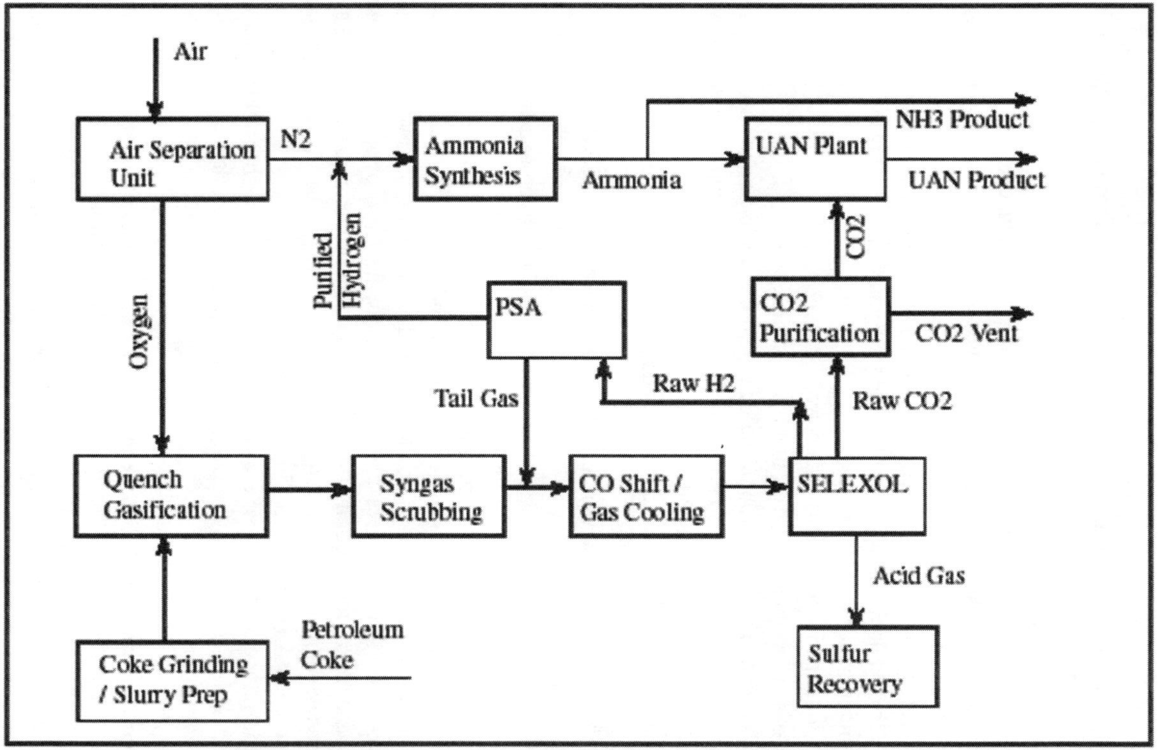

Figure 5.27. Process details of Farmland's petroleum coke -to-ammonia plant in Coffeyville, Kansas. (Reproduced by permission of UOP and Lawrence Reid Gas Conditioning Conference)

5.2.4. By-Product Hydrogen

Large volumes of by-product hydrogen are generated from a variety of production processes. Some of these processes are[40]:

- Petroleum gases from platforming or catalytic reforming operations in refineries,
- Sodium chlorate production,
- Styrene production by dehydrogenation of ethylbenzene,
- Purge-gas streams in ammonia, cyclohexane and methanol production,
- Ethylene production,
- Acetone production by dehydrogenation of isopropyl alcohol,
- Formaldehyde production by dehydrogenation of methanol over a silver catalyst,
- Liquid petroleum gas (LPG) dehydrogenation,
- As a co-product of carbon black production,
- Pyridine production from acetaldehyde, formaldehyde and ammonia,
- 1,3 butadiene production by dehydrogenation of n-butanes or n-butenes
- Methanol production and
- Para-methylstyrene production by dehydrogenation of p-ethyltoluene.

Some of the processes that can be used to obtain purified by-product hydrogen are summarized below:

The HyTex (**Hy**drogen **Tex**aco) process makes pure hydrogen from waste gases in oil refineries in a three-step process. In the first step, partial combustion with oxygen in a noncatalytic reactor yields a mixture of carbon monoxide and hydrogen (i.e., syngas). In the second step, the shift reaction with steam converts the carbon monoxide to carbon dioxide and more hydrogen [see Eq. (5.2)]. The third stage uses PSA (Pressure Swing Absorption) to separate the hydrogen from the carbon dioxide and various other impurities[1]. The process was developed by Texaco and announced in 1991. The first commercial unit started up in Anacortes, Washington, USA in 1993.

Platforming operations make a gas that contains as much as 90 to 95% hydrogen. This gas is usually purified by low temperature fractionation or by washing with liquid nitrogen.

Kvaerner Oil & Gas - Norway has introduced a new process for manufacturing carbon black that generates substantial quantities of by-product hydrogen. This Plasma-Arc process operates at temperatures of ~1600°C and separates hydrocarbons into pure carbon and hydrogen. During 1998 the process was installed in Montreal, Québec (Canada) to produce 20,000 tonnes of carbon black and 50 million Nm^3 per year (5.2 million SCF per day) of hydrogen. The process appears to be economic only when the carbon black can also be sold as an end product[40, 161]. The 2002 cost estimate for a plant that is designed to produce 120 million Nm3 per year is $US 150 million[161].

Air Products and Chemicals, Inc. has been selected to supply a hydrocarbon and nitrogen recovery system for a new polyethylene manufacturing plant in Baytown, TX. The plant will be owned by Chevron Phillips Chemical Company and Solvay Polymers, Inc. The recovery system uses partial condensation in conjunction with Air Products' pressure swing adsorption technology to recover hydrocarbons in the polyolefin plants, and recycle nitrogen with a purity of greater than 99%[55].

The Air Products' technology will be able to recover almost 100% of the hydrocarbon off-gas from the slurry phase polyethylene process and more than 97% of the high-purity nitrogen that is used in degassing. Both of these gas streams will be recycled and re-used to reduce the VOC and NO_x environmental emissions that are typically associated with flaring or burning plant off-gasses[55].

A few ammonia plants have been located where a hydrogen off-gas stream is available from a nearby methanol or ethylene operation (e.g., Canadian plants at Kitimat, BC and Joffre, Alberta). Gas consumption at such operations range from 25 million to 27 million BTU per tonne of ammonia, depending on specific circumstances. Perhaps more important, the capital cost of such a plant is only about 50% of the cost of a conventional plant of similar capacity because only the synthesis portion of the ammonia plant is required. However, by-product carbon dioxide is not produced and downstream urea production is therefore not possible[56].

5.2.5. Electrolysis

In water electrolysis electrical energy is applied to the water, and the result is the production of hydrogen and oxygen:

$$2\,H_2O + \text{Energy} \rightarrow 2\,H_2 + O_2 \qquad (5.14)$$

This is the opposite reaction to hydrogen oxidation that occurs in a fuel cell. To achieve the desired hydrogen capacity, several electrolysers can be connected[173].

Electrolytic hydrogen production has many advantages. It yields the highest-purity hydrogen (up to 99.999%), benefits from wide-spread raw materials availability (electricity and water), boasts simple system architecture and can be scaled economically to serve applications ranging from the smallest hydrogen uses to the larger-volume uses. The main disadvantage of water electrolysis is that electricity is an expensive "fuel". As a result the technology generally is practical only for systems with hydrogen requirements of approximately 2,000 SCF per hour or less[48].

Some of the common electrolysers are Alkaline Electrolysers, Polymer Electrolyte Membrane (PEM) Electrolysers (also known as Proton Exchange Membrane electrolysers), and Steam Electrolysers. In alkaline electrolysers a liquid electrolyte, such as a 25% potassium hydroxide solution, is used. At

Norsk Hydro in Norway, this technology was used from 1928 until 1988 to produce hydrogen that was used in an ammonia plant[173].

In 2002 Norsk Hydro Electrolysers (NHE) was a leading producer of alkaline electrolysers. Some of NHE's electrolysers have an efficiency of over 80% (based on high heating value). Efficiency is an important factor in electrolysis because the use of energy (~4.5 kWh/Nm^3 of H_2) makes up a significant portion of the operating costs. NHE estimates that electrical costs make up about 65% of the operating costs. Electrolysers are most effective when they operate at a low production rate with a low current density. Optimum economy of operation will depend on current density, cost of production materials and the demands for hydrogen production[173]. These electrolysers are best suited for larger systems that are connected to a power grid[173].

Another type of electrolyser uses polymer membranes to both support the electrolysis reaction and to separate the gases. Efficiency factors for PEM electrolysers are predicted to reach 94%, but this is only theoretical in 2002. These electrolysers are best suited for small plants that have a variable output of hydrogen[173].

Production of hydrogen in a PEM electrolyser involves four steps: Water Dissociation, Proton Exchange, Electron Conduction and Hydrogen Evolution (see Figure 5.28). Step One is Water Dissociation and Oxygen Evolution. The water electrolysis reaction occurs at the anode where water, in the presence of an electric field and a catalyst, is split into protons (4 H^+), electrons (4 e^-) and gaseous oxygen (O_2). In Step Two, the Proton Exchange takes place. The 4 H^+ protons are drawn into the Proton Exchange Membrane (PEM) and driven directly through the solid proton exchange membrane by the potential difference that is applied to the cell. In Step Three, electron conduction occurs. The electrons (4 e^-) are conducted through the external power supply. In Step Four, Hydrogen Evolution occurs. At the cathode the protons (4 H^+) combine with the electrons (4 e^-) to make gaseous hydrogen (2 H_2)[188].

The proton exchange membrane conducts H^+ ions (protons) directly through its structure. The membrane is a highly stable, completely inert, solid polymer. In some electrolyzers it is comprised of a Teflon® backbone with sulfonic acid groups attached. The protons pass from one sulfonic acid group to the next under the influence of an electromotive field. When the protons reach the cathode, they evolve as hydrogen gas. As hydrogen evolves, pressure is achieved by simply allowing hydrogen to accumulate within a confined volume until it reaches the desired pressure. Most PEM materials have no known bubble pressure so they are able to withstand thousands of pounds of pressure when they are properly supported[188].

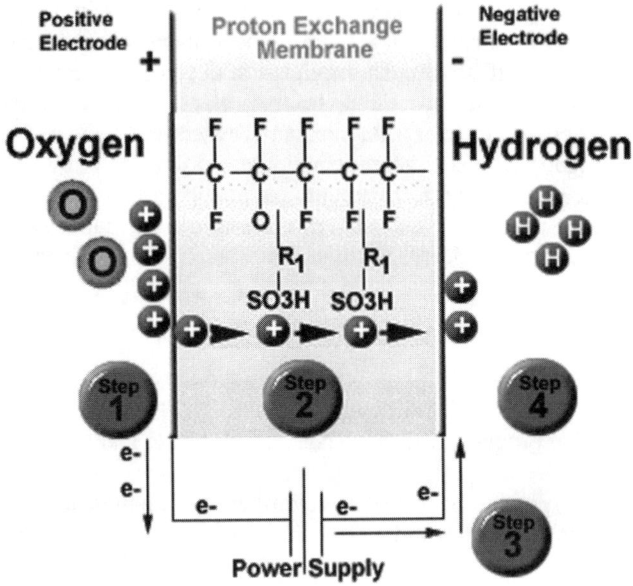

Figure 5.28. Production of hydrogen in a Proton Exchange Membrane (PEM) water electrolyzer [188]. (Reproduced by permission of Proton Energy Systems)

A third type of electrolyser is the Steam electrolyser. These electrolysers use a ceramic ion-conducting electrolyte. Steam electrolysers can reach a very high efficiency factor, but in 2002 they are not commercially feasible. A tubular steam electrolyser is under development at Lawrence Livermore National Laboratory. Another type of steam electrolyser is the German "Hot Elly". This system can reach an efficiency of 92%[173].

In 2002 several manufacturers can provide advanced water electrolysis systems that are standardized, compact in size, need minimal operator intervention and require little maintenance. The basic electrolysis reaction has not changed. However new cell designs, materials of construction, standardized designs and manufacturing techniques have enabled manufacturers to decrease dramatically the fixed costs per unit of capacity for electrolysis technology. In addition, these new systems operate automatically and require very little maintenance, which reduces personnel costs[48].

An example of the automatic systems is Honda's solar-powered hydrogen production and fueling station in Torrance, CA. The station's operation starts with solar cells that use the sun's energy to produce electricity. The electricity then is used to extract hydrogen from water. A compressor pressurizes the extracted hydrogen, and it is stored in tanks at the station. The station also has back-up power to increase the hydrogen production capacity[58].

Developing regions (such as China and other countries in the Far East) are the largest market for electrolysis systems. Electrolysis currently accounts for a very small portion of the hydrogen generated in developed countries that have a commercial hydrogen infrastructure. However electrolysis can be economic for small-scale generation in areas with inexpensive electricity, and a few plants still produce hydrogen for small-volume ammonia production[57]. In 1997, Messer-MG Industries announced two hydrogen gas plants based on water electrolysis in the United States. In Canada, a portion of the capacity installed at HydrogenAl's merchant hydrogen plant in Becancour, Québec is based on electrolytic cells[40].

5.2.6. Other Processes

Several other processes that have been developed to make hydrogen are: [1]:

- The Hypro process that makes hydrogen by catalytically decomposing hydrocarbons to carbon and hydrogen. The carbon is burned to provide the heat for the reaction. This process was developed by UOP.
- The Lane process makes hydrogen by passing steam over sponge iron at approximately 650°C. The iron is converted to magnetite.
- The MRH (**M**ethanol **R**eformer **H**ydrogen) process for generating hydrogen from methanol and separating it by PSA. This process was developed by the Marutani CPE Company.
- The Proximol process makes hydrogen by reforming methanol. This technology is offered by Lurgi.
- The SBA-HT (**S**ociété **B**elge de l' **A**zote-**H**aldor **T**opsøe) process is a combination of both steam reforming and partial oxidation. The process converts liquid petroleum gas (LPG) to syngas that is rich in hydrogen. Two cracking processes are conducted in two zones of one reactor. In the first zone, the LPG is autothermally cracked with steam and oxygen. In the second, the products from the first zone are catalytically cracked. This process was operated in France and Belgium in the 1960's.

5.2.7. New Developments

Several research organizations are developing low-cost methods of hydrogen production. Much of this work is sponsored by the U.S. Department of Energy as a result of the Matsunaga Hydrogen Research, Development and Demonstration Act of 1990 and the Hydrogen Futures Act of 1996.

Some of the research programs are listed in the following tables:

Table 5.26. Fossil-Fuel Based Hydrogen Production Research Programs - U.S. Department of Energy - 2000[59]

Research Program	Lead Organization or Company
ITM Syngas Process	Air Products
Sorption-Enhanced Reaction Process for Production of Hydrogen	Air Products
Liquid Fuel-Reformer Development	Argonne National Laboratory
Integrated Ceramic Membrane System for Hydrogen Production	Praxair
Novel Catalytic Fuel Reforming	InnovaTek
Production of Hydrogen by Superadiabatic Decomposition of H_2S	Institute of Gas Technology
Thermo-Catalytic CO_2-Free Production of Hydrogen Using Hydrocarbon Fuels	Florida Solar Energy Center
Thermal Dissociation of Methane using a Solar Coupled Aerosol Flow Reactor	University of Colorado/NREL
Hydrogen Membrane Separation	Sandia National Laboratories
Separation Membrane Development	Savannah River Tech. Center

Table 5.27. Biomass-Based Hydrogen Production Research Programs - U.S. Department of Energy - 2000[59]

Research Program	Lead Organization or Company
Biomass-to-Hydrogen via Fast Pyrolysis & Catalytic Steam Reforming	National Renewable Energy Laboratory
Biomass Pyrolysis for Hydrogen Production	Jet Propulsion Laboratory
Integrated Hydrogen Production for Agricultural Residues for Urban Transportation	Clark Atlanta University
Supercritical Water: Partial Oxidation	General Atomics
Biohydrogen Production from Renewable Organic Wastes	Iowa State University

Table 5.28. Photo-Electro-Chemical / Biological Hydrogen Production Research Programs - U.S. DOE - 2000[59]

Research Program	Lead Organization or Company
Photoelectrochemical Hydrogen Production	University of Hawaii
Photoelectrochemical-Based Direct Conversion Systems for Hydrogen Production	National Renewable Energy Laboratory
Solar Photocatalytic Hydrogen Production from Water Using a Dual-Bed Photo System	Florida Solar Energy Center
Development of an Efficient Algal Hydrogen Producing System	National Renewable Energy Laboratory
Two-Phase Photobiological Algal Hydrogen Production System	National Renewable Energy Laboratory
Hydrogen Production by Photosynthetic Water Splitting	Oak Ridge National Laboratory
Maximizing Photosynthetic Efficiencies and Hydrogen Production by Microalgal Cultures	University of California - Berkeley
Production of Hydrogen from Glucose	Oak Ridge National Laboratory
Biological Hydrogen from Fuel Gases and Water	National Renewable Energy Laboratory
Bioreactor Development for Biological Hydrogen Production	National Renewable Energy Laboratory

The ITM Syngas Process that is listed in Table 5.26 involves the direct conversion of methane to synthesis gas (see Figure 5.29). The process utilizes a mixed, conducting ceramic membrane and partial oxidation to produce the synthesis gas[40]. The goal of this advanced reformer technology is to reduce the cost of hydrogen production by over 25%[59]. Ion Transport Membranes (ITM's) are ceramic membranes that are non-porous, multi-component metallic oxides. They operate at high temperatures and have exceptionally high oxygen flux and selectivity.

A key part of this process is a novel ceramic membrane that has been developed by Argonne National Laboratory. Researchers at Argonne National Laboratory developed a ceramic membrane in tubular form based on the Sr-Fe-Co-O system. These tubes have operated without failure for over 1,000 hours at ~900°C with a methane conversion efficiency of over 98%. This membrane allows the use of ambient air for syngas production and eliminates the need for an oxygen plant or for cryogenic separation. Air flows on one side of the membrane while methane flows on the other side. Because the membrane is permeable to oxygen but not to nitrogen, oxygen diffuses through the membrane and reacts with methane to generate the product syngas. No external electrical source or circuitry is required. The membrane is a mixed ionic-electronic conductor that is selectively permeable to specific gases[79].

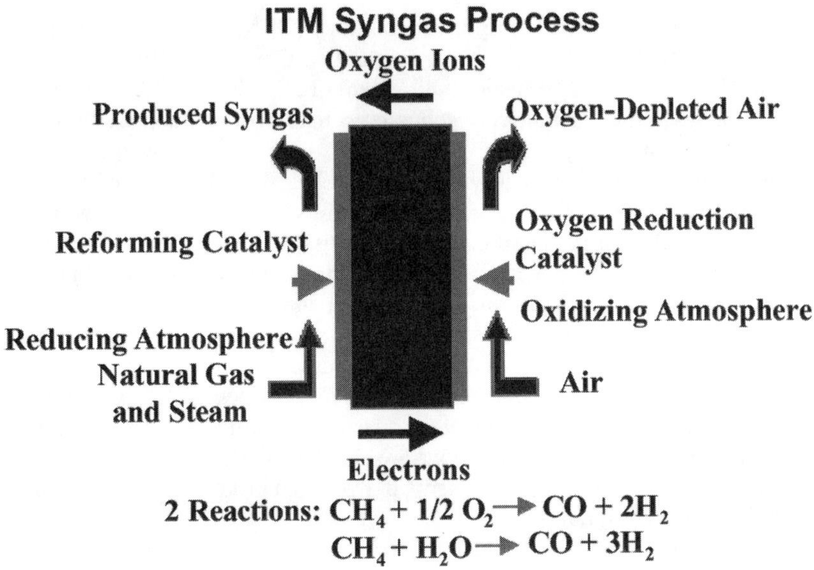

Figure 5.29. ITM syngas process[59].

The economics for an ITM Hydrogen process were developed for a plant that would produce 760 Million SCFD at 100 bar and 14,000 tonnes per day of CO_2 at 80 bar. For this application the IMT process was compared to a conventional oxygen-blown Autothermal Reformer (ATR) with a cryogenic air separation unit (ASU) to supply oxygen. The ITM process showed the potential for over 30% capital cost savings in the syngas process area and over 20% capital cost savings for the overall syngas/H_2/CO_2 process. The ITM process also has a higher thermal efficiency of 74% compared to 71% for the ATR process. The capital cost and efficiency advantages for ITM are due to the combination of oxygen separation and high temperature syngas production into a single unit operation as well as the ability to use low-pressure air as the oxidant source[189].

Argonne has also developed a ceramic membrane – based on another mixed ionic-electronic conductor – to separate hydrogen from gas streams. As with the membrane used to generate syngas, this membrane separates hydrogen without electrodes or other electrical circuitry[79]. The membranes contain $BaCe_{1-x}Y_xO_{3-d}$ and have operated in a temperature range of 600 to 900°C.

Two projects were announced in the late 1990's to develop the ITM Process and another related technology. Air Products will lead an $8^1/_2$-year, $US 90 million research project that is supposed to culminate in the construction of a 500 million SCFD Process Development Unit in 2005. Members of the team include McDermott, Ceramatec, Eltron Research, Battelle, Penn State University, ChevronTexaco, Norsk Hydro, University of Alaska - Fairbanks and the University of Pennsylvania.

A separate Oxygen Transport Membrane (OTM) Syngas Alliance was formed in 1997 and includes Amoco Production Company, British Petroleum, Praxair, Statoil, Foster Wheeler and Sasol Technology. This alliance is developing ceramic membrane technology that will economically convert natural gas to synthesis gas[190].

Another process that is listed in Table 39 is the CO_2-free production of hydrogen via thermocatalytic decomposition of hydrocarbon fuels. The process involves a single-step decomposition (pyrolysis) of hydrocarbons over carbon catalysts in an air- and water-free environment. This approach eliminates the need for a water-gas shift reactor, CO_2 removal and catalyst regeneration, which significantly simplifies the process[60].

Preliminary assessments of the process indicated that hydrogen could be produced at a cost of $US 5.00 per million BTU (if carbon sold at $US 100/ton). This production cost is less than that for a steam reforming process coupled with CO_2 sequestration. The CO_2 emissions are also lower[60].

Some additional new developments in hydrogen production are summarized in the following paragraphs.

The ICAR (**I**ntermediate **C**atalytic **A**ccumulation of **I**onizing **R**adiation **E**nergy) process converts nuclear energy to chemical energy. Energy from a nuclear reactor is used to promote the catalytic reforming of methane to syngas. This process was proposed by Yu. A. Aristov in 1993[1].

A related process is referred to as EVA-ADAM (**E**inzelrohr-**V**erzuchs-**A**nlage / **A**nlage mit **D**rei **A**diabaten **M**ethanator). This is a cyclic process that is based on the exothermic conversion of CO/H_2 to CH_4. In this circulation process, methane is steam reformed (an endothermic step) into a CO/H_2 mixture using helium-transported heat from a nuclear reactor in the EVA – a single tube experimental unit. The gas mixture is then supplied to the consumer by pipeline. There it is methanated in the ADAM (an exothermic step). The test unit consisted of three adiabatic methanation reactors. The methane that is formed can then be fed back to the EVA reformer. This process was developed by Kernforschungsanlage (KFA) in Jülich, Germany. In 1979, an EVA-ADAM pilot plant was brought on line in KFA-Jülich. Then in 1981, it was expanded to a capacity of 10 MW[1,46].

The U. S. Department of Energy is sponsoring a research project on the High Efficiency Generation of Hydrogen Fuels Using Nuclear Energy. This project identified a Sulfur-Iodine Thermochemical Water-Splitting Cycle in conjunction with a Modular Helium Nuclear Reactor as a way to economically produce hydrogen from nuclear energy. (The hydrogen production costs are estimated to be $US 1.00 to $US $1.50 per Kg of hydrogen or $US 2.36 to $US 3.54 per 1000 SCF). This process should be able to produce hydrogen at an efficiency of 45% to 55%[191].

NGK Insulators of Japan were issued US. Patent 5,741,474 in 1998 for a Process for Production of High Purity Hydrogen. This process is a combination of reforming, partial oxidation and membrane separation.

Northwest Power Systems obtained U.S. Patent 5,997,594 in 1999 for a Steam Reformer with Internal Hydrogen Purification. This process is based on a steam reformer that has a membrane separation system and a methanation system in close proximity to the area in which the reforming reaction occurs. Several potential equipment configurations are described in the patent.

BP and Kvaerner Process were finalizing the demonstration of their Compact Reformer Technology in 2001. This technology involves the integration of combustion, heat transfer and catalytic reaction within a simple tubular module. This reformer is about 30% lighter (in weight) than a conventional reformer of equivalent capacity[90].

Alchemix has developed the HydroMax® technology to produce hydrogen. This is a two-step process. First, steam contacts a molten metal to form a metal oxide and produce hydrogen. In the second step, the metal oxide is reduced with a carbon source into metal again. Both steps occur in the same reactor at about 1300°C, but at different times. A production plant requires at least two furnaces operating in tandem to produce hydrogen continuously. Metal smelting furnaces that are commercially available can be adapted for use as the reactor vessel[192].

Metal is not consumed in the process. It simply acts as a carrier for the oxygen from one part of the process to the other. A mixture of iron and tin is used. Iron is the primary oxygen transfer agent while tin is used to remove sulfur from the carbon-based feed materials. A variety of carbon sources can be used such as petroleum coke, automobile tires, high sulfur coal, municipal waste, biomass and sewage sludge. As of 2002, the process had been demonstrated in a laboratory reactor, and a pilot scale reactor (0.3 meter) was in operation at CSIRO (Commonwealth Scientific & Industrial Research Organization) in Melbourne, Australia. Hydrogen production costs are estimated to be about $US 0.02 per pound of hydrogen ($US 0.11 per 1000 SCF). Operating costs are actually provide an overall credit of - $US 0.03/lb when co-product credits from electricity, steam and ammonium sulfate are included. Depreciation (i.e., capital recovery) accounts for $0.05/lb. The HydroMax® process that is used to make ammonia is shown in Figure 5.30, and the overall process diagram is illustrated in Figure 5.31[192].

Synergy Technologies Corporation (www.synergytechnologies.com) of Conroe, TX is developing a technology called SynGen. It employs the proprietary GlidArc process to convert natural gas to syngas. GlidArc (or "gliding arc") is a cold plasma process featuring at least two diverging knife-shaped electrodes that operate under high voltage. The electric discharge (or cold plasma) converts natural gas to syngas via partial oxidation and is capable of processing feeds with up to 35% CO_2[224].

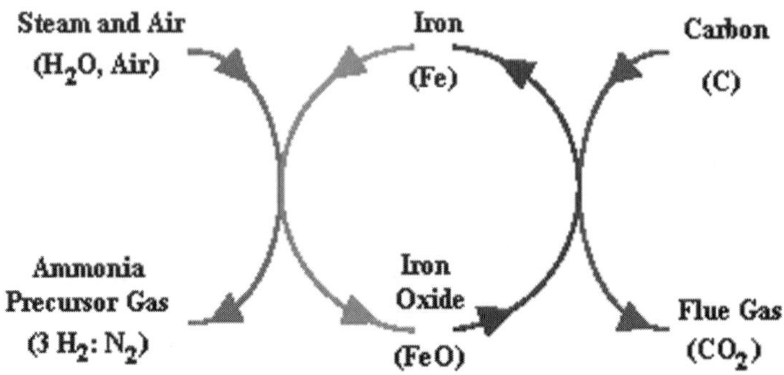

Figure 5.30. HydroMax® ammonia process steps[192].
(Reproduced with permission of Alchemix Corporation)

The electric discharge is similar to that in a neon sign so the power consumption is only a fraction of what is needed for thermal-plasma processes. Operating and maintenance costs are said to be low because GlidArc can be automated and has no moving parts. Furthermore, the electrodes are self-cleaning and the high-voltage discharge prevents electrode corrosion and erosion, which results in long electrode life. Syngen's capital costs are estimated to be about 40% less than conventional partial-oxidation and/or steam-reforming processes[224].

In 2003 Beta testing of the electric arc process was underway to convert natural gas to syngas. The SynGen technology has already been demonstrated using a 4-inch diameter reactor. After beta testing is completed, the process will be scaled up using 90 to 156-inch-diameter units for high-flowrate gas-to-liquids conversion (e.g., a 10,000-bbl/day gas-to-liquids plant might have four, 90-inch-diameter SynGen reactors operating in parallel). The company expects to be ready to license the technology in 2004[224].

Battelle Pacific Northwest National Laboratories (PNNL, Richland, WA) are developing microreactors that produce synthesis gas. These reactors can be mass-produced to yield efficient, compact and cost-effective systems, and they have been made from copper, aluminum, stainless steel, high-temperature alloys, plastics and ceramics. Conventional technologies cannot take full advantage of the intrinsically rapid surface reactions involved in the catalytic conversion of hydrocarbon fuels, but microreactors with integrated catalyst structures can[61].

A related effort is FORSiM (**F**ast **O**xidation **R**eaction in **Si**-technology-based **M**icroreactors) which is funded by the Dutch Technology Foundation and is a cooperative venture between the University of Twente and the Technical University of Eindhoven. The objective of this work is to build and operate the first microreactor for catalytic partial oxidation for small-scale and on-demand hydrogen production[61].

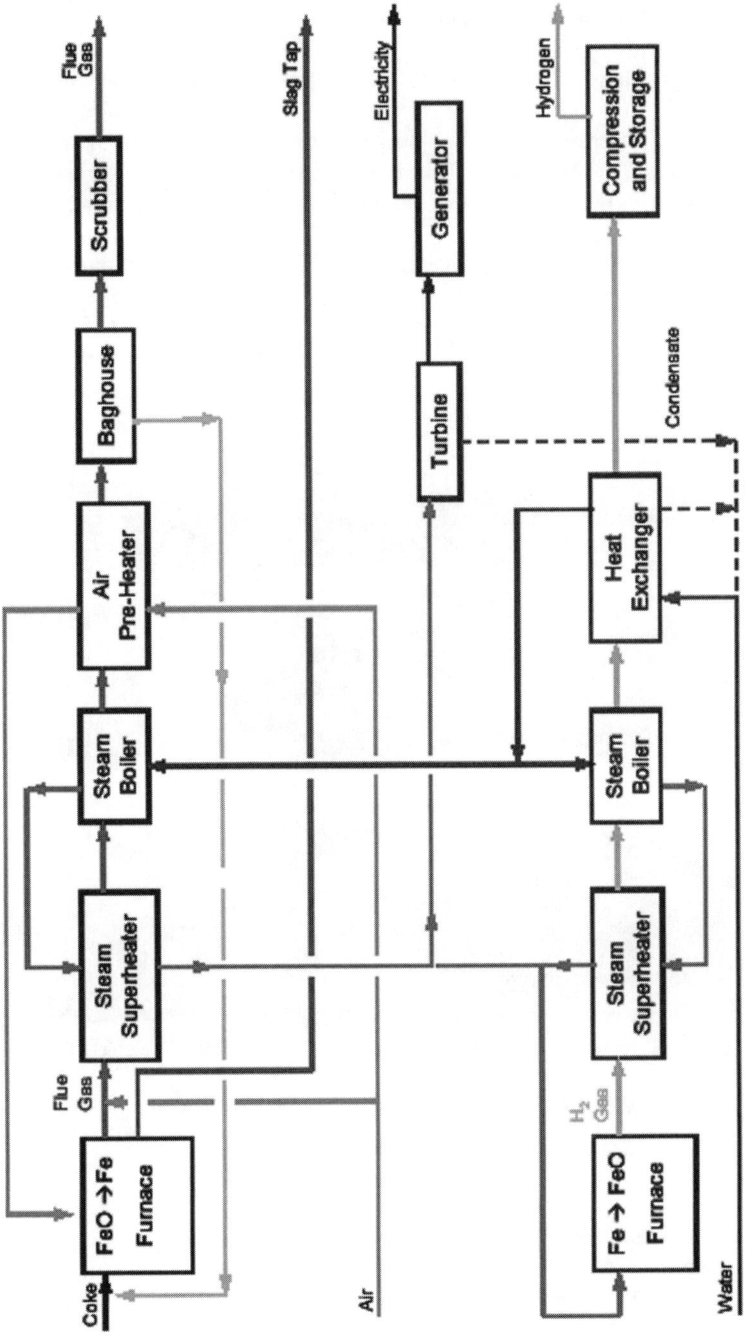

Figure 5.31. HydroMax® Process Diagram.
(Reproduced by permission of Alchemix Corporation)

5.2.8. Initial Purification of Synthesis Gas

Synthesis gas that is produced from the gasification of fossil fuels must be purified by removing several undesirable gaseous compounds that would affect its further use in different ways. The purification steps depend on the type of syngas process and the feedstock. This is illustrated by Figure 5.2 (Hydrogen Manufacturing Steps), Figure 5.3 (Conversion Routes to Synthesis Gas), Figure 5.4 (Steam Reforming), Figures 5.17 through 5.21 (Partial Oxidation) and Figures 5.24 to 5.27 (Coal Gasification). In Figure 5.32 the purification capabilities of different processes are illustrated. Table 5.22 illustrates how the synthesis gas stream can vary with different feedstocks. Figure 5.33 also illustrates the different process steps that may occur before ammonia synthesis takes place[74].

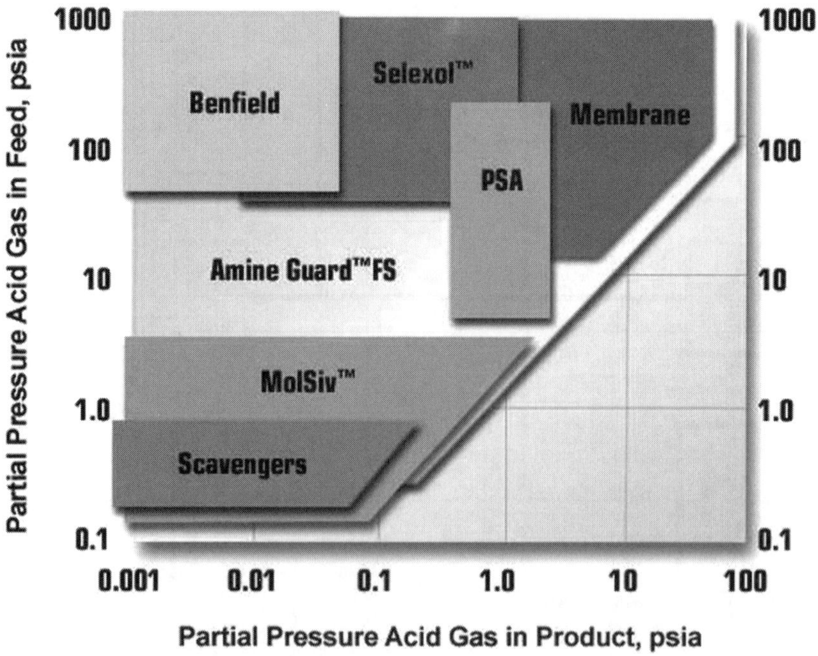

Figure 5.32. Operating conditions for different purification processes.

If sulfur is present as H_2S or COS, it is a poison for many catalysts and will partly or completely inhibit the catalyst activity[46]. Carbon monoxide (CO) and carbon dioxide (CO_2) can poison the ammonia synthesis catalyst so both of these compounds must be removed[53].

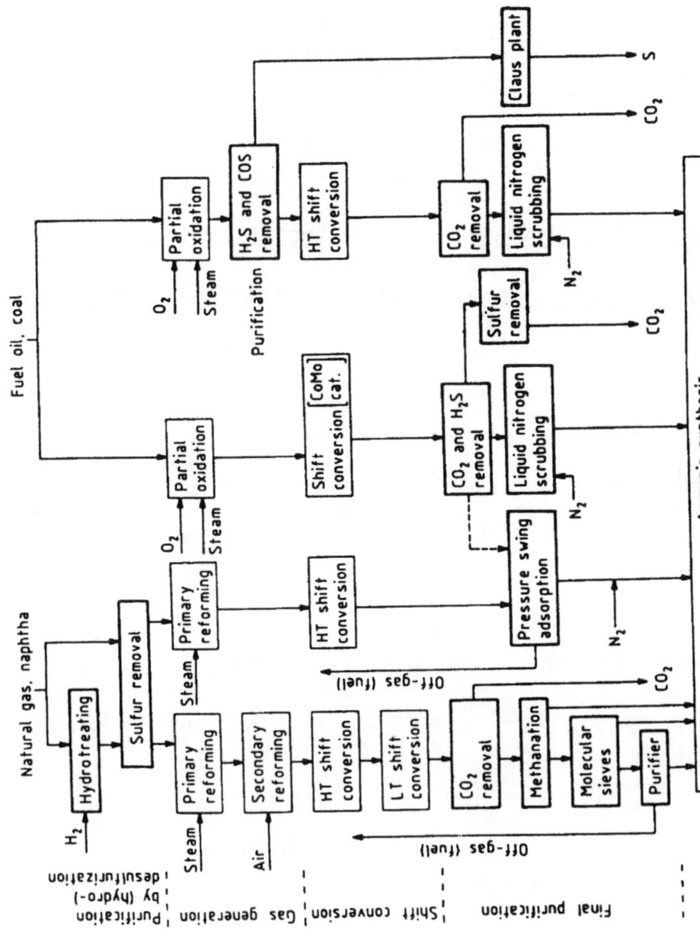

Figure 5.33. Alternative process steps for generation and purification of synthesis gas. (Reproduced by permission of Wiley-VHC)

After the sulfur is removed, most traditional ammonia processes have employed the purification steps that are shown in Figure 5.4 and Figure 5.17. However, ammonia plants have been built that use hydrogen purification via PSA, membrane separation with polymeric membranes, and cryogenic separation. PSA achieves the greatest product purities (about 99.999% pure hydrogen), but it is capital-intensive. Polymeric membranes are the least capital-intensive, but they achieve the lowest purity (about 96% to 98%). In 1998 PSA was considered the standard process for high-purity hydrogen recovery from raw synthesis gas. It has replaced the traditional steps of the shift conversion of carbon monoxide followed by carbon dioxide removal by absorption and then final purification by methanation. Characteristics of these three purification technologies are outlined in Table 5.29[40].

Table 5.29. Hydrogen recovery technology characteristics[40,64,172]

Characteristic	Membrane	Absorption	Cryogenics
Hydrogen Purity (%)	90% to 98%	99.9+%	90% to 96%
Hydrogen Recovery (%)	85% to 95%	75% to 92%	90% to 98%
Hydrogen Product Pressure	<< Feed Pressure	Feed Pressure	Feed / Low Pressure
By-Products Available	No	No	Yes: Liquid Hydrocarbons
Feed Pressure (psig)	300 to 2300	150 to 600	> 75 to 1100
Feed H_2 Content	> 25% to 50%	> 40%	> 10%
Hydrogen Capacity (MM SCFD)	1 to 50+	1 to 200	10 to 75+
Pretreatment Requirements	Minimum	None	CO_2 & H_2O Removal
Capital Cost	Low	Medium	Higher
Scale Economics (Ease of Expansion)	Modular	Moderate	Good

Besides the economic benefits, the PSA-based hydrogen plant has a number of technical advantages compared to traditional Steam Methane Reforming with solvent extraction and methanation. These advantages include[172]:

- The reformer can be operated at higher pressures and temperatures,
- There is a lower steam to carbon ratio of 2.8 to 3.0 compared to 3.5 to 6.0.
- No low temperature shift unit is needed.
- Hydrogen is available at higher purity.
- No methane molecules are lost in the hydrogen stream.
- Greater steam export is available at 650 psig.

5.2.8.1. Purification with PSA and Polymeric Membranes

The PSA process is based on the selective adsorption of gaseous compounds on a fixed bed of solid adsorbent in a series of identical adsorption beds. The adsorbent is an active carbon or a carbon-molecular sieve. Each bed undergoes a repetitive cycle of adsorption and regeneration steps. Feed gas is introduced to the bed at an elevated pressure, and it is sequentially directed to each bed so that a continuous flow of purified product is maintained. As the feed gas passes through a clean bed, the impurities are adsorbed and purified hydrogen exits the bed. The contaminated beds are regenerated and repressurized with a portion of the product hydrogen before they receive fresh feed. Impurities and unrecovered hydrogen are rejected at low pressure as fuel[161, 193].

PSA provides the hydrogen at essentially the same pressure as the feed, but recoveries are typically lower than from other technologies. Feed impurities are less of a problem than for membranes. However, heavy hydrocarbons in the feed can be irreversibly adsorbed, which can cause performance problems. The purges gas is often produced at or near atmospheric pressure and may require recompression to be used[170].

The first purification plant that used PSA was developed by Union Carbide Corporation (UCC) and was built at the Yokkaichi Plant of Mitsubishi Petrochemical Industries in 1971[4]. This process used several adsorbent beds and a complex valving system to produce high purity hydrogen. The process is now licensed by UOP and more than 400 units were operating worldwide in 1992[1].

The HYSEC Process was developed by Mitsubishi Kakoki K. Ltd. and The Kansai Coke & Chemicals Company. It has basically the same PSA unit as the UCC Process. It has prefilter beds with activated carbon that remove dirty components. After the main PSA beds, trace amounts of remaining oxygen are removed by a deoxo catalytic converter followed by a zeolitic dehumidifier. A Ni-La_2O_3-Rh catalyst, supported on silica, could lower the reaction temperature to about 30°C[4].

The LO-FIN (**L**ast **O**ut – **F**irst **In**) is another process that separates hydrogen from other gases. It includes a unique gas-retaining vessel that preserves the concentration gradient in one stream before using it to repressurize another bed. This process was jointly developed by Toyo Engineering and Essex Corporation and is shown in Figure 5.34[4].

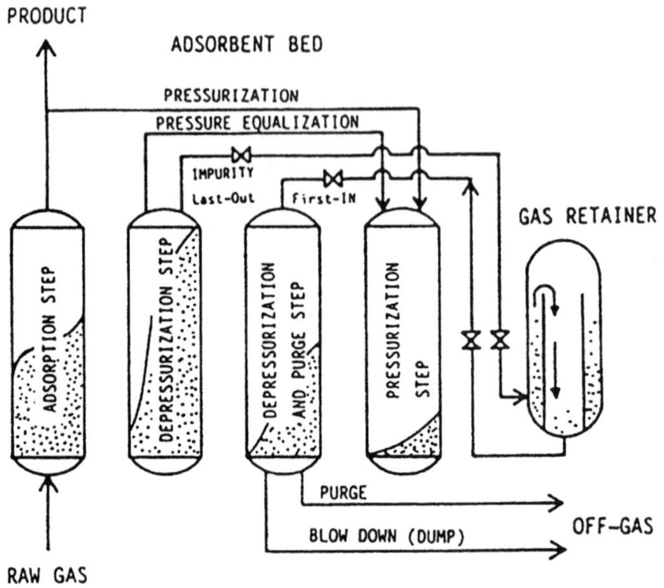

Figure 5.34. LO-FIN process for hydrogen purification[4].
(Reproduced by permission of AIChE. Copyright© 1998 AIChE – all rights reserved))

The Sumitomo-BF PSA process uses carbon molecular sieves (CMS) as the selective adsorbent. CMS has a higher capacity of adsorption than zeolites for methane and oxygen, and it is considered to be advantageous for hydrogen purification. If dirty raw gases are fed to this process, minor amounts of heavy hydrocarbon components such as aromatics are likely to cause deterioration of the adsorbents. To remove the heavy hydrocarbons, prefilter columns that contain activated carbon are placed upstream of the main CMS adsorbent beds[4].

5.2.8.1a. PSA Retrofits. A number of options are available to increase hydrogen production as a result of modifications to or around the PSA unit. PSA-related options include intermediate hydrogen off-gas feed, adsorbent replacement, cycle time adjustment, operating at atmospheric purge gas pressure and purge gas recycle[172].

Adding a carbon dioxide (CO_2) wash system to a PSA hydrogen plant will improve the hydrogen recovery of the PSA. Hydrogen recovery can be improved by 3% to 5% because both the total feed gas volume and the CO_2 content of the feed gas are reduced. The incremental hydrogen benefit is not sufficient to make this option economically attractive unless the recovered CO_2 has value. A liquid or gaseous CO_2 market must be available[172].

Adding a carbon dioxide wash unit and a methanator and eliminating the PSA can increase hydrogen production quite significantly - without an increase in feed rate. The drawbacks are a reduction in hydrogen purity from 99.99% to 97% (or lower). This option has limited application since most hydrogen users in 2002 are interested in high purity hydrogen[172].

The PSA Retrofits are summarized in Table 5.30[172].

Table 5.30. Summary of PSA Retrofit Options[172]

Debottleneck Option	Incremental Hydrogen	Steam Rate	Feed Rate	Fuel Rate	Cost	Issues
Low Temperature Shift	2-5%	1-3% Increase	No Change	Increase	Medium	Steam Purity
Improved PSA Recovery	1-2%	Same	Same	Increase	Low	Low Purge Gas Pressure
Addition of CO_2 Recovery	3-5%	6-10% Decrease	Same	Decrease	Low / High	Specific Market
Replace PSA with CO_2 Recovery	15-20%	4-10% Decrease	Same	Increase	Low / High	Purity Benefits in Refinery Units

5.2.8.2. Membrane Processes

Monsanto and Ube (Japan) developed the membrane process for purification of hydrogen gas mixtures. This process is based on the selective diffusion of hydrogen through semi-permeable membranes in the form of hollow fibers. Each gas has a characteristic permeation rate that is a function of its ability to dissolve in, diffuse through and dissolve out of the hollow fiber membrane. These separators are comprised of thousands of hollow fiber membranes that selectively separate "fast" gases (like hydrogen) from "slow" gases (like nitrogen and hydrocarbons) based on their relative permeation rates. Monsanto uses a polysulfone fiber while Ube uses an aromatic polyimide fiber[46, 193].

In the Monsanto process (referred to as PRISM® separators), the gas mixture flows around an array of hollow fibers, and only hydrogen and helium can diffuse into the fibers through the semi-permeable, polysulfone-coated walls. Separation from CH_4, CO, O_2 and N_2 can be accomplished in this way[46]. The PRISM® technology is now owned and marketed by Air Products.

The maximum operating temperature is 195°F (90°C). The maximum feed pressure is only limited by the mechanical design pressure of the flanges on the pressure vessel. The maximum differential pressure across the membrane depends on the operating temperature and can be as high as 1650 psi (113 bar) at 105°F (40°C) [193].

Optimum design and performance depend on several factors such as feed gas composition, feed pressure and the required hydrogen product pressure. In addition, a tradeoff always exists between the rate of hydrogen recovery and it purity. Several factors determine the required pretreatment. A feed heater is usually required to maintain a constant operating temperature and constant membrane performance. If the feed gas contains entrained liquids, a mist eliminator vessel with a high-efficiency coalescing element is required. Particulate filtration of the feed gas is not required. The membrane separators are designed so that the feed stream enters from the shell-side. Therefore particulates such as rust, pipe scale and others do not plug the separator. The control system can be designed to provide turndown to as low as 30% of the design feed flow rate[193].

5.2.8.3. New Membrane Technologies

Conventional polymeric hydrogen separation membranes yield hydrogen at low pressure. Air Products and Chemicals has demonstrated a carbon membrane on an alumina support that removes hydrocarbons from hydrogen/hydrocarbon mixtures and leaves the hydrogen at high pressure[40].

Los Alamos National Laboratory has developed a membrane based on polybenzimidazole (PBI) combined with a porous metallic support that can function at temperatures as high as 370°C - significantly higher than the 150°C limit of most polymer membranes. This PBI thin-film composite membrane outperforms other high-temperature membranes in terms of selectivity for the separation of H_2 from CO_2, has the highest demonstrated operating temperature of polymer-based membranes; is effective at higher pressures than conventional membranes, is chemically resistant and is easily processed[195].

For ultra-high-purity hydrogen (up to 99.9999% pure), palladium membranes are used. The hydrogen is separated from other gases by passing the gas mixture over a heated metal membrane such as palladium or palladium alloys at high pressure. The hydrogen molecule is first adsorbed on to a palladium site, and then it is dissociated into hydrogen atoms. The hydrogen atoms then diffuse through the membrane at a speed determined by the hydrogen front-to-back partial pressures and the metal temperature. Eventually these atoms recombine to form hydrogen molecules at the back surface of the metal membrane. As long as no leaks occur through or around the membrane, the hydrogen is 100% pure - since no other gas can pass through a solid metal membrane[194].

For effective mass transfer through the membrane, gas pressures far greater than those for PSA are needed. Metal membrane gas cleanup appears to be feasible only with a reformer system that is operating at a pressure of 20 atm[194].

Johnson Matthey is believed to be the leading supplier of these membranes. In 1996, Generex (Tryon, NC) licensed newly developed palladium membrane technology from Los Alamos National Laboratories that it plans to commercialize. In 1998 Wah Chang, an Allegheny Teledyne Company, purchased all patents and rights to the hydrogen metal membrane technology that Bend Research developed, and they plan to commercialize this technology[40].

US Patent 6,183,542 was issued in 2001 for a palladium membrane process. This process provides an apparatus that can handle high flow rates of gas, per unit area of membrane, while using a minimal amount of hydrogen-permeable material. This is accomplished by using stainless steel mesh elements to reinforce the thin-walled, palladium or palladium alloy membranes. This process also provides the ability to withstand large pressure gradients in opposite directions and thus will make it easier to clean membranes that have been clogged with contaminants.

Proton Energy Systems (Rocky Hill, CT) received US Patent 6,168,705 in 2001 for a system that can purify and simultaneously compress hydrogen from a dirty gas stream without relying on moving parts. This electrochemical gas purifier system can purify and produce hydrogen gas at pressures exceeding 2,000 psig.

5.2.9. Carbon Monoxide Shift

As shown in Figure 5.4 and Figure 5.17 and as described for each of the major processes that produce synthesis gas, the Water Gas Shift Conversion or the Carbon Monoxide Shift reaction is one of the traditional purification steps that will still be found in many ammonia plants. The CO must be removed because it acts as a poison to the catalyst that is used in ammonia synthesis.

The Carbon Monoxide Shift removes most of the carbon monoxide (CO) from the synthesis gas, and [as shown by Eq. (5.2)] it also produces more hydrogen.

$$CO + H_2O \leftrightarrow H_2 + CO_2 \quad \Delta H = -10 \text{ kcal/mol or } 42 \text{ kJ/mol} \qquad (5.2)$$

Homogenous Water Gas Reaction (Water Gas Shift)

The "shift" from carbon monoxide to carbon dioxide generally occurs in two steps - first a High Temperature Shift Conversion and then a Low Temperature shift conversion. In some cases the two steps may be combined in one isothermal or adiabatic step called Medium Temperature Shift Conversion. When the feed gas to the CO conversion is not desulfurized, the CO conversion is called Sour Gas Shift and a special type of sulfur-resistant catalyst is used[166].

As shown above, Eq. (5.2) is exothermic, and high temperatures are unfavorable for complete conversion. In the High Temperature Shift (HTS) conversion, the synthesis gas is passed through a bed of iron oxide/chromium oxide catalyst at around 400°C and a pressure of 25 to 28 bar. The CO content of the gas is reduced to about 3% (on a dry gas basis), and this is limited by the shift equilibrium at the actual operating temperature. The catalyst is not particularly sensitive to sulfur although high concentrations have been found to be detrimental[53]. Some HTS reactors are able to operate with an inlet temperature of 350°C, and they lower the CO level from 10-15 mole % (dry) to 1-2 mole % (dry). Ideally, the catalyst takes the reaction to equilibrium at as low a temperature as possible because this favors hydrogen production. Copper has been added to some iron-chrome catalyst formulations, and this has resulted in enhanced activity[70].

The material of construction in HTS reactors is usually low-alloy steel containing 1% Cr and 0.5% Mo. If CO_2 and H_2S are in the steam condensate that is fed to the shift conversion reactors, they should be made of 304 stainless steel to avoid corrosion problems[88].

Adding a LT-shift (adiabatic reactor, inlet temperature 200 °C) to a HT-shift reactor (adiabatic reactor, inlet temperature 340 °C) would have the following results[225]:

- Reduction of feed by 4 %.
- Increase in fuel by 19 %.
- Marginal increase in feed and fuel by 0.6 %.
- Increase in steam export by 3 %.
- Reduction of absorbed duty (i.e. reformer size) by 4 %.

If a Low Temperature Shift (LTS) converter is installed (see Figure 5.35), the gas from the HTS is cooled to increase the conversion, and then it is passed through the LTS converter. The LTS converter is filled with a catalyst containing copper oxide, zinc oxide, and aluminum oxide. It operates at about 200-220°C. The residual CO content in the converted gas is about 0.2% to 0.4% (on a dry gas basis)[53]. Some LTS reactors operate with an inlet temperature of 190-210°C and reduce the CO level to 0.1 to 0.2 mole % (dry). Again, the catalyst takes the reaction to equilibrium at as low a temperature as possible because this favors the hydrogen production[70].

The copper-based LTS catalyst has operational limits. First, the inlet gas temperature must be above its dew-point by a reasonable margin because water condensation damages the catalyst. This limits the minimum inlet temperature to about 190°C. Second, the LTS catalyst is affected by traces of poisons such as sulphur and chloride that do not affect the HTS catalysts[70].

Substituting the HT and LT shift with an isothermal Medium Temperature shift with a typical exit temperature of 260 °C would produce results similar to those provided by the combination of HTS and LTS. It is obvious that only a relatively expensive feed and a cheaper fuel would justify the additional investment needed to supplement the HT shift with a LT shift or install an isothermal MT shift in place of both of them[225].

The Medium Temperature Shift conversion normally takes place in an isothermal system, but it may also be accomplished in an adiabatic reactor with an exit temperature around 300°C. A copper-zinc catalyst is used for this

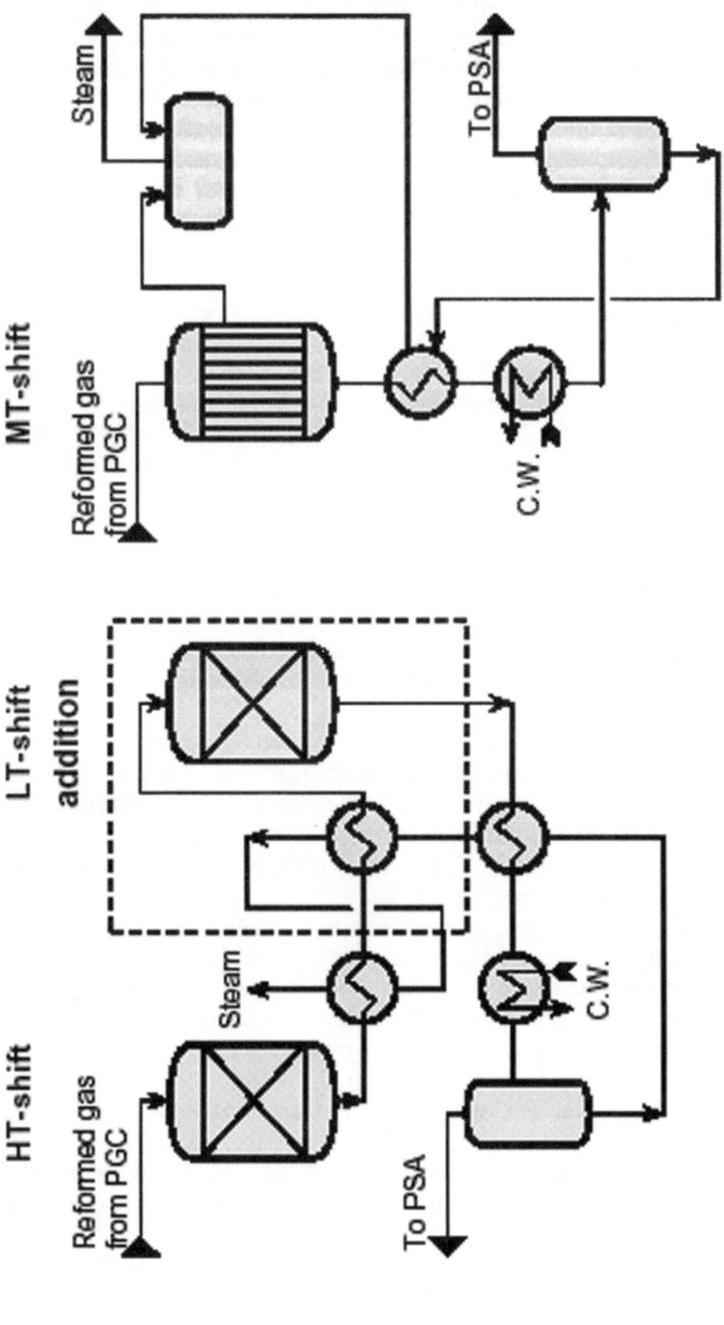

Figure 5.35. CO Shift Reaction Options.

reaction. If a Sour Gas Shift is part of the process, the catalyst will be cobalt-molybdenum on stabilized alumina. This catalyst can process a stream that contains high levels of sulfur and traces of unsaturated hydrocarbons[166].

The use of Au/α-Fe_2O_3 in the catalysis of the water-gas shift reaction has been reported by Andreeva. The gold-on-iron-oxide catalyst was active at low temperatures, and its performance overall compares favorably with the $CuO/ZnO/Al_2O_3$ catalyst used in LTS[83]. Sakurai also employed gold catalysts prepared by deposition-precipitation or co-precipitation, and supported by TiO_2, Fe_2O_3, Al_2O_3 or ZnO, for the forward and reverse water-gas shift reaction. Gold catalysts gave results comparable to $CuO/ZnO/Al_2O_3$ catalysts[84].

Experience in operation of the large single-train ammonia plants has shown that performance of the LTS catalyst is critical to good operating continuity. This catalyst contains copper and zinc and is very sensitive to sulfur and halogens[62]. Therefore, a guard catalyst frequently is used to protect it. The use of a small guard vessel ahead of the main LTS catalyst bed is valuable. The following operating practices have been found to be helpful[63]:

1. Dry reduction of the catalyst using natural gas or nitrogen as the carrier instead of steam,
2. The use of a low-temperature shift catalyst instead of zinc oxide as a guard catalyst,
3. Elimination of known sources of poisons,
4. A separate guard vessel,
5. Regular replacement of the guard catalyst, and
6. Dry inlet temperature control rather than quenching.

The steam requirements in an ammonia unit can be reduced by lowering the steam-to-carbon ratio to the primary reformer. However a number of drawbacks can exist downstream in the HTS and LTS reactors. The drawbacks include: By-product formation in the HTS, Pressure drop buildup in the HTS, Reversible poisoning of the LTS catalyst, and Higher CO equilibrium concentrations exiting the HTS and LTS reactors.

The simplest solution to this problem is to modify the classic iron-based HTS catalyst by adding a moderate amount of copper to reduce the amounts of byproducts formed and to give a much higher catalyst activity for the shift reaction. A second and more radical solution, is the use of an iron-and chromium-free HTS catalyst that is copper-based[73].

5.2.10. Removal of Sulfur Compounds and Carbon Dioxide

If sulfur is present as H_2S or COS or if CO_2 is present, any of these compounds will be a poison for many catalysts and will partly or completely inhibit catalyst activity. As shown in Figures 5.4, 5.5, 5.13 and 5.17, the point at which sulfur removal is employed depends on the synthesis gas process that is used. Table 5.31 lists many of the processes that are available[46].

Table 5.31. Processes for removal of carbon dioxide and sulfur compounds from synthesis gas[46]

Process Name	Key Chemicals
Alkazid Process	Alkali Salts of Amino Acids (N-Methylaminopropionic Acid)
aMDEA®	Activated Methyl Diethanolamine
Benfield Process	Hot Potassium Carbonate
Carsol Process	Potassium Carbonate
Catacarb Process	Hot Potassium Carbonate
Fluor Solvent Process	Propylene Carbonate
Giammarco-Vetrocoke	Process Potassium Carbonate with Arsenite
Hi Pure Process	Two-Stage Hot Potassium Carbonate Washing
Purisol Process from Lurgi	N-Methyl-2-Pyrrolidone (NMP)
Rectisol Process	Low Temperature Methanol
Selexol Process	Dimethyl Ethers of Polyethylene Glycol
Shell Sulfinol Process	Di-isopropanolamine Dissolved in Sulfolane & Water
Pressurized Washing	Monoethanolamine (MEA) or Diglycolamine

The Amine Guard is a corrosion inhibitor that was developed by Union Carbide. It is added to the MEA solvent and has the further advantage that MEA concentration can be increased. The higher MEA concentration leads to a lower circulation flow rate and a lower energy demand during regeneration[74].

The heat requirements for solvent regeneration in some of the processes listed in Table 5.31 are given in Table 5.32:

Table 5.32. Regeneration heat requirement for CO_2 removal solvents[74]

Process Description	Regeneration Heat (MJ/kmol CO_2)
MEA (before Amine Guard)	209
MEA (Amine Guard II)	140
MEA (Amine Guard III)	116
Benfield-Amine (Single-Stage Regeneration)	107
MEA (Amine Guard IV)	88
Benfield-Amine (Single-Stage Regeneration, Lean Solution Flash, Steam Ejectors)	88
Benfield-Amine (Two-Stage Regeneration, Lean Solution Flash, Steam Ejectors)	76
Benfield-Amine (Two-Stage Regeneration, Lean Solution Flash, Mechanical Heat Pump)	63
Activated MDEA (Two-Stage Regeneration)	42.5

As shown in Figures 5.4, 5.17 and 5.32, the CO_2 Removal step is normally after the Shift Conversion step. The process gas from the LTS converter contains mainly hydrogen, nitrogen, CO_2 and excess process steam. The gas is cooled and most of the excess steam is condensed before it enters the CO_2

removal system. This steam condensate normally contains 1500 to 2000 ppm of ammonia and 800 to 1200 ppm of methanol. Therefore it should be stripped or recycled. However it should be noted that some LTS catalysts are available that are designed to produce a minimum amount of methanol by-product[166]. The heat that is released during the cooling/condensation can be used: to regenerate CO_2 scrubbing solution (if needed), to drive an absorption refrigeration unit, or to preheat boiler feedwater.

The amount of heat released depends on the process steam to carbon ratio (see Tables 5.9, 5.15 and 5.22). If all this low-level heat is used for CO_2 removal or absorption refrigeration, the high-level heat has to be used for the boiler feedwater heating system. An energy-efficient process should therefore have a CO_2 removal system with a low heat demand[53].

The CO_2 is removed in a chemical, a physical or a hybrid absorption process. Residual CO_2 contents are usually in the range of 50 to 1000 ppmv – depending on the type and design of the removal unit. The typical range of heat consumption in a chemical absorption process is 30 to 60 million Joules/kmol CO_2. The physical absorption processes may be designed for zero heat consumption, but for comparison with the chemical processes, the mechanical energy requirements have to be considered[53]. All the processes operate in essentially the same manner. They scrub the feed gas in absorption towers to collect the CO_2 and then regenerate the solvent and release the CO_2[199].

In the chemical absorption process, the CO_2 reacts with chemical solvents to form a weakly-bonded intermediate compound that is then broken down by the application of heat. The heat regenerates the original solvent and produces a CO_2 stream. Typical solvents are amine- or carbonate-based. Examples are MEA, diethanolamine (DEA), ammonia and hot potassium carbonate. These processes can be used at low CO_2 partial pressures, but the feed gas must be free of SO_2, O_2, hydrocarbons and particulates. Hydrocarbons and particulates cause operating problems in the absorber[199].

In the physical absorption process, the CO_2 is absorbed in a solvent according to Henry's Law and then regenerated using heat, pressure reduction or both heat and pressure reduction. Typical solvents are Selexol (dimethylether of polyethylene glycol) and Rectisol (cold methanol) which are applied at high pressure. At lower pressures, the chemical absorption processes are more economical. The Selexol physical solvent process is frequently specified for coal gasification applications[199].

Hybrid solvents combine the best characteristics of both the chemical and physical solvents and are usually composed of a number of complementary solvents. Hybrid solvents have out-performed existing solvents and those developed for one application have been easily adapted for to others. However this is not always desirable and the tendency has been to develop tailor-made solvents using complementary solvents where the proportions are varied to suit the applications. Typical solvents are aMDEA, Purisol, Sulfinol and UCARSOL[199].

5.2.10.1. Alkazid Process

The Alkazid process removes sulfur compounds from gas streams. All the sulfur compounds are first catalytically hydrogenated to hydrogen sulfide (H_2S) using a cobalt/molybdena catalyst. The H_2S is then absorbed in an aqueous solution of potassium salt of either methylamino propionic acid ("Alkazid M"), or dimethylamino acetic acid ("Alkazid DIK"). This solution is heated to regenerate the hydrogen sulfide as a concentrate. This concentrate is then treated by the Claus process to recover the sulfur[1].

The Claus process has two stages. In the first stage, one third of the H_2S is oxidized with air to produce sulfur dioxide. In the second step, this sulfur dioxide stream is blended with the remainder of the H_2S stream and passed over an iron oxide catalyst at approximately 300°C. The resulting sulfur vapor is condensed to liquid sulfur[1].

5.2.10.2. aMDEA Process®

The aMDEA® (**A**ctivated **M**ethyl **D**iethanol**a**mine) process removes CO_2, H_2S and trace sulfur compounds from natural gas and syngas via a pressurized wash with activated diethanolamine. This process was developed by BASF. The first unit was started up in 1971 in Germany and in 2002 it was used in more than 140 plants with 20 more under construction[1]. About 75 of these plants involve another process technology (MEA, DEA, Benfield, etc.) that was modified to use the aMDEA® technology.

The aMDEA® solvents are non-corrosive due to their chemical structure. This allows a high sour gas loading and therefore reduced solvent circulation rates, which results in low investment and low operating costs.

5.2.10.3. Benfield Process

The Benfield (**Ben**son and **Field**) process removes carbon dioxide, H_2S, and other acid gases from industrial gas streams by scrubbing with hot aqueous potassium carbonate that contains activators. The chemical reactions are[1,46]:

$$K_2CO_3 + CO_2 + H_2O \leftrightarrow 2\,KHCO_3 \qquad (5.15)$$

$$K_2CO_3 + H_2S \leftrightarrow KHS + KHCO_3 \qquad (5.16)$$

In one variation of this process that removes CO_2 from the gas stream, the aqueous scrubbing solution contains 40% K_2CO_3 and is circulated at 118°C in the absorber. The CO_2 is released in the stripper at atmospheric pressure by heating the solution to 230°C. An inhibitor generally is needed to reduce corrosion and stainless steel equipment is used in many places. This system is characterized by its low steam requirements and its compatibility with catalysts[65]. Typical heat consumption is 30,000 to 40,000 BTU per lb-mole of CO_2 removed.

In potassium carbonate processes, arsenic and vanadium salts are used to suppress corrosion. A passive layer of magnetite on steel surfaces is maintained by adding a small amount of air[88].

The process was invented by H.E. Benson in 1952 and then developed with J.H. Field at the U.S. Bureau of Mines. It was first licensed by the Benfield Corporation. Subsequently it was acquired by Union Carbide, and it is now licensed by UOP. The UOP version has offered different solution activators (diethanolamine - DEA, MEA and since 1992, ACT-1™) and incorporates zeolites or membrane processes for complete separation of acid gases and minimal loss of product gases[1]. More than 700 plants were operating in January, 2000[200].

A typical Benfield process is shown in Figure 5.36.

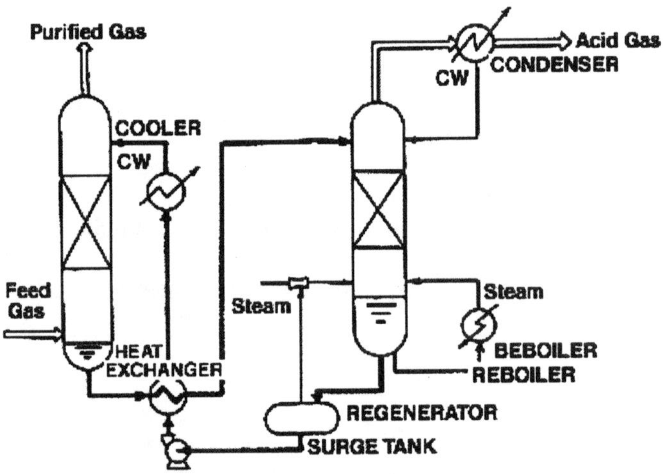

Figure 5.36. Benfield Process for Acid Gas Purification. [196]

Potassium formate and a few other carboxylic acid salts are one result of the breakdown of the DEA molecule. These salts are benign at low concentrations. However, when at concentrations of 5% or more, they interfere with operations by altering the physical properties of the carbonate solution. Some amine degradation compounds are even considered to be corrosion accelerators. They may solubilize iron, keep it in solution and prevent it from forming the passivation layer on the pipe and equipment. UOP's ACT-1 activator is an amine that contains a more-stable molecule that is more resistant to degradation[206].

Typical activator concentrations in plant solutions are 0.3 to 1.0 wt. % ACT-1 compared to about 3 wt. % for DEA. The overall activator consumption costs are estimated to be about the same for ACT-1 and DEA. However CO_2 Slip with ACT-1 is about half of that achieved with DEA[206].

In Table 5.33 the Benfield process options with different packings and activators are compared. This table assumes this equipment will be used in a 1,500 tonne per day ammonia plant. The costs assume the plant was built in January, 1997[206].

Table 5.33. Benfield Process Design Comparisons[206]

	Case 1	Case 2	Case 3
Packing Type	Pall Rings	IMTP	IMTP
Activator	DEA	DEA	ACT-1
		Major Equipment Sizes	
Absorber Tower Top			
Diameter (ft)	8.75	7.75	7.75
Packed Height (ft)	2 x 27	3 x 22	2 x 24
Packing Volume (ft3)	3,247	3,113	2,264
Packing Details	2 inch Metal	#50 Metal	#50 Metal
Absorber Tower Bottom			
Diameter (ft)	14.0	12.5	12.5
Packed Height (ft)	2 x 21	2 x 26	2 x 22
Packing Volume (ft3)	6,465	6,381	5,400
Packing Details	2 inch Metal	#50 Metal	#50 Metal
Regenerator Tower			
Diameter (ft)	17.75	15.75	15.50
Packed Height (ft)	3 x 27	4 x 25	3 x 26
Packing Volume (ft3)	20,043	19,483	14,718
Packing Details	2 inch Metal	#50 Metal	#50 Metal
Utilities			
Net Heating Duty:			
MM BTU/hr	130.3	130.6	123.6
BTU/lb mol CO_2	33,400	33,500	31,700
Lean Pump (gpm)	5,935	5,935	5,890
Electricity (kWh)	1,920	1,920	1,830
Cooling Water (gpm)	12,180	12,180	11,640
Capital Investment (+30%)			
Purchased Equipment Cost ($ Million)	4.1	4.0	3.6
Installed Equipment Cost ($ Million)	8.3	8.2	7.3

(Reproduced by permission of UOP)

As shown in Table 5.33, the combination of IMTP packing and the ACT-1 activator results in smaller columns, less packing and lower capital costs. Packing volumes are 25% to 35% smaller, and capital costs are reduced by about 10%[206].

Some packings that provide more efficient CO_2 absorption than Pall rings in hot carbonate service are[206]:

- Norton Co. IMTP Packings
- Glitsch Co. Mini-Ring Packings
- Nutter Engineering Co. Nutter Ring Packings
- Koch Engineering Co. Fleximax Packing
- Sulzer Flexipac Structured Packings

5.2.10.4. Benfield Hybrid LoHeat Process

The Benfield Hybrid LoHeat process is similar in design to the standard 4-stage ejector, lean solution flash tank design of the Benfield LoHeat Process. It is a low-energy CO_2 removal process can achieve a net thermal energy consumption of only 650 kcal/Nm^3 of CO_2 removed. The conventional LoHeat process reduces energy consumption from about 1,200 kcal/Nm^3 for Benfield units without LoHeat, down to about 800 kcal/Nm^3 of CO_2 removed. This energy recovery is accomplished by flashing the reboiled solution to generate steam, and then steam ejectors are used to compress the flashed steam for injection back to the regenerator column[206].

To improve the LoHeat process, a fifth stage of solution flashing was added along with mechanical vapor recompression (MVR) to boost the fifth-stage flashed steam back to the pressure of the regenerator column. The combination of ejectors plus MVR for multi-stage heat recovery is referred to as the Benfield Hybrid LoHeat Process. If this improvement is combined with ACT-1 activator the energy consumption can be reduced to 600 kcal/Nm^3 of CO_2 removed.

The installed costs (in January, 1997 dollars) for a Benfield Hybrid LoHeat Process in a 1,500 tonne per day ammonia plant are shown in Table 5.34. Table 5.35 lists the energy use in different Benfield processes that could be installed in a 1,500 tonne per day ammonia plant[206].

Table 5.34. Installed Costs for Benfield Hydrid Process[206]

Equipment	Cost ($US Million)
Towers & Vessels	2.820
Tower Packing	0.343
Flash Tank	0.455
Pumps	1.037
Compressor	0.618
Turbine	0.462
Exchangers	2.428
Total Installed Costs	8.163

(Reproduced by permission of UOP)

Table 5.35. Comparison of Energy Use in Different Benfield Processes[206]

	Different Benfield Processes		
Energy Consumption	No LoHeat	LoHeat	Hybrid LoHeat
Thermal Energy			
MM kcal/hr	47.91	34.01	23.90
kcal / Nm^3 of CO_2	1297	920	647
Electrical Power			
kWh	1826	1826	2365
MM kcal/hr Equivalent	6.23	6.23	8.07
Net Total Energy			
MM kcal/hr	54.14	40.24	31.97
kcal / Nm^3 of CO_2	1465	1089	865

(Reproduced by permission of UOP)

5.2.10.5. Carsol Process

The Carsol Process is used to remove carbon dioxide from gas streams by scrubbing with aqueous potassium carbonate.

5.2.10.6. Catacarb Process

The Catacarb (**Cata**lyzed Removal of **Carb**on Dioxide) Process removes carbon dioxide and hydrogen sulfide from gas streams by adsorption in a hot potassium carbonate solution that contains a proprietary catalyst. The process was developed and licensed by Eickmeyer and Associates based on work at the U.S. Bureau of Mines in the 1950's. More than 100 plants were operating in 1997[1].

5.2.10.7. Fluor Solvent Process

The Fluor Solvent Process removes CO_2 from natural gas and various industrial gas streams by dissolving the CO_2 in a propylene carbonate solvent. The process is based on the principle that CO_2 is much more soluble in low-temperature propylene carbonate than other common gases. It should be noted that the process cannot be used when H_2S is present[1].

The absorber operates at a relatively low temperature as a result of the refrigerating effect of CO_2 being released, but some auxiliary refrigeration is needed. The CO_2 is released during the regeneration of the solvent when staged pressure reductions with flashing (rather than heating) are used to regenerate the solvent. Some of the flashed gases are also used to drive turbines for electrical power recovery. However if the CO_2 is needed for urea production, it can be released at 100 to 200 psig[67, 68].

The process was invented in 1958 by the Fluor Corporation and can be licensed from Fluor Daniel. The first plant was built for the Terrell County Treating plant in El Paso, TX in 1960. By 1985, 13 plants were operating[1].

5.2.10.8. Giammarco-Vetrocoke Process

Two processes are known by the Giammarco-Vetrocoke name. Both processes use an aqueous solution of sodium or potassium carbonate and arsenite to absorb acid gases. Some variations of the process use glycine instead of arsenite to activate the potassium carbonate solution. In one process, the solution is used to extract carbon dioxide from natural gas or synthesis gas. In the other, hydrogen sulfide is extracted from coke-oven or synthesis gas and yields elemental sulfur from a complex sequence of reactions. In 1992, more than 200 plants were operating[1].

By varying the process conditions, the CO_2 level can be reduced as low as necessary, and CO_2 with a purity of 99% can be produced. The process is reported to be non-corrosive to carbon steel[66].

5.2.10.9. Hi Pure Process

The Hi Pure Process is a variation of the Benfield process. (Compare Figure 5.36 to Figure 5.37[196].) It uses two stages of scrubbing by hot potassium carbonate solution to reduce the CO_2 content of gases to very low levels[1].

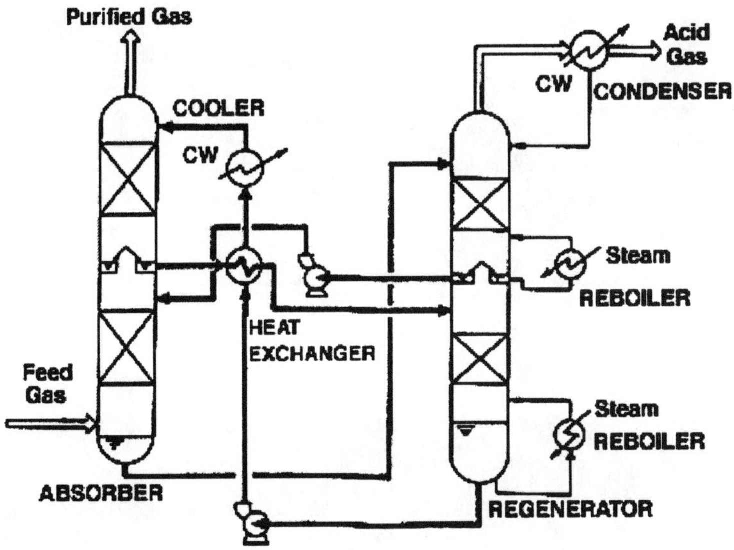

Figure 5.37. Benfield Hi Pure process sketch[196].

5.2.10.10. Purisol Process

The Purisol Process removes H_2S from gases by selective absorption in N-methyl-2-pyrrolidone (NMP). It was developed and licensed by Lurgi, particularly for desulfurizing waste gases from IGCC (Integrated Gasification Combined Cycle) plants. IGCC plants are coal gasification processes that use hot gases from combustion to generate electrical power. However the technology might be applied to some synthesis gas processes. In 1996, seven Purisol units were either in operation or under construction[1].

5.2.10.11. Rectisol Process

The Rectisol Process (See Figure 5.38[197]) was originally developed jointly by Lurgi and Linde. It is designed to remove sulfur and acid gas components compounds from gas mixtures that are produced from the partial oxidation of hydrocarbons. It is based on pressurized washing with low-temperature (below 0° C) methanol, which results in the physical absorption of the sulfur compounds in the methanol. Lurgi applied the process for the first time in 1953 at the SASOL coal gasification plant in South Africa. The process was further developed by Linde, and they installed it for Texaco in the purification of a gas from oil gasification near Los Angeles in 1968.

In 2003 the Rectisol process is offered for licensing by both companies for the removal of CO_2, H_2S, HCN, C_6H_6 (benzene), COS and gum-forming hydrocarbons from syngas and fuel gas. In 2003, over 70 units were in operation or under construction[1].

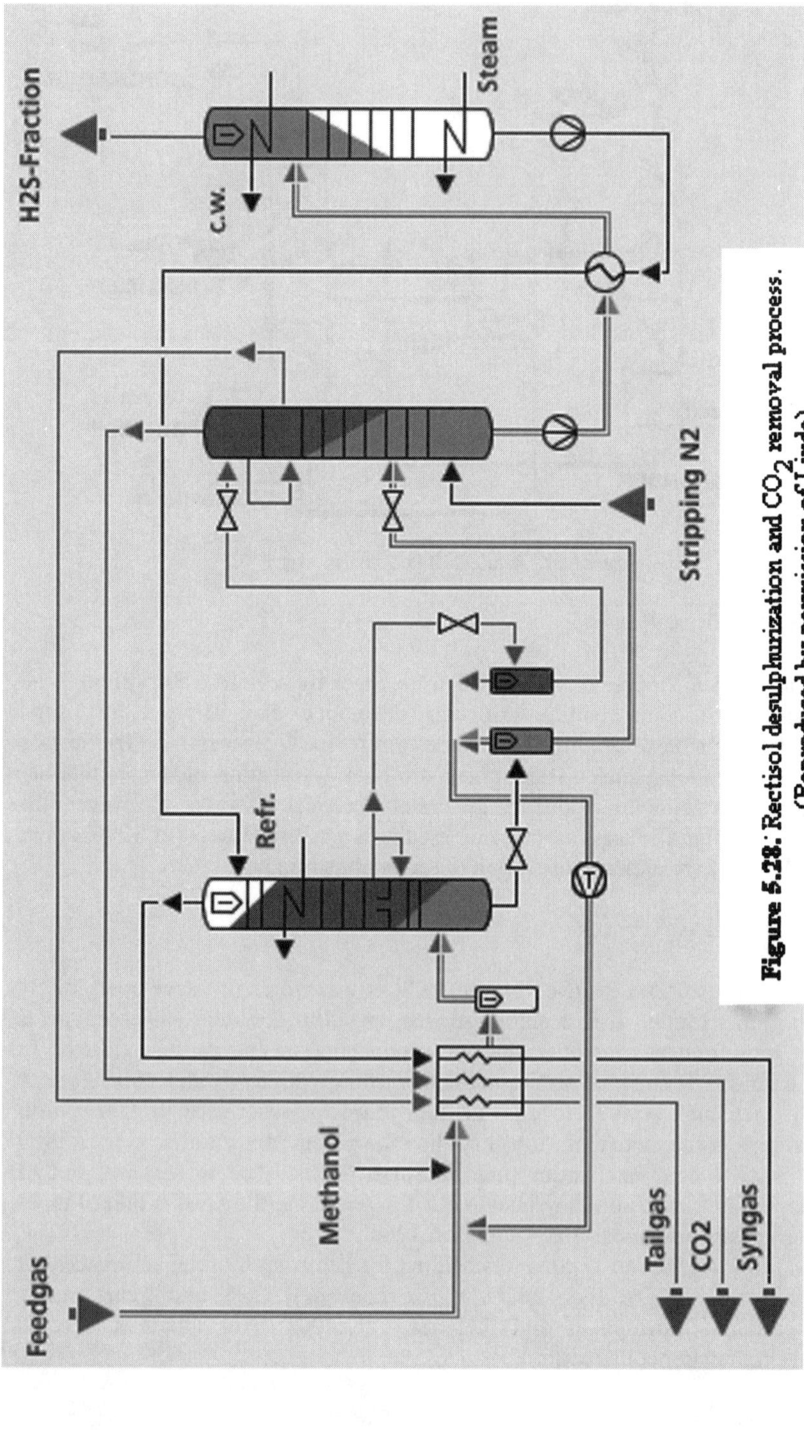

Figure 5.28: Rectisol desulphurization and CO_2 removal process. (Reproduced by permission of Linde)

5.2.10.12. Selexol Process

The Selexol Process (see Figure 5.39) removes acid gases from hydrocarbon gas streams by selective absorption in polyethylene glycol dimethyl ether (DMPEG). It exploits the pressure-dependent solubility of the acidic gases in DMPEG. It absorbs H_2S, CO_2, COS and mercaptans. Absorption takes place in a counter-current extraction column that is under pressure. The solvent is regenerated by "flashing" (evaporation) or by "stripping" (passing an inert gas through it). The process has also been used to remove CO_2 from syngas, natural gas and coal gas. It was developed by Allied Chemical in the 1960's and acquired by Union Carbide (UCC) in 1990. Dow acquired UCC (and their subsidiary UOP) in 2000. In 2003 the Selexol process is available for license exclusively from UOP. As of May 2002, 55 Selexol units were operating[1, 198].

The DMPEG is chemically inert and is not subject to degradation. Due to its non-aqueous nature and inert chemical characteristics, the most of the Selexol process equipment can be made of carbon steel. Acid gas partial pressure is the key driving force for the Selexol process. Typical feed conditions range between 300 and 2000 psia with acid gas composition (CO_2 + H_2S) from 5% to more than 60% by volume. The product specifications depend on the application and can range from ppmv up to percent levels of acid gas[198].

5.2.10.13. Shell Sulfinol Process

The Shell Sulfinol process removes H_2S, CO_2, COS (carbonyl sulfide), and organic sulfur compounds from natural gas by scrubbing with di-isopropanolamine dissolved in a mixture of sulfolane ($C_4H_8SO_2$) and water. It was developed in the 1960's by Shell. In 1997, over 200 commercial units were operating or under construction[1].

The Sulfinol process is a mixed solvent process that in 2002 is offered with the traditional Sulfinol-D solvent, which is formulated with the secondary amine DIPA (di-isopropanolamine), or with selective Sulfinol-M, which is formulated with the tertiary amine MDEA (methyldiethanolamine). As a mixed solvent system, the solvent formulation can be tailored to obtain good treating economy, single-step treating for sweetening and organic sulfur removal, high acid gas loading or selective treating[260].

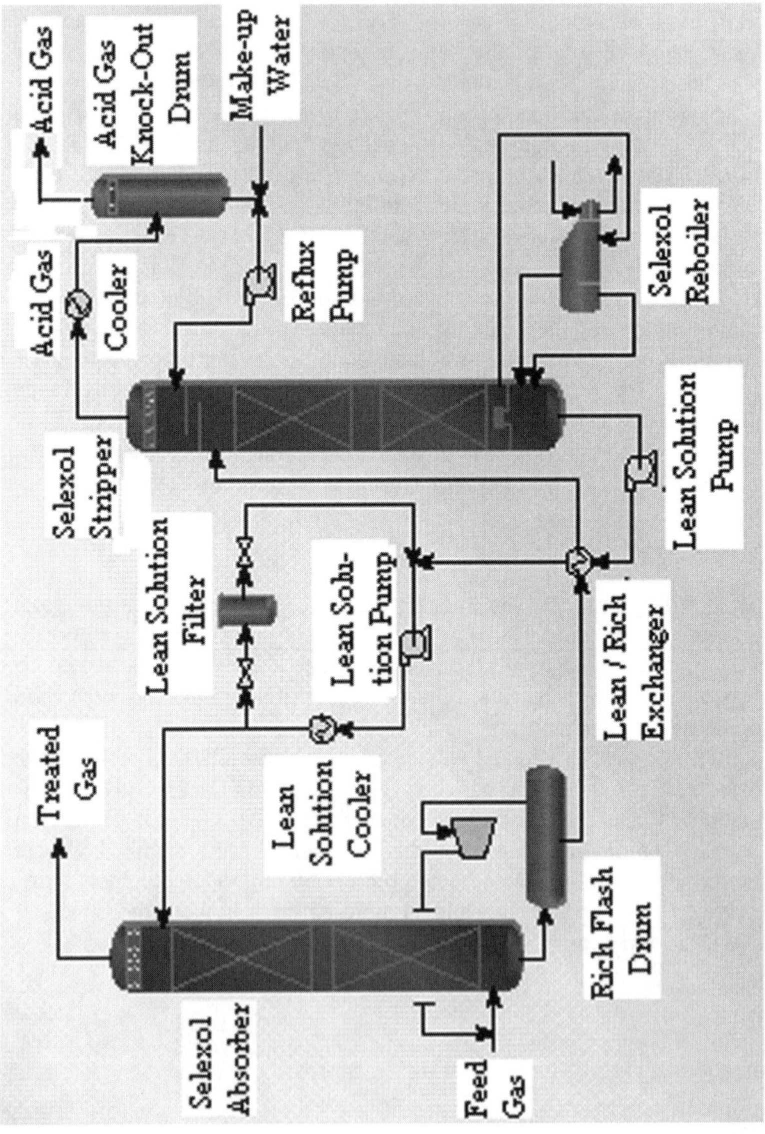

Figure 5.39. Selexol Process for removal of acid gases. (Reproduced by permission of UOP)

The advantages of the Sulfinol process over conventional amine treating systems include[260]:

- Reduced solvent circulation for higher acid gas loading
- Lower energy requirements for regeneration
- Tailored process to meet treating needs exactly
- Selective removal of H_2S
- Lower corrosion rates
- Lower foaming tendency and
- A single process for complete removal of acid gas and organic sulfur species.

Capital costs for a Sulfinol unit are lower than alkanolamine units that have the same capacity. This is because the equipment is smaller as a result of less foaming and lower circulation rates. However, certain treating applications may not be ideal for the Sulfinol process. A reclaimer may be needed to remove DIPA degradation products when CO_2 partial pressure is high. Also, aromatic and hydrocarbon co-absorption occurs when solvent -to-gas ratios are high[260].

This process is characterized by its high CO_2 retention under pressure and by low steam requirements for regeneration. According to U.S. Patent 3,347,621, a sidestream regenerator is required to remove by-products that build up in the system. Mild steel is suitable in much of the process, but stainless steel is preferred where CO_2 concentrations are high. The process must be carefully engineered to protect the downstream nickel catalysts from sulfur that may carry-over from the absorber. Likewise, the CO_2 that is produced must be protected from entrainment of the solvent and contamination with sulfur[260].

Typical operating data are shown in Table 5.34.

Table 5.34. Shell Sulfinol Process: Typical Operating Data[260]

		Feed Gas	Treated Gas
Feed Gas Flowrate	MM scfd	100	
H_2S	Volume %	1.25	1.0
CO_2	Volume %	8.0	0.05
COS	ppm (vol)	200	3.0
RSH	ppm (vol)	-	5.0
CS_2	ppm (vol)	-	1.0
Other S	ppm (vol)	-	3.0
Total S	ppm (vol)	-	5.0
Absorber Pressure	psia	600	
Absorber Temperature	°F	115	
Low Pressure Steam	M lb/hr	40	
Electrical Power	KW	530	

5.2.10.14. Pressure Washing with Monoethanolamine (MEA)

In this process a 15% to 30% solution of MEA in water is used to absorb the CO_2 under pressure. The solution is then regenerated by heating it in a stripper to release the CO_2. This process is characterized by good CO_2 absorption properties at low pressure. However the high regeneration energy consumption limits its use.

A corrosion inhibitor is used with this process. With the Amine Guard inhibitor, the MEA concentration in the circulating solution can be increased to 30% from a normal 20%. Hence, the circulating rate can be decreased by 33% and the heat requirements are decreased by 43%.

Stainless steel is used in critical areas where hot carbon dioxide-laden solutions are present. It is reported that antifoam agents (silicones) are used in some cases with this process[69]. Some pressurized wash processes also use diglycolamine to remove sulfur compounds and CO_2[46].

Two examples of this technology are the Amine Guard FS Process and the Amine Guard II Process. A comparison of the two processes is given in Table 5.35[200].

The Amine Guard FS flowsheet is shown in Figure 5.40, and the Amine Guard II flowsheet is shown in Figure 5.41. Both of these technologies are available from UOP[200].

Table 5.35. Comparison of Amine Guard FS and Amine Guard II Processes[200]

	Amine Guard FS	Amine Guard II
Key Driving Force	Acid Gas Partial Pressure	CO_2 Partial Pressure
Minimum Feed Pressure (psia)	150	50
Maximum Feed Pressure (psia)	1800	1800
Minimum Acid Gas Concentration (vol %)	5	5
Maximum Acid Gas Concentration (vol %)	35	30
Heat Consumption (Btu/lb-mole CO_2)	Not Available	60,000 to 70,000
Year of Technology Introduction	1983	1973
Units Installed (as of 1/2000)	> 400	> 175
Impurities Removed	CO_2, H_2S, Other Sulfur Compounds	CO_2
Solvent	UCARSOL	Alkanolamine Solution with Metal-Passivating Corrosion Inhibitors

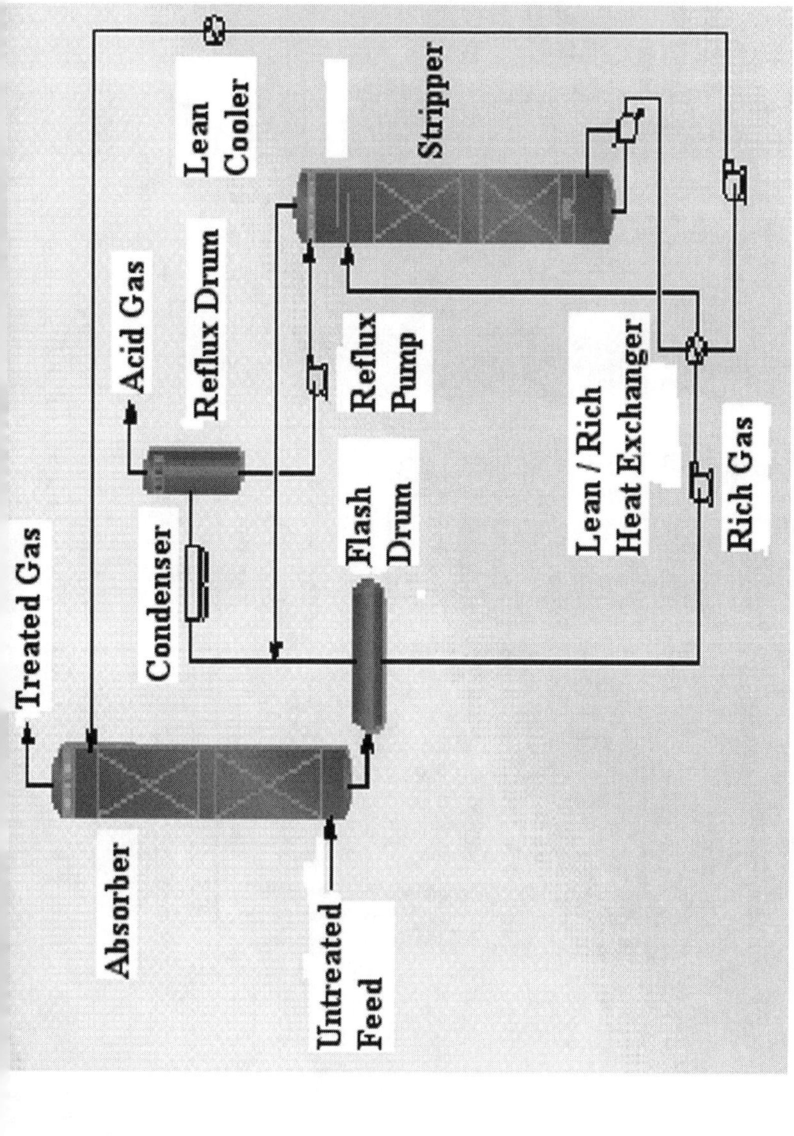

Figure 5.40. Amine Guard FS Process.
(Reproduced by permission of UOP)

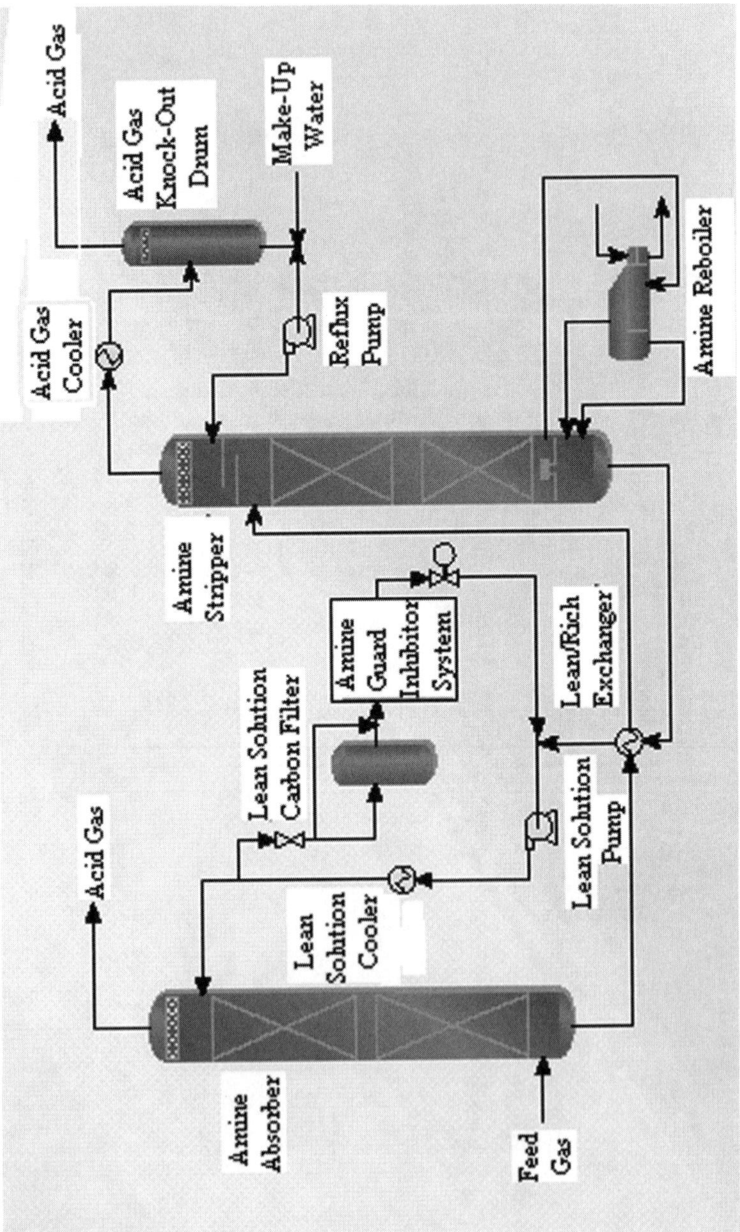

Figure 5.41. Amine Guard II Process. (Reproduced by permission of UOP)

5.2.10.15. Retrofits of the CO_2 Removal System

In a CO_2 removal system that uses wet scrubbing, the existing towers are the major limit to more capacity because they are expensive to replace. Therefore a strategy must be developed to increase CO_2 absorption in the existing towers.

Separation equipment tends to have less flexible design limits and can be costly to modify. For example, the CO_2 absorber and solvent regenerator towers in a wet scrubbing unit are difficult to replace. Fortunately, it is often possible to make changes to tower internals, or to change the solvent composition, to provide more capacity[86].

In an amine system, one way to increase absorption is to increase amine concentration. A higher amine concentration requires 1) more extensive filtration to keep the solution clean and 2) the addition of corrosion inhibitors and to minimize corrosion. Another option is to change from monoethanolamine (MEA) to methyl diethanolamine (MDEA)[86].

In a potassium carbonate system, different additives can be used to increase the CO_2 absorption rate. In any wet scrubbing system, improved tower packing can be used. A change from random to structured packing can lead to higher solvent circulation rates and improved mass transfer[86].

If more heat transfer is needed, air coolers in a CO_2 removal unit often need to be modified by adding surface area or increasing fan horsepower[86].

5.2.11. Final Purification of Synthesis Gas

Before the synthesis gas enters the ammonia synthesis loop, essentially all of the oxygen compounds must be completely removed. This step is required to 1) avoid poisoning the ammonia synthesis catalyst and 2) keep CO_2 from forming carbamates and ammonium carbonate in the synthesis loop. Typical oxygen compounds include CO, CO_2, O_2 and H_2O. It is also advantageous to remove the inert gases (methane, argon, etc.) to achieve a higher synthesis conversion per pass[74].

5.2.11.1. Selectoxo Process

The Selectoxo (**Select**ive Catalytic **Ox**idati**o**n) Process reduces 1) the hydrogen consumption of the methanation system and 2) the inert gas content of the purified synthesis gas that is fed to the ammonia synthesis loop[74].

In the early 1960's Engelhard developed and commercialized the Selectoxo™ catalyst and process for H_2 plants. The heart of this technology is a highly selective catalyst, which oxidizes up to 10,000 ppm CO without significantly oxidizing the 70% H_2 (dry) in the feed stream. CO levels were reduced to less than 5 ppm under steady state conditions (50°C, 10,000 h^{-1} space velocity and 200-400 psig). The Selectoxo™ catalyst contains 0.5% platinum (Pt) supported on γ -alumina, $^1/_8$ inch tablets promoted with a base metal oxide.

The alumina is impregnated with salts of Pt and base metal oxide, dried, and then calcined[202].

The Selectoxo process provides good energy efficiency because it minimizes carbon monoxide "slip" (only about 0.03%), improved process flexibility, and increased productivity in revamps when compared to other oxidation options. The Selectoxo unit is capable of increasing a plant's capacity by 1.5-2.0%[202].

The Selectoxo unit can also help in a grass root plant by maintaining carbon dioxide/ammonia production ratios which is favorable for full conversion of ammonia to urea. The economics of this option are to be considered against the extra cost of carbon dioxide production by other means (either from the flue gas of the primary reformer or through back burning of extra synthesis gas) [202].

5.2.11.2. Methanation

In a steam reforming process that includes CO shift conversion and CO_2 removal, the synthesis gas still contains 0.1% to 0.2 mole % CO and 100 to 1000 ppmv of CO_2. The simplest method for eliminating these small concentrations of oxygen compounds is shown by the reactions in Eqs. (5.13) and (5.17) (that are the reverse of the methane reforming process). The methane that is formed does not cause any problems in the downstream ammonia process. The methane simply acts as an inert diluent[70].

$$CO + 3\,H_2 \leftrightarrow H_2O + CH_4 \quad \Delta H = -49.27 \text{ kcal/mol or } 206 \text{ kJ/mol} \qquad (5.13)$$

$$CO_2 + 4\,H_2 \leftrightarrow 2\,H_2O + CH_4 \quad \Delta H = -39.44 \text{ kcal/mol or } 165 \text{ kJ/mol} \qquad (5.17)$$

Methanation

The normal methanation operating temperature is 250°C to 300°C, and a large excess of hydrogen is present. Typical space velocities in the reactor are 5,000 to 9,000 hr^{-1}. The equilibrium lies far to the right side of the above reactions so the CO and CO_2 impurities can be reduced to about 5 ppm[70]. The advantages of methanation (simplicity and low cost) more than outweigh the disadvantages (hydrogen consumption and production of additional inerts in the ammonia synthesis feed gas)[74]. Some methanation catalysts, that have been pre-reduced, can be used at temperatures down to 190°C[75].

The two methanation reactions are strongly exothermic. The temperature rise for typical methanator gas compositions in hydrogen plants is about 74°C (133°F) for each 1% of carbon monoxide converted and 60°C (108°F) for each 1% of carbon dioxide converted. At higher temperatures, the intrinsic rates of both methanation reactions can become sufficiently fast for diffusion effects to become important as shown in Figure 5.42. Under these conditions, film diffusion controls the overall rate of reaction. Diffusion limitations can be overcome to some extent by using a catalyst with a smaller particle size (3.1mm diameter by 3.6 mm long compared to regular catalyst dimensions of 5.4 mm by

3.6 mm. The smaller catalyst 1) provides a higher geometric surface area and 2) enhances mass transfer[70].

The catalyst is formulated on a calcium oxide / aluminate base with the active nickel incorporated in a NiO/MgO solid solution. This results in negligible nickel sintering (i.e., crystallite growth) during operation. The NiO content from different vendors varies from 25% to 43% by weight. The NiO content that is chosen will depend on the feed conditions and the product purity requirements of the application. Deactivation results mostly from solvent carryover from the upstream CO_2 removal section. A catalyst lifetime of over 10 years is common with careful operation. Catalyst life has been extended by employing back-washing techniques when the catalyst is contaminated by solvent carry-over. Further enhanced performance has been achieved from pre-reduced catalysts and smaller pellet sizes that allow operation at lower temperatures and higher space velocities[70]. Pre-reduced catalysts are only recommended for applications in which the inlet operating temperature is extremely low, or if start-up procedures do not enable adequately high inlet temperatures during the reduction procedure. A partial loading of pre-reduced catalyst may also be considered to shorten the reactor start-up process[166].

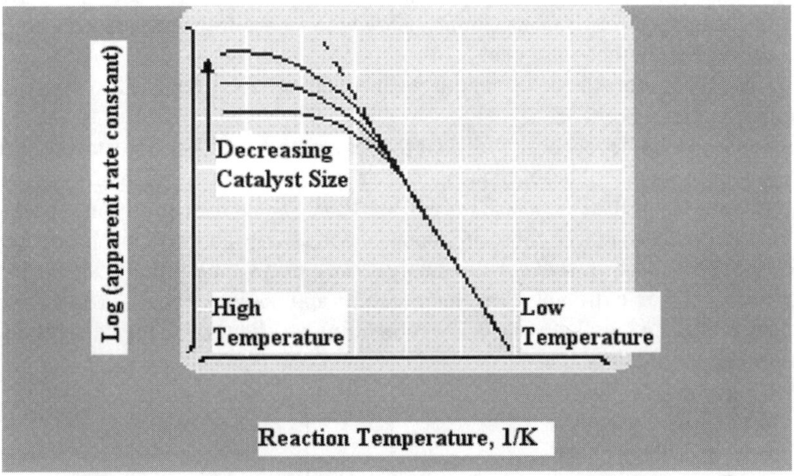

Figure 5.42. Effect of diffusion and reaction temperature on methanation rate [70].
(Reproduced by permission of Johnson Matthey Catalysts. Copyright Johnson Matthey PLC)

Methanation catalysts are not usually deactivated by thermal sintering. The principal reason for any loss of activity is poisoning. Sulfur compounds will poison methanation catalysts, but sulfur is not present unless the low temperature shift catalyst is by-passed. The poisons most likely to occur under normal operating conditions are those originating from the carbon dioxide removal system that precedes the methanator. Carry-over of a small amount of liquid into the methanator is not serious. Large volumes of liquid will have a serious effect, particularly if the liquid contains arsenic or sulfur, because irreversible loss of catalyst activity results[70].

When only potassium carbonate or organic solvents are used, the effects are less important. Potassium carbonate blocks the catalyst pores, and can be removed by washing with water to restore normal performance. Methanation catalysts can be protected from poisons by installing a guard bed of zinc oxide absorbent. This will remove traces of sulfur and droplets of liquid from the carbon dioxide removal system[70].

Since the methanation reaction is strongly exothermic, a sharp temperature rise can be measured across the reaction zone in the catalyst bed. Most methanation reactors are designed with a number of thermocouples that monitor the position of the exotherm. A strong indicator of the amount and rate of methanation catalyst deactivation is the position of the temperature profile in the catalyst bed and its rate of movement over time. A record of the temperature profile should be kept to detect any movement during the first one to two years of operation. An estimate of future life can then be made[70].

If a breakthrough of CO occurs from the LTS converter or a breakthrough of CO_2 from the absorption system occurs, the intensely exothermic methanation reaction can reach temperatures that exceed 500°C. Controls should be installed and other measures taken to avoid these high temperatures because the catalyst may be damaged or the maximum allowable operating temperature of the pressure vessel may be exceeded[74].

The poisoning effects of liquids from different CO_2 removal systems are summarized in Table 5.36[70].

Some papers have been published that examine Ru/SiO_2 as a catalyst in the Methanation step. These papers looked at the effects of hydrogen and temperature as well as how a Cl-modified Ru/SiO_2 catalyst performs. The Cl decreased catalytic activity, but it enhanced selectivity for methane formation - even though it was present on the catalyst only during the initial stages of the reaction[77,78]. For extremely low temperature applications (i.e., $<$ 180°C), one company offers a catalyst with 0.3% ruthenium on alumina. This catalyst does not contain any NiO or CaO[166].

Table 5.36. Poisoning Effects of Liquids on Methanation Catalysts[70]

Process	Chemical	Effect
Benfield	Aqueous Potassium Carbonate	Blocks pores of catalyst by evaporation of K_2CO_3 solution.
Benfield DEA	Aqueous Potassium Carbonate with 3% Diethanolamine	Blocks pores of catalyst by evaporation of K_2CO_3 solution (DEA does not affect the process).
Vetrocoke	Aqueous Potassium Carbonate plus Arsenious Oxide	- Blocks pores of catalyst by evaporation of K_2CO_3 solution. - As_2O_3 is also a poison - 0.5% of As on the catalyst will reduce its activity by 50%.
Sulphinol	Sulpholane, Water, Di-2-Propanolamine	Sulpholane will decompose and cause sulfur poisoning.
MEA, DEA	Mono- or Di-Ethanolamine in Aqueous Solution	None
MDEA	Aqueous solution of Methyl-Di-Ethanolamine and Activators	None
Rectisol	Methanol	None
Catacarb	Aqueous Potassium Carbonate with Borate Additive	Blocks pores of catalyst by evaporation of K_2CO_3 solution.
Selexol	Dimethyl Ether of Polyethylene Glycol	None

5.2.11.3. Cryogenic Purification

In the cryogenic purifier all the methane and the excess nitrogen are removed from the synthesis gas as well as a part of the argon. The cooling is produced by depressurization and no external supply is needed. The purified syngas is then practically free of all impurities, except for a small amount of argon. The cryogenic unit also receives the purge from the ammonia synthesis section and delivers an off-gas for fuel. Removal of essentially all impurities from the make-up synthesis gas is a significant improvement – compared to the conventional purification by methanation only. The combination of higher conversion per pass and reduced purge flow result in a more efficient process[53].

The KBR Purifier, which is located between the methanation and compression steps, is a cryogenic process that removes excess N_2, CH_4, most of the argon and other impurities from the synthesis gas by condensing them at low temperatures. The resulting makeup gas to the loop has an inert concentration of less than 0.3%. The result is a highly pure makeup gas to the synthesis loop, which reduces the required synthesis pressure and loop purge rate. The small loop purge is recycled to the Purifier for hydrogen recovery, thus eliminating the purge gas recovery system. The H_2 to N_2 ratio of the loop makeup gas is controlled at the Purifier, which simplifies the operation of the plant[81]. A KBR Purifier is shown in Figure 5.43[201].

Excess N_2, which is needed to make the cryogenic separation work, is provided by using excess process air in the secondary reformer. This design has the further advantage of transferring a significant amount of the reforming duty from the primary reformer to the secondary reformer. As a result, the size of the primary reformer is reduced by about 35%. Both the primary and secondary reformers have outlet temperatures that are about 100°C lower than those found in conventional ammonia plants[81].

Fifteen Purifier plants have been built since 1966. These plants range in capacity from 680 to 1750 tonnes per day, and as of 2001 all fifteen plants were still operating[81].

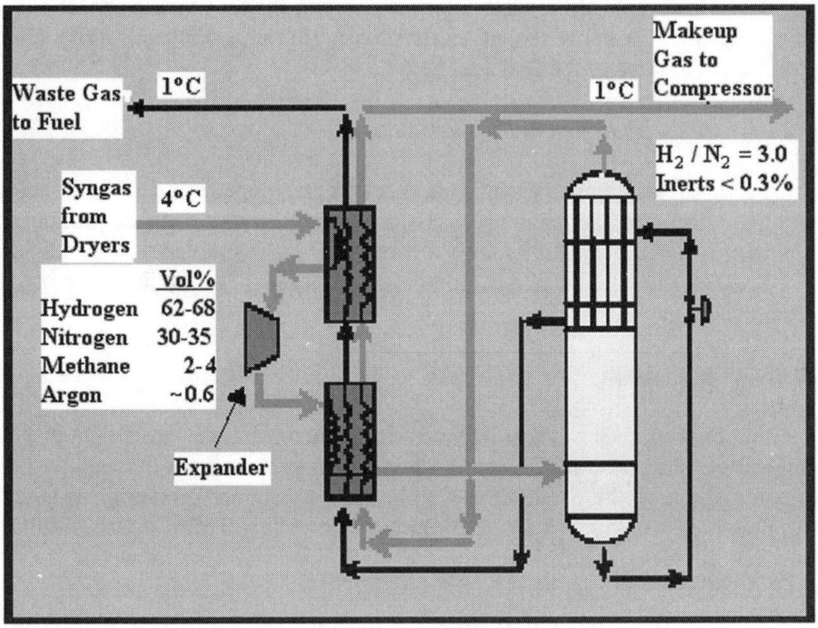

Figure 5.43. KBR Purifier[81].
(Reproduced by permission of Kellogg Brown & Root, Inc.)

5.2.11.4. Dehydration

If the makeup gas to the ammonia synthesis loop is absolutely free of catalyst poisons, such as H_2O and CO_2, it can flow directly to the ammonia synthesis converter. This leads to the most favorable arrangement from a minimum energy point of view. This can be accomplished by allowing the gas that leaves the methanation step to pass through beds of molecular sieves to remove water and traces of CO_2[74].

5.2.11.5. Liquid Nitrogen Wash

In many partial oxidation syngas processes, liquid nitrogen scrubbing is used to remove the carbon monoxide that remains after the shift conversion step. The CO content may be as high as 3% to 5% in plants that have only a HTS conversion. The Liquid Nitrogen Wash 1) delivers a gas to the ammonia synthesis loop that is free of all impurities (including inert gases) and 2) adds all or part of the nitrogen that is required for ammonia synthesis[74].

The nitrogen is obtained from an air-separation plant that provides the oxygen for the partial oxidation process (see Figure 5.17). Before the nitrogen wash operation, the gas must be thoroughly dried and the CO_2 content reduced to less than 5 ppm. Then the gas is cooled to the point where it can be scrubbed with liquid nitrogen. Otherwise, the moisture and the CO_2 would freeze out in the equipment. Therefore the gas is first passed through silica gel dryers and a caustic wash to remove the H_2O and CO_2[74].

5.2.11.6. Adjust Hydrogen to Nitrogen Ratio

The optimum ammonia synthesis reaction rate depends on several factors including pressure, temperature, H_2-to-N_2 molar ratio, concentration of impurities and catalyst activity. Therefore the H_2-to-N_2 molar ratio is adjusted to suit the requirements in the ammonia synthesis. This adjustment occurs before the compression step.

5.2.12. Compression

Ammonia synthesis is normally carried out at a pressure higher than that for synthesis gas preparation. Therefore the purified synthesis gas that is fed to the ammonia synthesis loop must be compressed to a higher pressure. Synthesis loop pressures employed industrially range from 8 to 45 MPa (80 to 450 bar). However, the great majority of ammonia plants have synthesis loops that operate in the range of 15 to 25 MPa (150 to 250 bar)[74].

Due to several major developments in ammonia process technology that began in the late 1960's, ammonia plants with 1,000 to 1,500 tonnes per day capacities became the industry standard for new plant construction. In 2001 plants as large as 2,000 tonnes per day have become common. These plants have much lower average production costs than the earlier generation of smaller plants. A major factor has been the switch from electrically-driven, reciprocating compressors to the use of natural gas-fired, steam-driven centrifugal compressors[57]. Some compressors are also driven by steam turbines using high-pressure steam generated in the plant waste heat boilers. This steam is generated by utilizing the process gas leaving the secondary reformer and the flue gases from the primary reformer. The steam turbine drives also avoid the energy losses associated with generation and transmission of electric power[74].

For plants with a capacity of less than 600 tonnes per day, reciprocating compressors are generally used. Horizontally opposed construction is the preferred reciprocating compressor layout – principally because good dynamic balance can be readily achieved. Depending on the required compression ratio, the units are designed with three to six compression stages. In some plants, the various compression services (air compression, natural gas compression, synthesis gas compression and recycle, and ammonia compression for refrigeration) are apportioned among the crankshaft "throws" in such a manner that a single compressor frame can perform all process compression services. Usually synchronous-type electric motors are used to drive reciprocating compressors[74].

6

AMMONIA SYNTHESIS

The ammonia synthesis reaction is:

$$N_2 + 3\,H_2 \leftrightarrow 2\,NH_3 \qquad \Delta H_{298} = 11.04 \text{ kcal/g-mol} \tag{6.1}$$

In most processes the reaction takes place on an iron catalyst. The reaction pressure is normally in the range of 150 to 250 bar, and temperatures are in the range of 350°C to 550°C. At the usual commercial converter operating conditions, the conversion achieved per pass is only 20% to 30%[53]. In most commercial ammonia plants, the Haber recycle loop process is still used to give substantially complete conversion of the synthesis gas. In the Haber process the ammonia is separated from the recycle gas by cooling and condensation. Next the unconverted synthesis gas is supplemented with fresh makeup gas, and returned as feed to the ammonia synthesis converter[74].

Synthesis loop arrangements differ with respect to 1) the points in the loop at which the make-up gas is delivered, 2) where the ammonia is taken out and 3) where the purge gas is taken out[53] (see Figure 6.1). The best arrangement is shown in Figure 6.1 (A). After the gas leaves the ammonia synthesis converter, ammonia is condensed/removed by cooling and the recycle gas is returned to the recycle compressor. This represents the most favorable arrangement from a minimum energy point of view. It results in the lowest ammonia content at the entrance to the converter and the highest ammonia concentration for condensation[74].

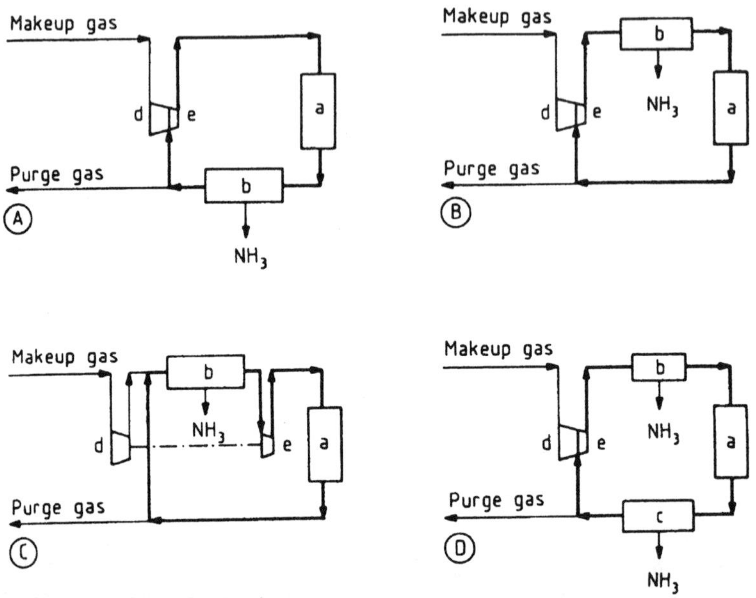

A) Synthesis Loop for Pure and Dry Makeup Gas
B) Product recovery After Recycle Compression
C) Product Recovery Before Recycle Compression (Four-Nozzle Compressor Design)
D) Two Stages of Product Condensation

a) Ammonia Converter with Heat Exchangers
b) Ammonia Recovery by Chilling and Condensation
c) Ammonia Recovery by Condensation at Ambient Temperature
d) Synthesis Gas Compressor
e) Recycle Compressor

Figure 6.1. Schematic flow diagrams of typical ammonia synthesis loops.[74] (Reproduced by permission of Wiley-VCH)

When the makeup gas contains water or carbon dioxide, the condensation stage is located partially or wholly between the makeup gas supply point and the converter [see Figure 6.1 (B)]. This arrangement is used because water and carbon dioxide are completely absorbed by condensing ammonia. The disadvantages of this arrangement are 1) the ammonia concentration for condensation is reduced by dilution with the makeup gas, 2) at equal condensing temperatures, a higher ammonia concentration exists at the inlet to the converter and 3) the ammonia produced in the vapor of the gas must be compressed with the recycle gas in the recycle compressor[74].

The design in Figure 6.1 (C) avoids the energy expenditure in Figure 6.1 (B). With this arrangement, recycle compression follows directly after the ammonia is condensed and separated. Therefore it is possible to cool the recycle gas using cooling water or air immediately before admixing the makeup gas (i.e., before diluting the recycle gas) and thereby reduce the energy expenditure for refrigerated cooling[74].

Figure 6.1 (D) splits the cooling step for ammonia condensation and also compresses recycle gas together with the makeup gas. This design is especially useful at synthesis pressures above 25 MPa (250 bar). At these pressures, a greater portion of the ammonia can be liquefied by cooling with cooling water or air[74].

The loop purge should be taken out after ammonia condensation and before make-up gas addition. This configuration depends on the make-up gas being treated in a drying step before it enters the loop. If the make-up gas contains traces of H_2O or CO_2, it must be added before ammonia condensation. However this addition point will have negative effects on both ammonia condensation and energy efficiency[53].

Conventional reforming with methanation as the final purification step produces a synthesis gas that contains inerts (CH_4 and argon) in quantities that do not dissolve in the condensed ammonia. Most of the inerts are removed by taking a purge stream out of the synthesis loop. The size of this purge stream controls the level of inerts in the loop at about 10% to 15%. The purge gas is scrubbed with water to remove ammonia and then it can be used as fuel or sent to hydrogen recovery.

6.1. REACTION RATE

Knowledge of the macrokinetics is important for solving the industrial problem of designing ammonia synthesis reactors, for determining the optimal operating conditions and for computer control of ammonia plants. Some of the considerations are:

- High pressure promotes a high rate of ammonia formation. This is based on Le Chatelier's principle that says a reaction that reduces the number of moles of gas (in this case from four moles of reactants to two moles of ammonia) will be favored by an increase in pressure (see Eq. 2.5);
- High ammonia concentration in the synthesis gas (recycle gas) restricts ammonia formation (see Figure 6.2);
- The rate of formation initially increases with rising temperature but then goes through a maximum as the system approaches thermodynamic equilibrium (see Figure 6.3); and
- With lower temperatures, the maximum rate shifts to a lower hydrogen-nitrogen ratio (see Figure 6.4)[74].

Figure 6.4. presents data from a commercial iron catalyst, Haldor Topsøe KMIR. The data shows a sharp drop in reaction rate with declining temperature at a 3:1 H_2-to-N_2 ratio in contrast to a 1:1 H_2-to-N_2 ratio. This may be attributed to a hindering effect by absorbed hydrogen at low temperature[74].

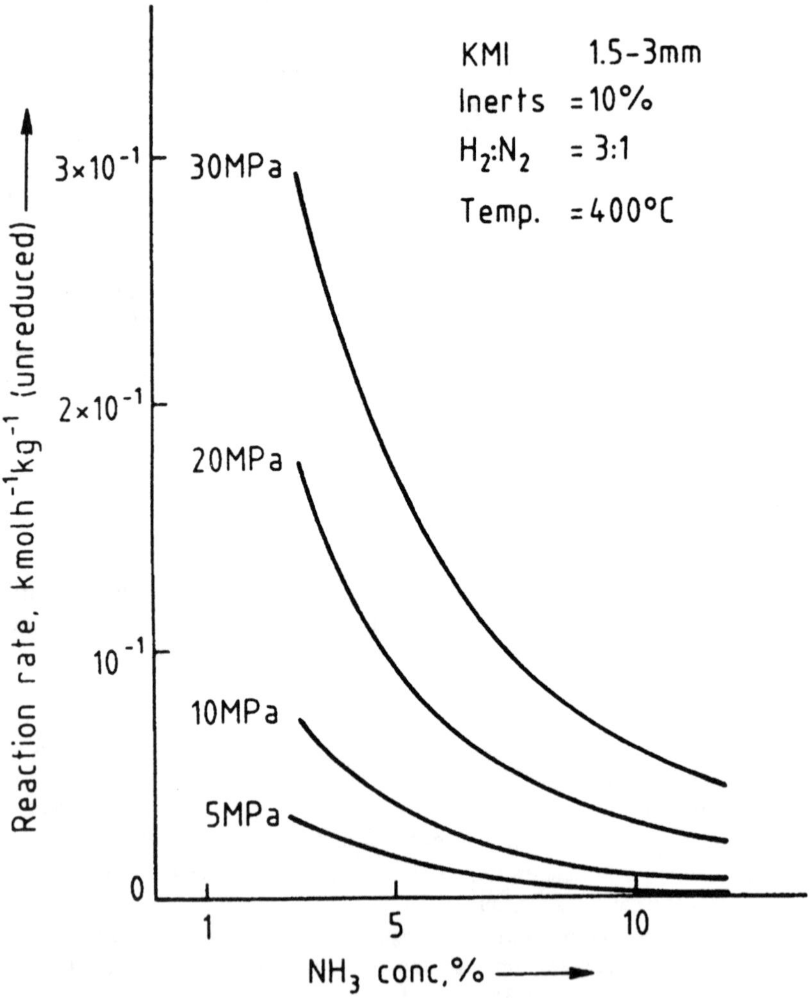

Figure 6.2. Reaction rate for NH_3 synthesis. Dependence on the ammonia concentration at various pressures.

(Reproduced by permission of Wiley-VCH)

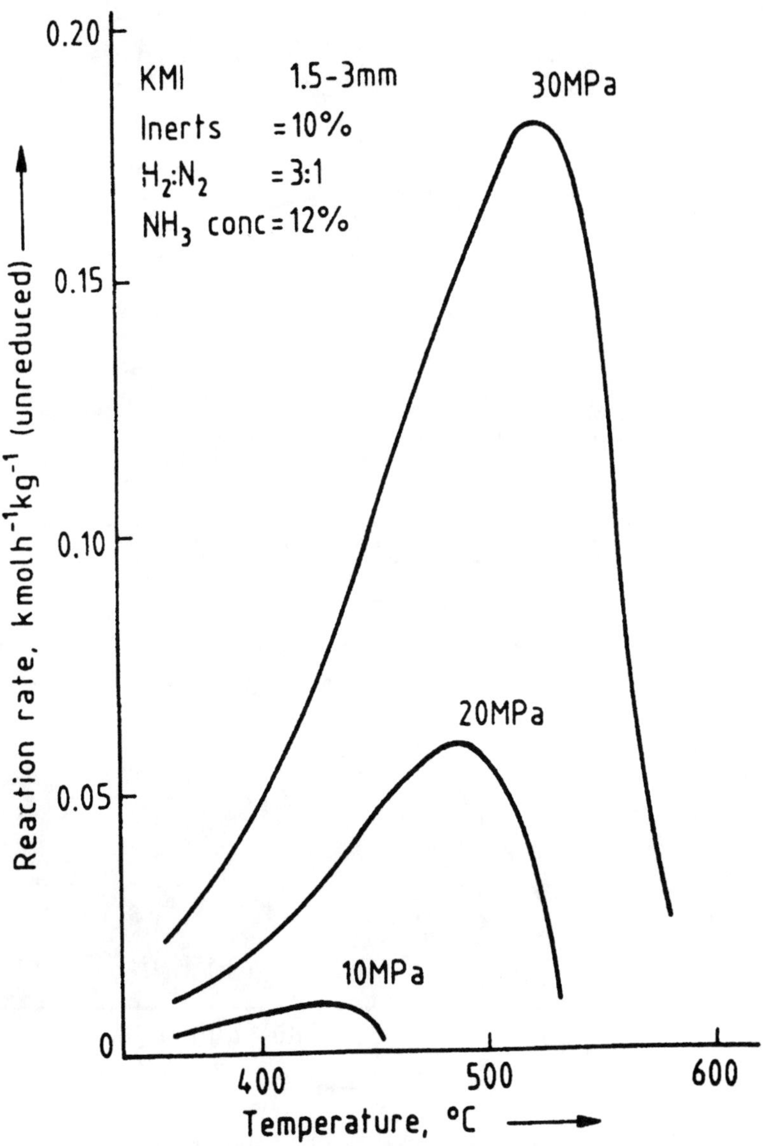

Figure 6.3. Reaction rate for NH_3 synthesis. Dependence on the temperature at various pressures. (Reproduced by permission of Wiley-VCH)

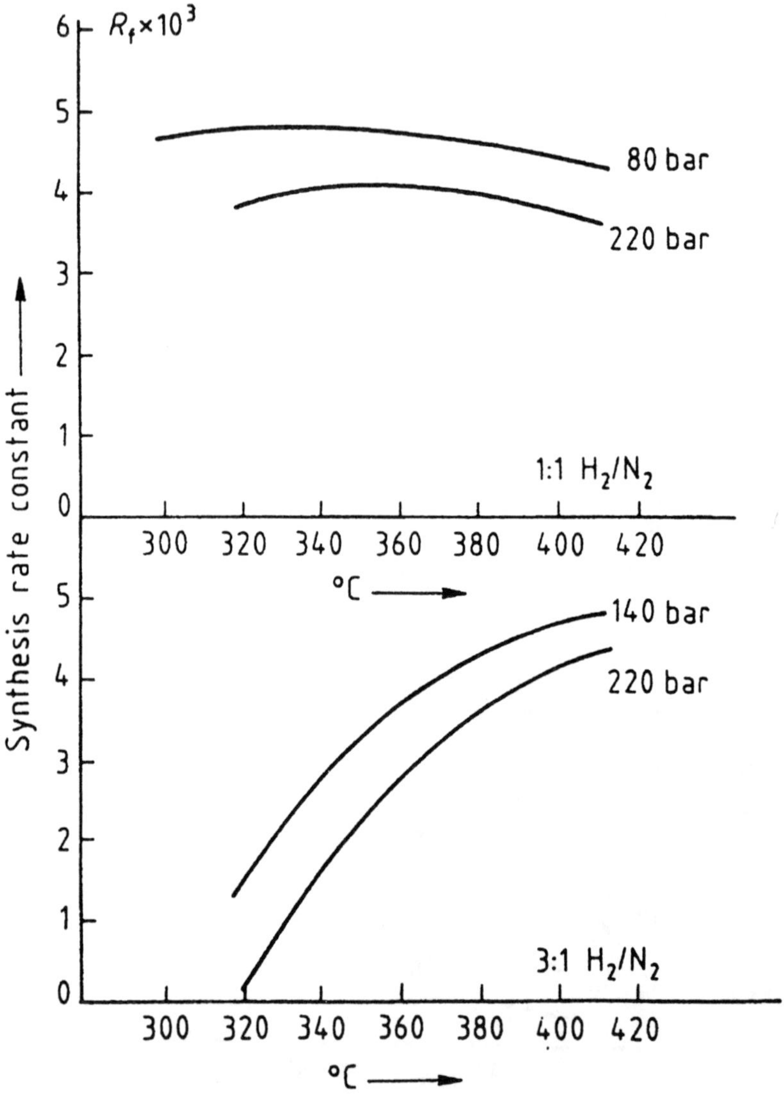

Figure 6.4. Ammonia synthesis rate constant dependence on hydrogen-nitrogen ratio. (Reproduced by permission of Wiley-VCH)

Rate equations for describing ammonia synthesis must therefore represent the influence of temperature, pressure, gas composition and equilibrium composition. Moreover, they must also take into consideration the dependence of the ammonia formation rate on the concentration of catalyst poisons and the influence of mass-transfer resistances. Since the beginning of commercial ammonia synthesis, a large number of different kinetic equations have been suggested. Each one is based on either a proposed reaction mechanism or an empirical evaluation. Contradictory data on the kinetics of ammonia synthesis may reflect a lack of attention to the influence of impurities in the gas. If oxygen compounds are present in the synthesis gas, reversible poisoning of the absorbing areas must be taken into account when developing rate equations[74].

In 2001 Hyprotech and Synetix announced an ammonia plant simulation that can be used for modeling, on-line monitoring and optimization of the plant. The simulation includes Synetix reactor models, customized thermodynamic data and information to simulate the performance of a range of catalysts. The reactor models in the simulation include Primary and Secondary Reformers, High Temperature Shift converter, Low Temperature Shift Converter, Methanator and Ammonia Synthesis Converter[80].

6.2. CATALYSTS

When ammonia is made from natural gas using the steam reforming process, several reaction stages are needed and catalysts are key to the economic operation of each stage. The chemistry for the different process steps and the basic materials that make up the catalysts are shown in Figure 6.5[82].

The ammonia synthesis catalyst may be viewed as the heart of an ammonia plant. For a given operating pressure and a desired production rate, the catalyst determines 1) the operating temperature range, 2) recycle gas flow and 3) refrigeration requirements. It also indirectly influences the makeup gas purity requirements. That means the catalyst affects the operating pressure as well as the capital and energy expense for synthesis gas production and purification[74].

Industrial catalysts for ammonia synthesis must satisfy the following requirements:

- High Catalyst Activity at the lowest possible reaction temperatures. This allows the process to take advantage of the favorable thermodynamic equilibrium that exists at low temperatures.
- The highest possible Insensitivity to Oxygen- and Chlorine-Containing Catalyst Poisons. Keep in mind that the effects of poisons (e.g., oxygen compounds) may become more severe as temperature declines.

- Long Life. This is determined essentially by resistance to a) thermal degradation and b) irreversible poisoning. In some plants, conventional iron catalysts have achieved service lives of up to 14 years.
- Mechanical Strength. Insufficient pressure and abrasion resistance may lead to an excessive increase in converter pressure drop. This eventually leads to a premature plant shutdown.

Metals or metal alloys are suitable as ammonia catalysts – especially those metals in the transition-metal group. Metals or metal compounds whose chemisorption energy of nitrogen is neither too high nor too low show the greatest effectiveness. Most catalysts are complex and contain other metal oxides that are hard to reduce. This promotes the activity of metallic iron[74].

Ammonia synthesis catalysts have traditionally been based on iron and have been made by the reduction of magnetite (Fe_3O_4). The difference between different commercially available products lies in optimized levels of metal oxide promoters that are included within the magnetite structure. These metal oxides promote activity and improve the thermal stability of the catalyst. Typical promoters are alumina (Al_2O_3), potassium oxide (K_2O), and calcium oxide (CaO). The interactions between the many components in the catalyst can radically affect 1) the initial reducibility, 2) the level of catalyst activity that is achieved, 3) the long-term catalyst performance and 4) the long-term catalyst stability[204].

Alumina functions as a "structural" promoter. It permits the production and preparation of an open and porous structure[204].

The potassium oxide is an "electronic" promoter. It increases the intrinsic activity of the high- surface-area iron particles that are produced when reduction occurs[204].

Calcium oxide, and other basic promoters, react with silica impurities in the raw materials to form glassy silicates. The glassy silicates enhance the thermal stability of the reduced iron. The main benefit is to minimize any neutralization of the K_2O promoter, which would reduce its effectiveness[204].

It should be noted that the KAAP process uses a ruthenium catalyst rather than an iron-based catalyst. The advantages of this catalyst and the KAAP process are discussed below[57]. In 2001 it was reported that Project and Development India Ltd. (PDIL) had a research program in place to produce ammonia at low temperature (100°C) and low pressure (20 to 40 kg/cm^2 g). The catalyst is based on cobalt and ruthenium[212].

The particle size and shape of commercial ammonia catalysts are determined mainly by two parameters: a) the catalyst performance and b) the pressure drop. From the standpoint of space-time yield, it is desirable to use the finest possible particle, which is about 1-2 mm. However, with decreasing particle size, the pressure drop increases and the danger of destructive movement of the catalyst increases[74].

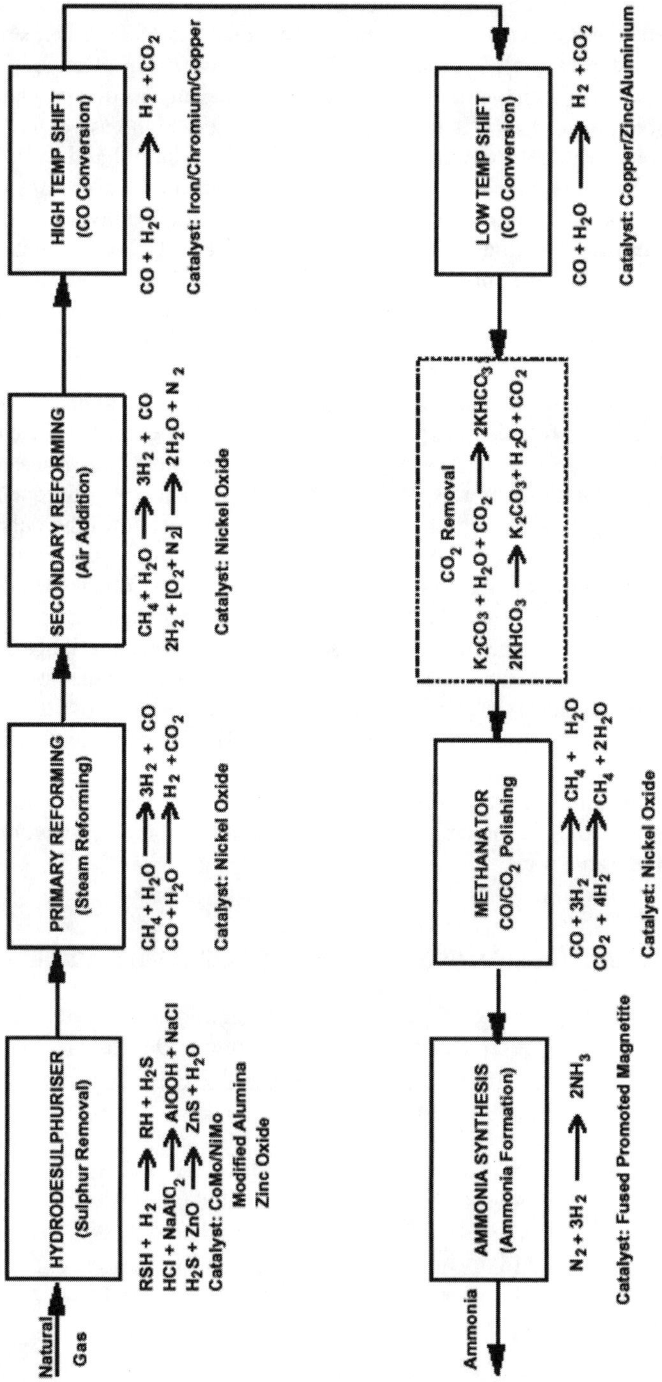

Figure 6.5. Basic catalyst materials and chemistry of ammonia process.
(Reproduced by permission of Johnson Matthey Catalysts. Copyright Johnson Matthey PLC)

For processes operating at pressures of 25 to 45 MPa (250 to 450 bar) and space velocities of 8,000 to 20,000 $m^3/m^3\ h^{-1}$ (with the gas at STP, i.e., standard temperature and pressure), a grain size of 6 to 10 mm is preferred. Larger granulations (8 to 15 mm or 14 to 20 mm), are used only in plants where the lowest possible pressure drop is essential because of very high gas velocities. In catalyst zones in which the ammonia formation rate is so high that the allowable temperature limits are exceeded, it may be advantageous as well to use coarse particles to suppress the reaction. The so-called radial converter and the horizontal cross-flow Kellogg converter operate at relatively low gas velocities. These converters can use small catalyst granulations (1.5 to 3.0 mm or 2 to 4 mm) and permit the optimal use of the converter volume. Fluidized bed processes were explored in the former Soviet Union in the 1960's, but no commercial scale processes are known to be operating[74].

Two effects cause the low production capacity of large-grained catalyst. First, large grain size retards transport of the ammonia formed inside the catalyst into the bulk gas stream. This is because the ammonia transport proceeds by slow diffusion through the pore system. The second effect is a consequence of the fact that a single catalyst grain in the oxide state reduces from the outside to the interior of the particle. The water vapor produced inside the catalyst by reduction comes into contact with already reduced catalyst on its way to the outer surface of the catalyst. This induces a severe recrystallization. As an example, if the particle size increases from about 1 to 8 mm, the inner surface decreases from 11 to 16 m^2/g to 3 to 8 m^2/g[74]. Therefore the choice of catalyst requires the optimization of 1) catalyst size versus catalyst activity, 2) catalyst size versus pressure drop across the converter and 3) the impact of 1 and 2 on operating costs[204].

The relationship between the catalyst size classification, the equivalent particle diameter and the percentage saving in catalyst or converter volume is summarized in Table 6.1[74].

Table 6.1. Effect of Catalyst Size on Savings in Catalyst Volume

Catalyst Size (mm)	Approximate Equivalent Particle Diameter[a] (mm)	Approximate Catalyst Volume Decrease (%)
6.0 to 10.0	7.5	-
4.5 to 8.0	5.5 to 6.5	4.8 to 7.4
3.6	4.5 to 4.7	10.0 to 11.5
1.5 to 3.0	2.0 to 2.2	18.0 to 20.0
1.0 to 1.5	1.2 to 1.3	21.0 to 22.3

[a] Effective Particle Diameter, d_{eff}, is defined as the ratio of equivalent diameter (Λ) and a form factor (ψ).

Ammonia synthesis catalysts are available in both 1) unreduced (i.e., oxidized) form and 2) pre-reduced and stabilized form. The pre-reduced catalysts give faster start-ups and minimize liquor disposal problems. The pre-reduced catalyst may be mixed with the unreduced catalyst, or a complete charge of pre-reduced catalyst may be used. The optimum proportion of pre-reduced catalyst depends on:

- The number of days typically taken to reduce the unreduced catalyst,
- Plant design and operation,
- Environmental considerations, and
- The value of extra ammonia production.

Typical reduction times with different catalyst mixtures for a three-bed quench converter are shown in Table 6.2[204].

Table 6.2. Typical Reduction Times with Different Mixtures of Unreduced and Pre-Reduced Catalysts in a Three-Bed Quench Converter

Composition of Catalyst in Converter	Typical Reduction Time (Days)
Complete Charge of Unreduced Catalyst	4 to 5
First Bed Charged with Pre-Reduced Catalyst	2.5 to 3.5
First & Second Beds Charged with Pre-Reduced Catalyst	2.0 to 2.5
Three Beds Charged with Pre-Reduced Catalyst	1.0 to 1.5

Oxygen compounds (CO, CO_2, etc.) are the most significant poisons of ammonia synthesis catalysts because they have a strong affinity for the catalyst and are preferentially adsorbed on its surface. Other poisons include sulfur, arsenic, phosphorus and chlorine. Some heavy metals (e.g., copper and nickel oxides as well as lead, tin and bismuth) are also considered poisons. Deposits from decomposition products of machine oil may cause macroscopic blocking of the catalyst surface while olefins in the makeup gas lead to carbon deposition that contaminates the catalyst. Under normal synthesis conditions methane is not a poison, but it concentrates in the ammonia recycle loop as an inert gas. This will reduce converter capacity by lowering the nitrogen-hydrogen partial pressure[74].

The activity of damaged catalysts may be almost completely restored by reduction with clean synthesis gas at a relatively low temperature. It has also been found that the degree of poisoning rises as the partial pressure ratio (p_{H2O} / p_{H2}) increases, and the degree of poisoning falls with increasing temperature. The effect of water vapor concentration on ammonia catalyst activity is shown in Figure 6.6. This figure shows the reversible effect of increasing water vapor concentrations in the synthesis gas on the activity of industrial ammonia catalysts[74].

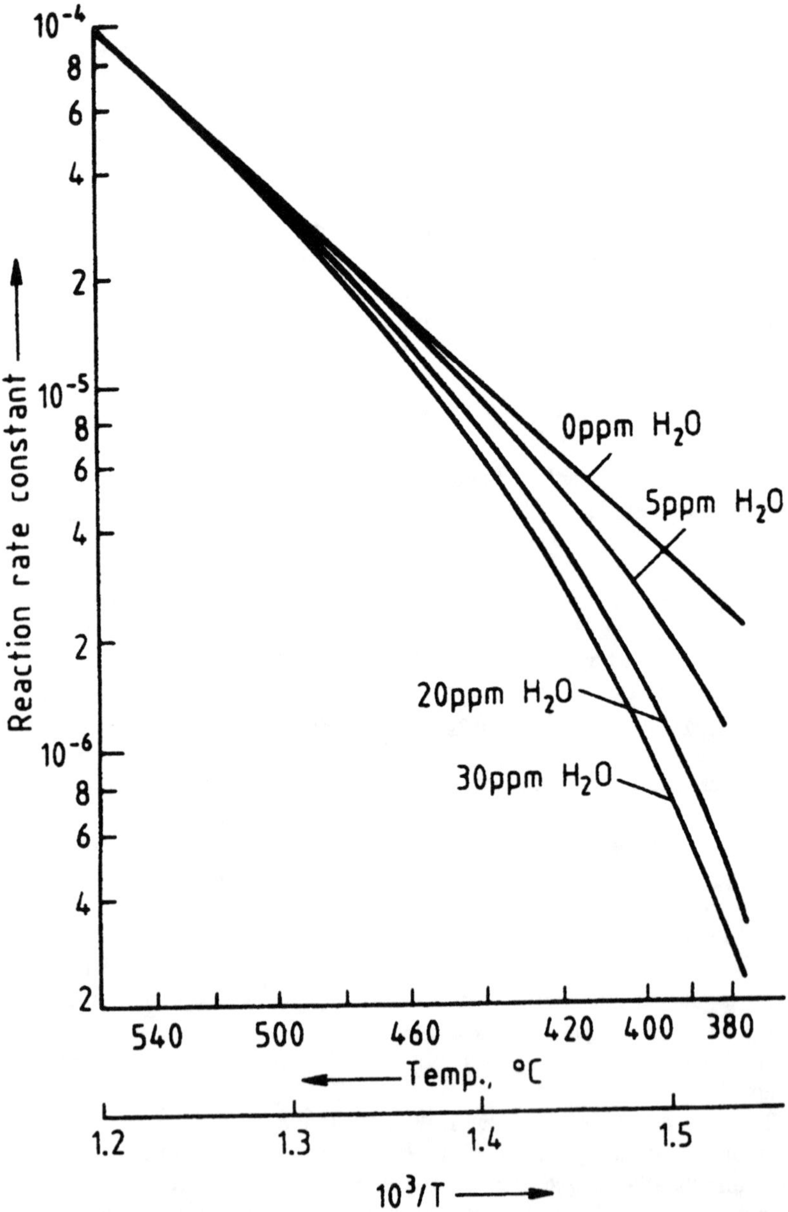

Figure 6.6. Reversible effect of increasing water vapor concentrations in the synthesis gas on the activity of industrial ammonia catalysts.

(Reproduced by permission of Wiley-VCH)

With continuing exposure to oxygen compounds, irreversible damage to catalyst activity may occur due to the growth of iron primary crystallites. This damage depends on the water vapor partial pressure and is especially serious at high temperatures.

6.3. ENERGY EFFICIENCY

Due to increased feedstock costs, some new ammonia plant designs use fuel more efficiently but their capital cost may be higher. The recovery of the hydrogen and ammonia from the synthesis purge gas by a cryogenic unit or a membrane system results in an ammonia capacity increase of about 5%[57].

Substantial improvements have been made in the energy efficiency of CO_2 removal systems. The first large-scale ammonia plants in the 1960's typically used monoethanolamine (MEA) as a solvent. Energy input was over 50,000 kcal/kg-mol of CO_2 removed. In 2001 plants use improved solvents and designs that can reduce the energy input to about 10,000 kcal/kg-mol of CO_2 removed[57].

A large ammonia plant in 2001 is more fuel-efficient than plants that were built in the 1970's and 1980's. A typical world-scale plant that was built in the 1970's consumed about 42 billion BTU of natural gas per tonne of ammonia produced. Retrofitting such a plant to improve fuel efficiency can reduce gas consumption to about 36 million BTU per tonne. Ammonia plants that were built in the late 1990's use only about 30 million BTU per tonne of ammonia, are easier to operate and have slightly lower conversion costs. Some new plants also recover more than one million BTU per tonne by generating electricity from waste heat[57].

6.4. AMMONIA PLANT DESIGN

A simplified flowsheet for an ammonia plant that processes natural gas via steam reforming is shown in Figure 6.7. A block diagram of this same plant is shown in Figure 6.8. This diagram lists typical stream compositions, typical operating conditions, catalyst types (recommended by Synetix) and catalyst volumes[82].

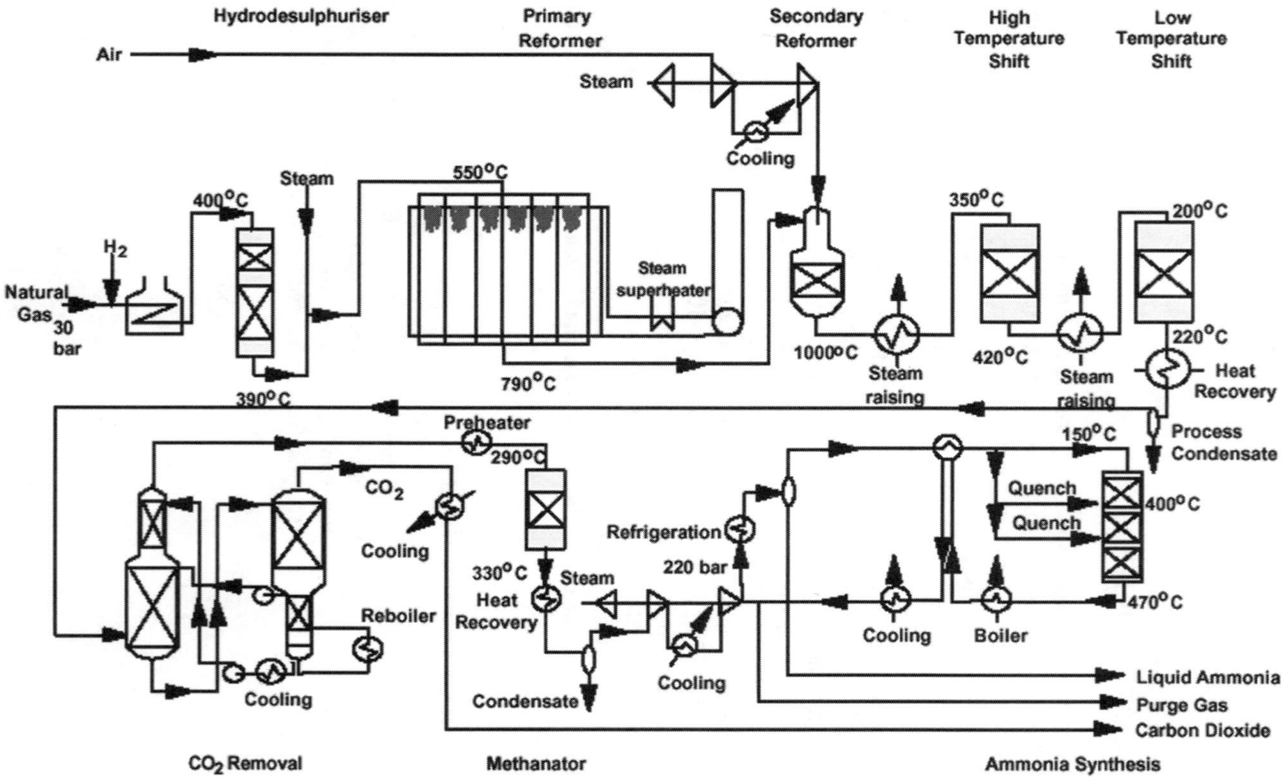

Figure 6.7. Simplified ammonia plant flowsheet.
(Reproduced by permission of Johnson Matthey Catalysts. Copyright Johnson Matthey PLC.)

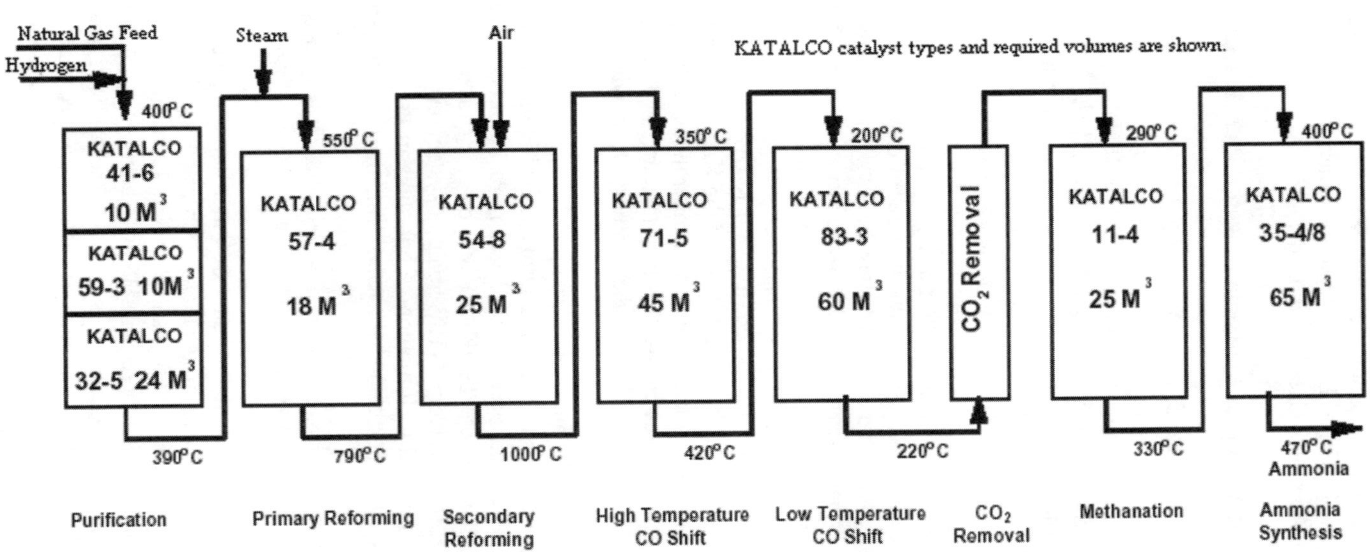

Exit Gas Composition						MUG	Loop
	Volume %	Volume %	Volume %	Volume %	Volume %	Volume %	Volume %
N_2	69.8	56.5	60.3	61.4	74.6	74.2	63.4
CO	9.3	12.9	3.0	0.3	0.4	-	-
CO_2	10.5	7.5	15.6	17.8	0.1	-	-
CH_4	10.4	0.3	0.3	0.2	0.3	0.8	9.5
N_2+ A	-	22.8	20.8	20.3	24.6	25.0	24.9
CO + CO_2	-	-	-	-	-	5ppm	<5ppm
							NH3 - 2.5

Figure 6.8. Block diagram of 1,000 tonne/day ammonia plant.
(Reproduced by permission of Johnson Matthey Catalysts. Copyright Johnson Matthey PLC.)

Some of the ammonia process technologies that are available include: KAAP*plus*™, Haldor Topsøe, LAC (or Linde Ammonia Concept), LCA (or Leading Concept Ammonia), Ammonia Casale, and Uhde.

6.4.1. Kellogg Brown & Root Advanced Ammonia Process – Plus (KAAP)

The KAAP*plus*™ (Kellogg Brown & Root Advanced Ammonia Process – Plus) is an ammonia process design made up of commercially proven technologies – KBR Reforming Exchanger System (KRES), KBR Purifier, and KBR Advanced Ammonia Process KAAP) for synthesis. KAAP*plus*™ uses KRES with excess process air fed to the autothermal reformer. Excess nitrogen, along with the methane, most of the argon, and other impurities are removed in the KBR Purifier. The resulting high-purity makeup synthesis gas is compressed to the KAAP synthesis loop pressure of about 90 bar[81]. This process uses a ruthenium-based, copromoted catalyst on a graphite support. It is as much as twenty (20) times more active than the iron-based catalysts. The new catalyst results in improved synthesis efficiency because the synthesis pressure is lowered from a typical pressure of 150 bar (2175 psig) to 90 bar (1305 psig)[57, 203]. Typical conversions with the ruthenium catalyst are 20% compared to 15% to 16% with conventional, iron-based catalysts.

The first application in 1992 used a two-bed, hot-wall KAAP reactor that featured a low pressure drop and radial flow. Because of the KAAP catalyst's high activity, thin beds are necessary to keep operating temperatures within the desired range[203]. In 2002 the KAAP reactor had evolved to a four-bed design. A magnetite catalyst is used in the first bed of the synthesis loop when the ammonia concentration is below 2% of the feed. Then the ruthenium catalyst is used in the next three beds to bring the ammonia level up to 18% or more[215].

In most radial-flow converters, the upper portion of the bed is sealed with excess, unused catalyst. This design prevents feed gas from by-passing the reaction section when the catalyst settles. The KAAP reactor uses a proprietary sealing system to overcome this problem. This sealing system avoids the catalyst maldistribution that can lead to formation of hot spots in the catalyst bed. The system also allows 100% of the loaded catalyst volume to be utilized for the ammonia conversion reaction[203].

For start-up the KAAP catalyst is loaded in the oxidized state. It is reduced by passing fresh synthesis gas over the catalyst beds and heating them to 300°C[203].

The KAAP catalyst generates a smaller amount of water during reduction because it contains much less metals than conventional iron catalysts. For this reason, the catalyst reduction proceeds swiftly and required 19 hours in one installation. (Iron catalyst reduction usually requires 36 to 48 hours.)[203]

Another item that can improve operation is a mass spectrometer that analyzes the KAAP streams for hydrogen and nitrogen. The mass spec enables closer control of the H/N ratio that enters the loop and, as a result, provides more consistent ammonia conversion[203].

Capital cost savings of 3% to 8% may be achieved with KAAP*plus*™ when compared with other designs. These savings are a result of the following design features:

- No air separation plant,
- No primary reformer,
- Single case synthesis gas compressor,
- Synthesis at 90 bar,
- No purge gas recovery unit and
- Reduced plot space.

The maintenance costs are lower because KAAPplus™ eliminates the primary reformer and greatly simplifies the syngas compressor – the two highest maintenance items in a conventional ammonia plant. KAAPplusTM also reduces the severity of the operating conditions. Reforming exchanger and autothermal reformer exit temperatures are about 100°C lower than in conventional plants. Synthesis pressure at 90 bar is substantially lower than other processes[81].

6.4.2. Haldor Topsøe

Haldor Topsøe's ammonia synthesis technology is based on the S-200 ammonia converter. This is a two-bed radial flow converter with indirect cooling between the beds. Features of the S-200 include efficient use of converter volume and low pressure drop (factors related to the use of small catalyst particles: 1.5 to 3.0 mm), and high conversion per pass due to indirect cooling[85].

The S-200 converter concept has been used extensively to upgrade existing converters (Haldor Topsøe or other designs) in modification projects to achieve higher capacity and/or better energy efficiency. Modifications may involve installation of a new S-200 converter, installation of a new S-200 basket in an existing pressure shell with full opening closure or an in situ modification of the existing internals of bottle-shaped converters. Modifications of this type will typically result in a reduction in specific energy consumption of about 0.3 Gcal/tonne of ammonia and/or a capacity increase of up to 20%[85].

Haldor Topsøe also offers the S-250 concept and the three-bed S-300 converter. The S-250 uses the S-200 converter followed by a one-bed S-50 converter with a steam boiler or steam superheater between the two converters[85]. The S-50 catalyst bed inlet temperature (which is equal to the converter inlet temperature) is controlled by the amount of steam superheating that occurs upstream of the S-50. The S-300 has an internal heat exchanger installed between the second and third catalyst beds. In 2001 a total of thirteen (13) S-300 converters were in operation or under construction. A cut-away view of an S-300 converter is shown in Figure 6.9[207].

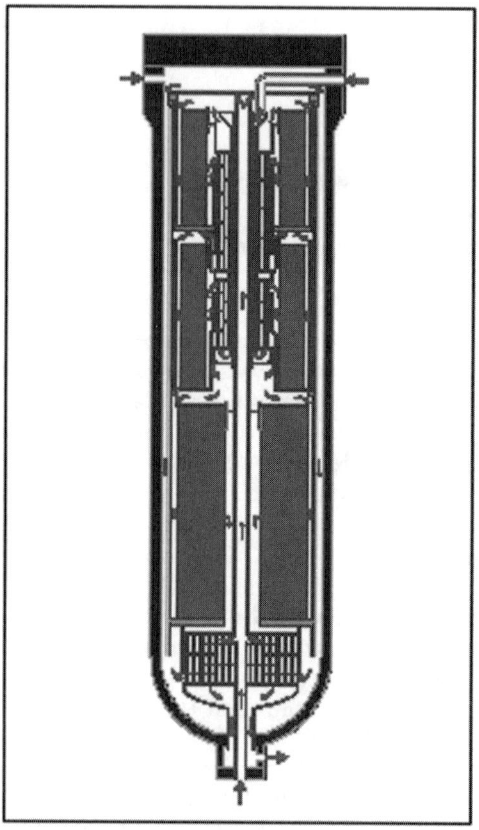

Figure 6.9. Haldor Topsøe S-300 Ammonia Converter.
(Reproduced by Permission of Haldor Topsøe)

6.4.3. Linde Ammonia Concept (LAC)

The Linde Ammonia Concept (LAC) plant consists of a modern hydrogen unit, a standard nitrogen unit and a high-efficiency ammonia synthesis loop based on the Ammonia Casale design. The first LAC plant, a 1,350 tonne per day ammonia plant, was designed and built for GSFC in Baroda, India. It started up in 1998. In 2001 two more LAC plants were under construction in Australia - a 230 tonne per day plant and a 600 tonne per day plant. Total energy needs are 7 Gcal per tonne or 25.2 MM BTU per short ton[208, 214].

The ammonia loop is based on the Ammonia Casale axial-radial three-bed converter with internal heat exchangers. Heat from the ammonia synthesis is used to 1) generate high-pressure steam and 2) preheat feed gas. The gas is then cooled and refrigerated to separate ammonia product. Unconverted gas is recycled to the syngas compressor[208, 214].

In Figure 6.10 the LAC technology is compared to a conventional ammonia plant[208].

6.4.4. Leading Concept Ammonia (LCA)

The Leading Concept Ammonia (LCA) process is designed to approach the capital cost and energy advantages of larger capacity (> 1,000 tonne per day) ammonia plants while producing only 400 to 600 tonnes per day of ammonia. This technology was developed by Imperial Chemical Industries (ICI), and the first plant started up in 1988.

- The main features of the process are:
- Milder Primary Reformer Conditions,
- A Gas-Heated Reformer (which receives heat directly from the secondary reformer process gas),
- A Single-Stage CO Shift Reactor,
- A Pressure Swing Absorption step to Replace the Conventional CO_2 Removal step, and
- A low pressure ammonia synthesis loop.

The gas-heated reformer (GHR) reduces the need for the high pressure steam system associated with the primary reformer and also minimizes or eliminates nitrogen oxide formation. The GHR is discussed in more detail in Chapter 5.

In the LCA process natural gas feed is mixed with recycle hydrogen, heated and desulfurized. It is then cooled by preheating the feed to the desulfurizer. Next it enters the feed gas saturator where it is contacted with circulating hot process condensate. The feed gas from the saturator is mixed with steam to give a steam-to-carbon ratio of about 2.5. Then it is preheated in the reformed gas stream. The reactants enter the primary reformer, which operates with an exit temperature of 700 to 750°F and a pressure of 30 to 45 bar. The gas mixture is then fed to a secondary reformer for further reforming with an excess of process air. The reformed gas is cooled by providing the heat for the primary reforming reaction and preheating the reactants. The cooled, reformed gas is shifted in an isothermal, water-cooled, low-temperature shift converter[17].

After shift conversion, the gas is cooled by direct contact with circulating process condensate and then fed to a pressure swing adsorption (PSA) unit to remove excess nitrogen, CO_2 and inerts. The CO_2 can be recovered from the pressure swing adsorption waste gas by using an aqueous solution of tertiary amine[17].

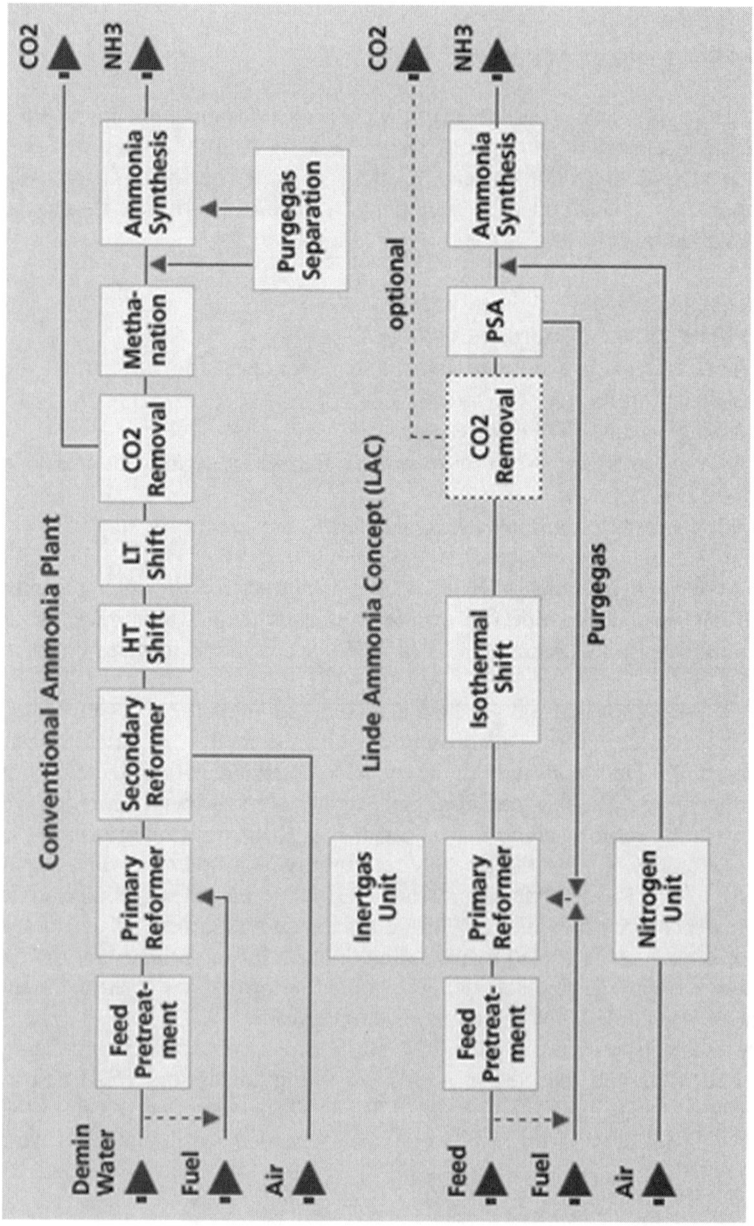

Figure 6.10. Linde Ammonia Concept (LAC) Plant compared to a conventional ammonia plant. (Reproduced by permission of Linde Engineering)

The gas that leaves the PSA unit is methanated, cooled and dried. The dried gas enters the ammonia synthesis loop at the circulator suction. In the synthesis loop, gas from the circulator is heated and passed over low pressure ammonia synthesis catalyst to produce ammonia[17].

The hot gas leaving the ammonia converter is cooled by generating 60 bar steam and by heating the feed gas to the converter. Ammonia is separated from the partially cooled gas by vaporizing ammonia. The unreacted gas is returned to the circulator. Argon and methane are removed from the synthesis loop by taking a purge and recycling back to the synthesis gas generation section as feed[17].

Energy consumption by the process is estimated to be 29.3 GigaJoules per tonne of ammonia for natural gas feed and 30.1 GigaJoules per tonne of ammonia for naphtha feed[16].

6.4.5. Ammonia Casale

Ammonia Casale of Lugano, Switzerland has been involved in synthetic ammonia production since 1921. Between 1985 and 2001 they were involved in 111 revamp projects and 17 new plants. Ammonia Casale sells an axial-radial reactor design (see Figure 6.11). In the design the gas flow pattern across the catalyst beds is axial-radial. Compared to an ammonia converter with axial flow, this design gives lower pressure drop and higher conversion per pass by 1) using a small catalyst (1.5 to 3.0 mm) and 2) ensuring better utilization of the catalyst volume. The lower pressure drop is possible because the overall synthesis loop flow is reduced. The reduced synthesis loop flow also leads to lower energy costs for compressing synthesis gas and lower energy costs for refrigeration[209, 210].

For revamps of ammonia synthesis converters, Ammonia Casale offers 1) an in-situ modification of bottle-shaped converters of the Kellogg type, and 2) a three-bed intercooled configuration. The intercooled design is similar in some ways to the Uhde design discussed below[213].

In an axial-radial catalyst bed, approximately 90% of the gas passes through the catalyst bed in a radial direction. The balance passes down through a top layer of catalysts in an axial-radial direction. This design eliminates the need for a top cover on the catalyst beds. The axial-radial design offers higher utilization of the available catalyst volume while the three-bed interchanger provides the most thermodynamically efficient cartridge configuration. Cooling is achieved by means of heat exchangers both between the first and second beds and between the second and third beds[213].

The Ammonia Casale design also includes catalyst-containing beds that are easy to handle and are inexpensive. Their design also uses a reverse bottom that increases the catalyst-filling efficiency[213].

Ammonia Casale does not have a proprietary catalyst. They design their equipment to work with the different catalysts that are commercially available[209].

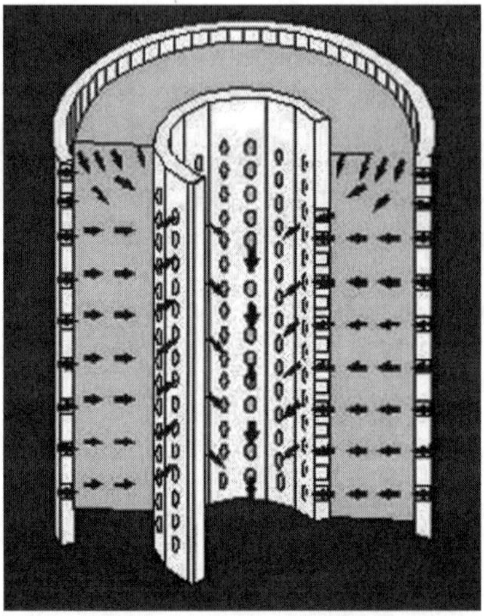

Figure 6.11. Ammonia Casale Axial-Radial Reactor Design.
(Reproduced by permission of Casale Group)

6.4.6. Uhde

Two Uhde design concepts are available and are compared in Table 6.3 and Figure 6.12[211]. They can provide 1) a single converter that contains three ammonia synthesis beds (see Figure 6.13) or 2) a two converter system that also contains three synthesis beds (Figure 6.14).

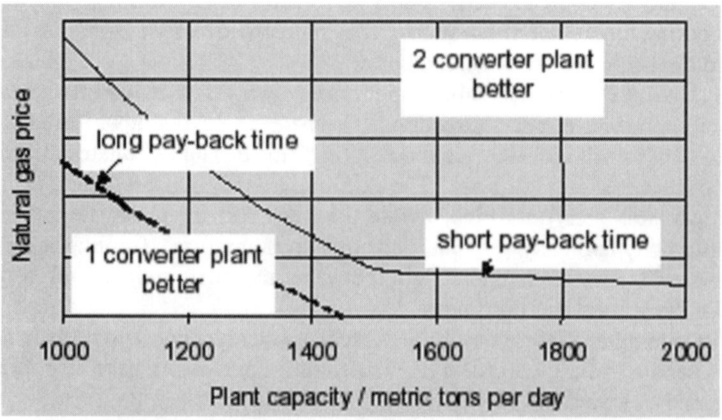

Figure 6.12. Ammonia converter concept selection.
(Reproduced by permission of Uhde GmbH and copyright by Uhde)

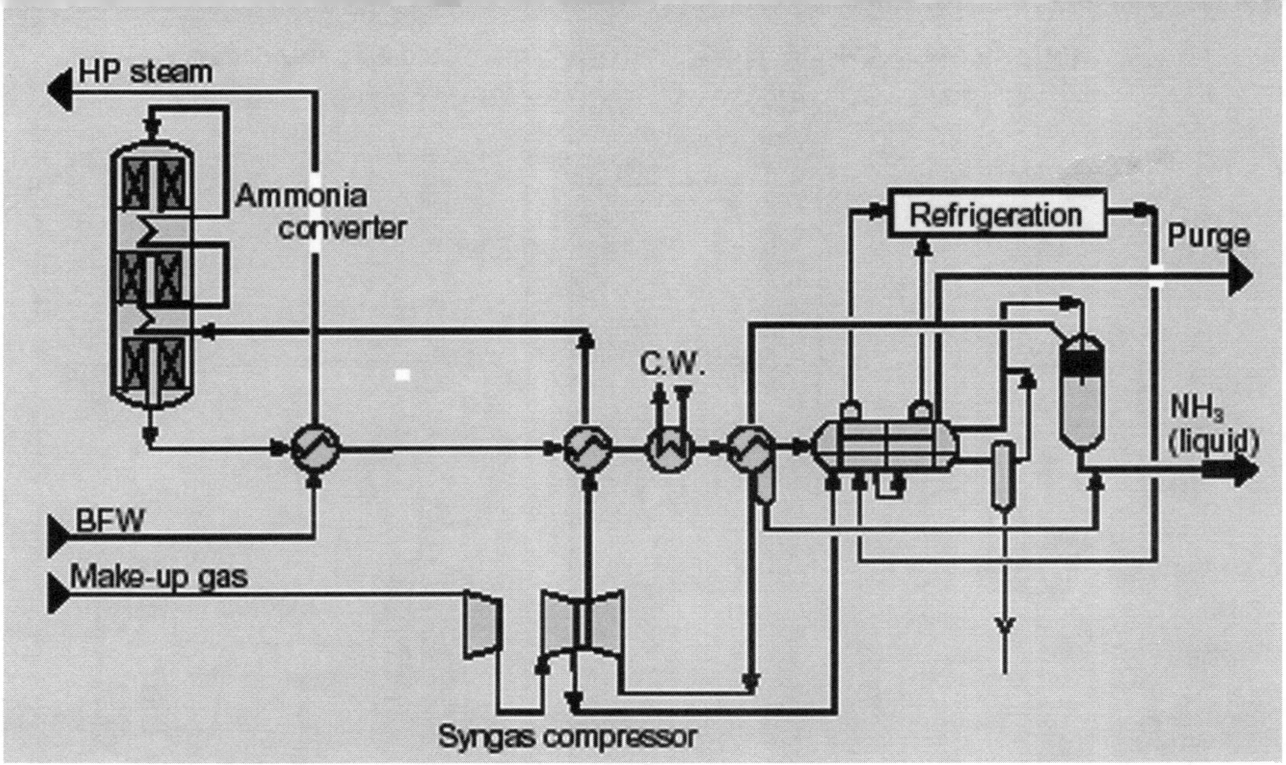

Figure 6.13. Single Ammonia Converter with Three Synthesis Beds.
(Reproduced by permission of Uhde GmbH and copyright by Uhde)

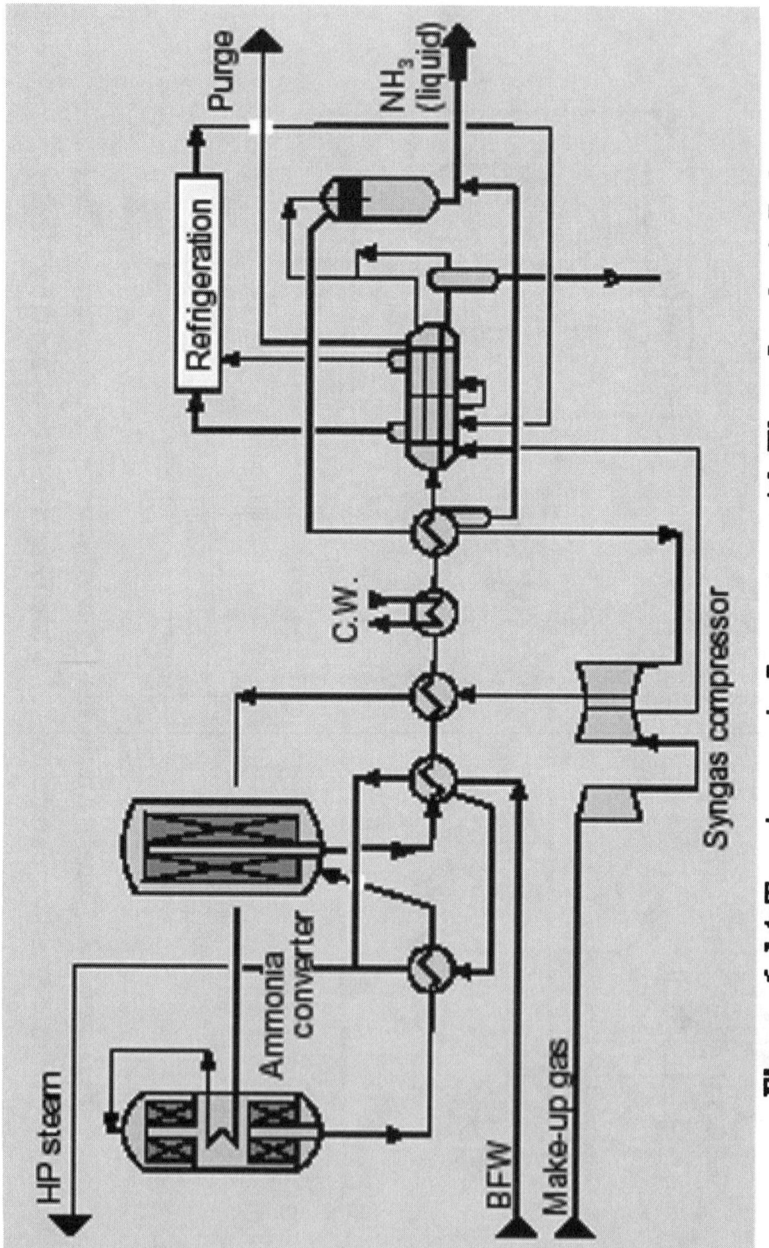

Figure 6.14. Two Ammonia Converters with Three Synthesis Beds. (Reproduced by permission of Uhde GmbH and copyright by Uhde)

Table 6.3. Comparison of Uhde Ammonia Converter Concepts

Item	Single Converter	Two Converters
Investment	Lower	Higher
Energy Efficiency	Reasonable	Best
Capacity	Up to 1800 tonnes per day	Very Large Capacities

6.4.7. Converter Comparison

As discussed above, several different types of ammonia converters are available. These types include axial quench converters (e.g., standard Kellogg reactors), tube cooled converters (e.g., TVA and Synetix designs), axial-radial designs (e.g., Ammonia Casale retrofit) and Kellogg's horizontal design. Typical operating data for different types of ammonia converters are shown in Table 6.4[204].

Table 6.4. Operating Data in Different Ammonia Converters

Converter Type	3-Bed Quench	4-Bed Quench	Axial-Radial	Tube Cooled
Design Output (tpd)	1,100	1,040	1,100	580
Catalyst Age (years)	4	5	1	8
Actual Output (tpd)	260	1,090	1,488	600
Catalyst Volume (m3)	44	71	51	16
Space Velocity (hr-1)	12,000	11,000	12,500	22,000
Pressure (bar)	220	145	197	320
Exit Ammonia (%)	17.3	10.3	17.7	14.9

6.5. AMMONIA SEPARATION

The removal of ammonia product is accomplished via mechanical refrigeration or absorption / distillation. The choice is made by examining the fixed and operating costs. Typically, refrigeration is more economical at synthesis pressures of 100 atmospheres (1470 psia) or greater. At lower pressures, absorption/distillation is usually favored. The two methods are compared in Table 6.5[212].

Table 6.5. Comparison of ammonia separation techniques

Condensation	Absorption
High Energy Costs at Lower Loop Pressures (Below 100 Atmospheres)	Almost constant energy costs that are independent of pressure. Less condensation separation below a pressure of 100 atmosphere.
Higher Fixed Costs at Pressures Below 100 Atmospheres	Almost constant fixed costs at a range of pressures. Less condensation separation at pressures below 100 atmospheres.
Economical at Higher Operating Pressures	Economical at lower synthesis pressures (above 100 atmospheres) compared to condensing process.
Energy consumption in refrigeration cycles.	Inefficient energy consumption in the distillation process.
Simple process with condensers and separators.	More complex process with absorber, distillation column, pumps, reboilers, condensers, and reflux accumulators. Associated instrumentation is also complex.
No change in potential for catalyst poisoning.	Increase in potential for catalyst poisoning due to oxygen in the absorbents.

Usually the ammonia concentration of the recycle is 3% to 4%, but reducing this amount to 1.5% can increase plant capacity by about 2.5%. However, the additional separation can often represent a significant addition to the capital cost of the plant and may not be economical for a retrofit (depending on operating pressure). However lower ammonia concentration in the recycle can be reviewed for a grass root project where capacity gains can be realized with an additional investment[212].

Decreasing the ammonia concentration in the recycle stream of existing plants is also hampered by the high energy cost required for water absorption. Norsk Hydro (Norway) developed a method to reduce the recycle ammonia concentration to about 0.5% via absorption in glycol (DEG or diethylene glycol). This process can be installed in a high pressure loop (> 100 atm) and in combination with a condensation unit. The installed cost may be lower than a comparable mechanical refrigeration system[212].

6.6. LARGE CAPACITY AMMONIA PLANTS

In 2002 the design and construction of ammonia plants with capacities of 3,000 tonnes per day or more were underway. These large capacity plants will be at least 50% bigger than the largest plants that are operating in 2002. The benefit of bigger plants is to realize economy of scale. Doubling the size of a single-train plant has the potential to reduce the capital-related cost of production by about 20%[215].

Qafco (in Qatar) is working with Uhde to build an ammonia plant with a capacity of 3,500 tonnes per day. Startup is scheduled for 2004[216]. The Saudi Arabian Fertilizer Co. (SAFCO) let a contract in 2002 for an ammonia plant with a capacity of 3,000 tonnes per day. Startup is scheduled for 2005[215].

In 2002 the world's largest single-train grassroots ammonia plant is a 2,500 tonne/day unit operated by Profertil S.A. (Buenos Aires, Argentina). It uses Haldor Topsøe technology, as does a 2,000 tonne/day plant owned by Kaltim Pasifik Amoniak (Bontang, Indonesia). BASF operates an upgraded plant in Antwerp, Belgium at 2,060 tonnes/day. This plant was built by Uhde GmbH and had an original capacity of 1,800 tonnes/day[215].

KBR (Houston, Texas) has built three 1,850 tonne/day plants in Point Lisas, Trinidad for Caribbean Nitrogen Co., and a fourth plant is under construction. The third plant has attained a production rate of up to 2,000 tonnes/day since it was started up in July, 2002. All these plants use KBR's KAAP technology[215].

Lurgi and Ammonia Casale have teamed up to develop the MEGAMMONIA® technology that is designed to produce 4,000 tonnes per day of ammonia or 1,400,000 tonnes per year[291].

6.6.1. KBR Design

KBR has designed a 4,000 tonne/day plant in which all the equipment is single-train except for the primary reformer and the ammonia converter, each of which consists of two identical, parallel units. The design uses the KAAP*plus*™ technology , which includes the KBR Reforming Exchanger system (KRES) and the KBR Purifier. These technologies are discussed earlier in this chapter and in Chapter 5[215].

The 4,000 tonne/day plant is designed to use two parallel KRES systems that have a capacity of 2,000 tonnes/day in each system. This choice should avoid scale-up problems[215].

6.6.2. Haldor Topsøe

The Haldor Topsøe design for large capacity ammonia plants will be based on their experience with the 2,050 tonne/day plant that Profertil operates in Argentina and on the 2,000 tonne/day plant that Kaltim Pasifik Amoniak operates in Indonesia.

In both plants the primary reformer uses a side-fired reformer design. Despite the large capacity of the plants, the number of reformer tubes is well below the maximum limit. By using a microalloy that contains titanium for the reformer tube material, a high heat flux can be used in the design. This will minimize the size of the primary reformer. The smaller size of the reformer box requires reformer burners with a higher heat release because the area available for burner installation is smaller than that in older reformers. These design practices show that a reformer for a 3,000 tonne/day, single-line ammonia plant can be built[217].

In the ammonia synthesis converter, a rate of 3,000 tonnes/day can be achieved by combining the higher loop pressure used at Profertil (195 kg/cm^2 g) with the S-250 ammonia synthesis loop (used at Kaltim Pasifik Amoniak)[217]. Further design studies have shown that their converter technology can use conventional a magnetite catalyst for plants that produce 4,000 to 4,500 tonnes/day. Such a plant could be built in a single train with no parallel pieces of equipment[215].

A ruthenium catalyst that can perform ammonia synthesis at 200 bar is being developed by Haldor Topsøe in 2002. Although the catalyst was not yet ready for commercial use, the combination of high activity with high pressure is very well suited for the design of plants with capacities greater than 4,000 tonnes/day. This is because high pressure is needed to keep equipment and piping at reasonable sizes in very large plants. The catalyst is able to handle higher pressures because it is on a boron nitride carrier[215].

6.6.3. Uhde

Uhde's solution to the challenge of building larger plants is a dual-pressure process that uses Synetix catalysts. A capacity of 3,300 tonnes/day can be designed with all proven high-pressure equipment for the synthesis loop and compressor. It is also possible to reach capacities of 4,000 tonnes/day if the casing of the synthesis gas compressor is increased to the next size, which is available[218].

A standard primary and secondary reforming front end of the plant will operate at proven process conditions. Scale-up of the reforming section is straight-forward when using a top-fired box reformer with a cold outlet manifold system. The size of the reformer is much smaller than reformers currently designed for large-scale methanol projects[218].

The ammonia synthesis takes place in two steps at different pressures (see Figure 6.15). In the first step, makeup gas is compressed to about 110 bar in a two-stage, intercooled compressor. The makeup gas is then fed to a three-bed, intercooled, once-through converter to produce about one-third of the ammonia. About 85% of the ammonia product is separated from the gas, which is then compressed to the standard pressure of up to 210 bar for the second-stage ammonia synthesis loop[215, 218].

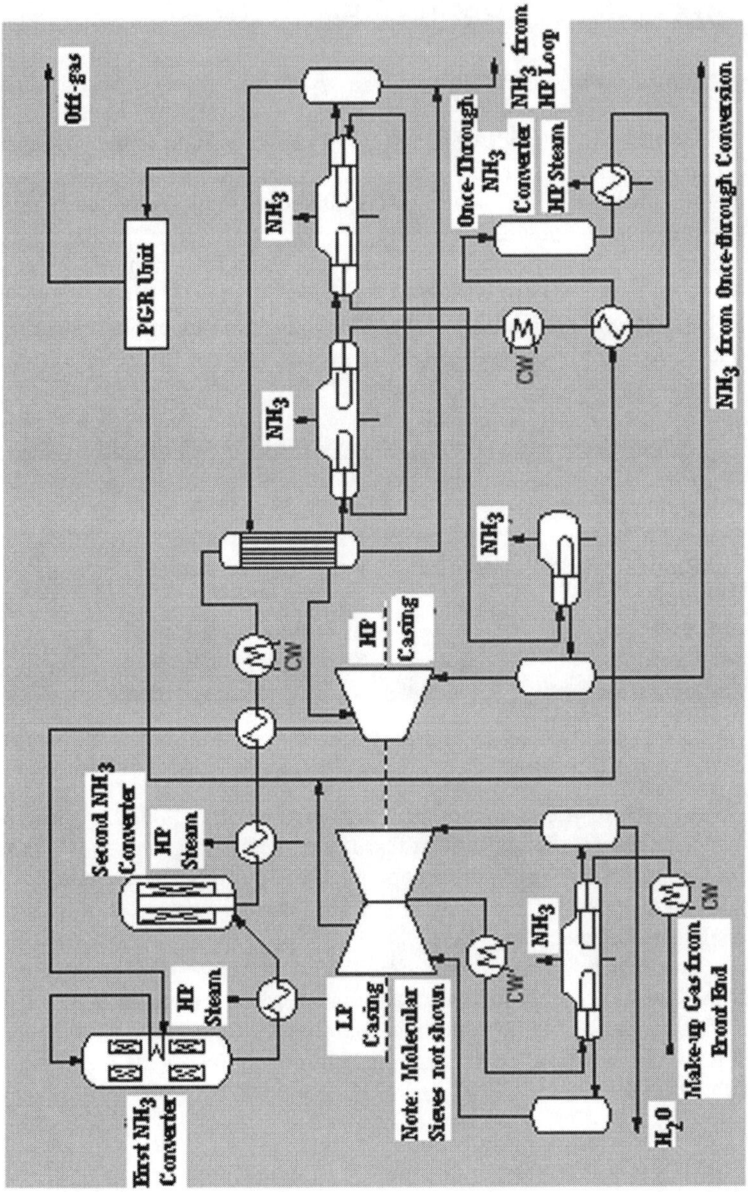

Figure 6.15. Uhde dual pressure process for large capacity ammonia plant. (Reproduced by permission of Uhde GmbH and copyright by Uhde)

The main benefit of the process is that the removal of ammonia after the first stage reduces the volume of the gas that has to be recompressed by about one-third. This effectively increases the capacity of the synthesis loop[215].

6.6.4. Lurgi and Ammonia Casale

The solution that Lurgi and Ammonia Casale have brought to the challenge of building larger plants is to replace air with oxygen and operate a single-line plant at higher pressures. A capacity of 4,000 tonnes/day is available with proven high-pressure equipment[291].

A comparison of the process steps in a conventional ammonia process (Figure 6.16) with the MEGAMMONIA® process (Figure 6.17) is given in Table 6.6. The MEGAMMONIA® process has one less process unit than the conventional process and two less catalytic reactors[291].

Table 6.6. Compare Conventional Process with MEGAMMONIA® Process

Step	Conventional	Step	MEGAMMONIA
1	Steam Reformer	1	Air Separation Unit (ASU)
2	Air Blow Secondary Reformer	2	Catalytic Partial Oxidation
3	CO Shift	3	CO Shift
4	aMDEA	4	Rectisol Wash
5	Methanation	5	Nitrogen Wash
6	Ammonia Synthesis	6	Ammonia Synthesis
7	Purge Gas Treatment		

At a plant size of 4,000 tonnes per day, the MEGAMMONIA® process offers a reduction in specific investment that is between 15% and 20% less than a scaled-up version of conventional technology. It also offers the potential for a reduction in cash costs that are 10% to 15% below the most advanced conventional ammonia technology[291].

If urea production is being considered, the recoverable CO_2 from the MEGAMMONIA® process is sufficient to convert 100% of the ammonia into urea, i.e., 4,000 tonnes per day of ammonia could be converted into 7,000 tonnes per day of urea. However in the conventional ammonia process the CO_2 supply is about 10% short[291].

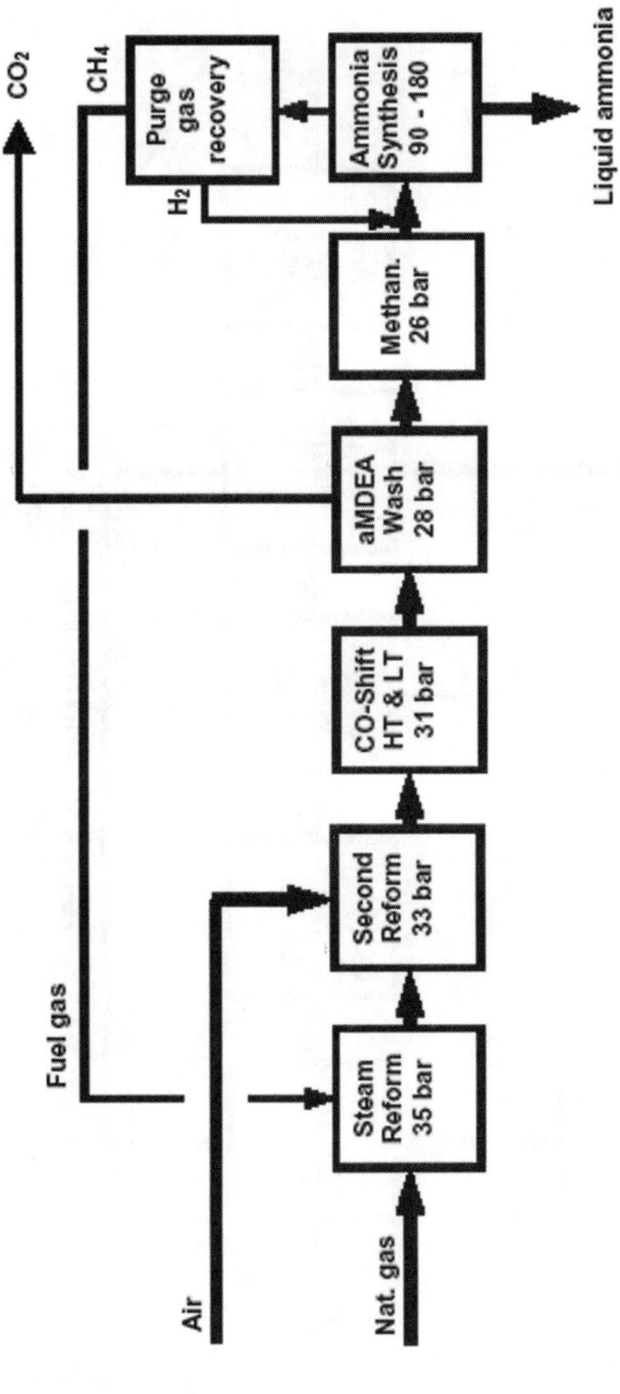

Figure 6.16. Conventional Ammonia Process.
(Reproduced by permission of Lurgi)

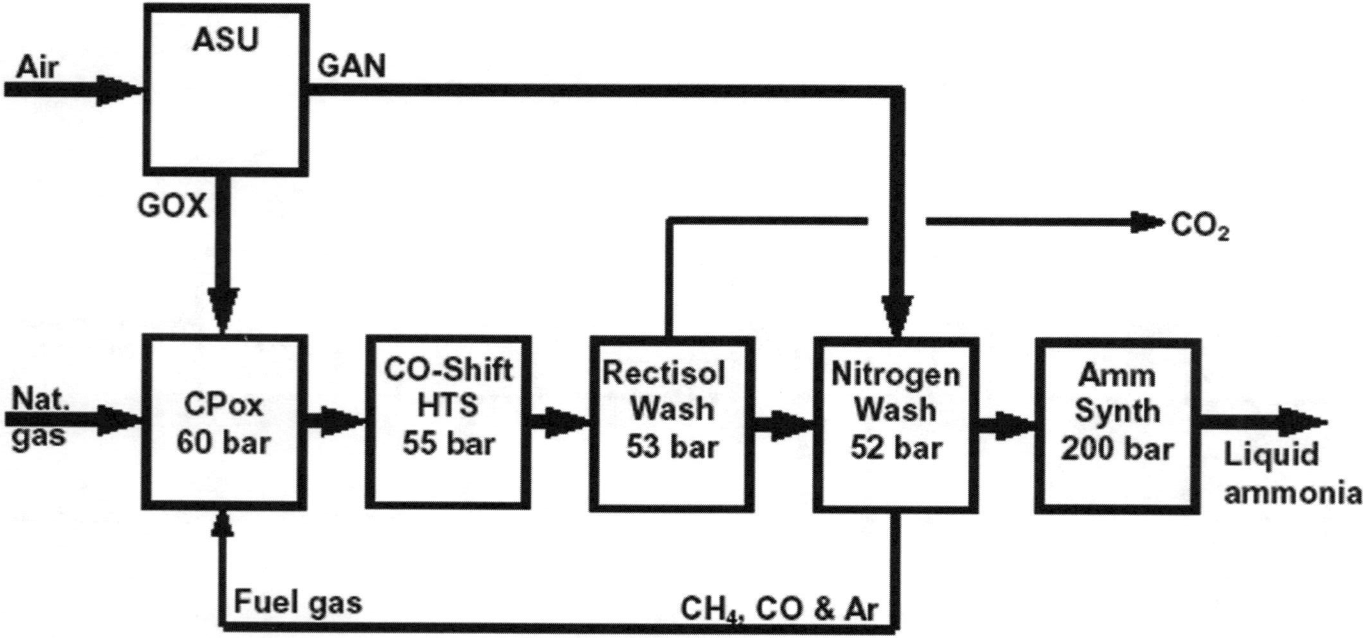

Figure 6.17. MEGAMMONIA® process.
(Reproduced by permission of Lurgi)

6.7. AMMONIA PRODUCTION COSTS

In the mid-1990's the ammonia industry accounted for about 5% of the worldwide natural gas consumption. For economic and environmental reasons, natural gas is the feedstock of choice. However, processes for ammonia production can use a wide range of energy sources. For example, in 2001 60% of China's nitrogen fertilizer production was based on coal. In 2002 natural gas is the most economic feedstock for the production of ammonia as shown in Table 6.7[36].

Table 6.7. Feedstock Economics for Ammonia Production

	Natural Gas	Heavy Oil	Coal
Energy Consumption	1.0	1.3	1.7
Investment Cost	1.0	1.4	2.4
Production Cost	1.9	1.2	1.7

Ammonia production costs are very dependent on feedstock cost – which is normally natural gas. Natural gas prices can vary significantly from region to region. They may also be impacted by the cycles in the world energy market. In addition, in regions where natural gas is owned or controlled by the government, social concerns (such as keeping people employed or generating hard currency from export sales) can have an impact on natural gas prices. Typical ammonia production costs are listed in the Table 6.7[57]:

Table 6.7. Approximate U.S. cash production costs for ammonia - 2000

Gas Price ($US / Million BTU)	$2	$4	$6
Gas Use (Million BTU / tonne)	36	36	36
Gas Cost ($US / tonne Ammonia)	$72	$144	$216
Conversion Cost ($US /tonne Ammonia)	$28	$28	$28
Total Cost ($US /tonne Ammonia)	$100	$172	$244

Figures 6.18 and 6.19 illustrate how production location and natural gas cost can have an impact on ammonia production costs:

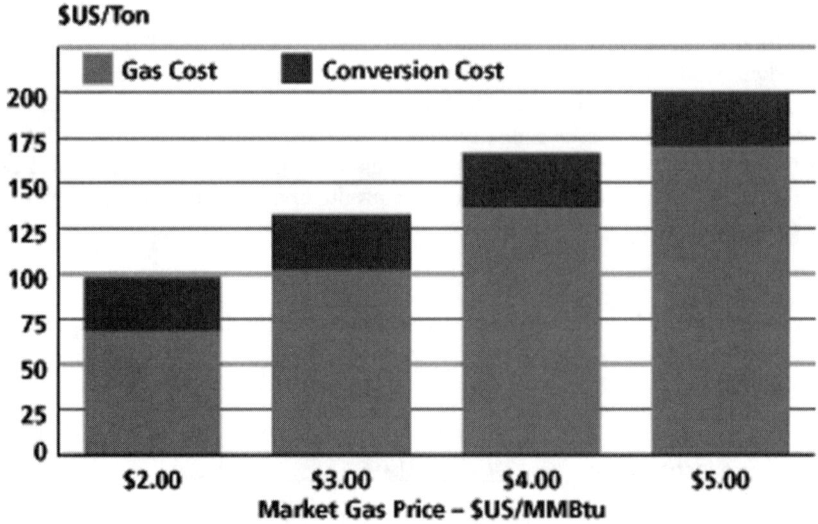

Figure 6.18. Ammonia production costs at different natural gas prices.[153]
(Reproduced by permission of Fertecon)

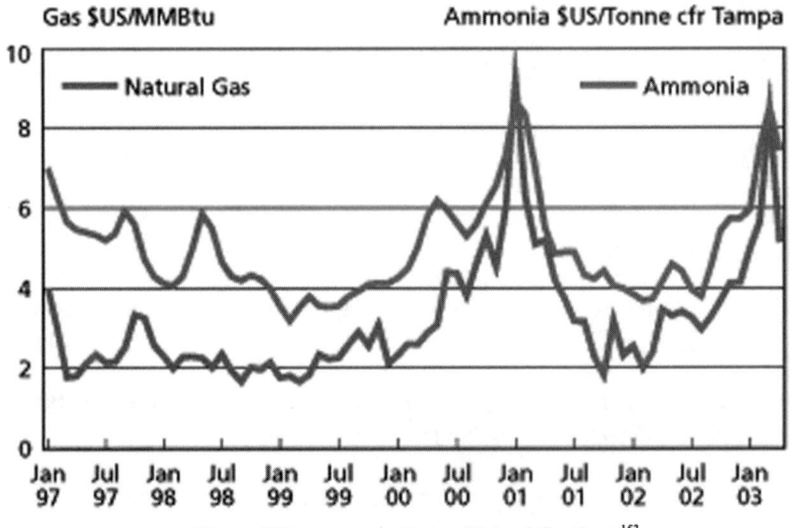

Figure 6.19. Ammonia Cost vs. Natural Gas Cost.[153]
(Reproduced by permission of Fertecon)

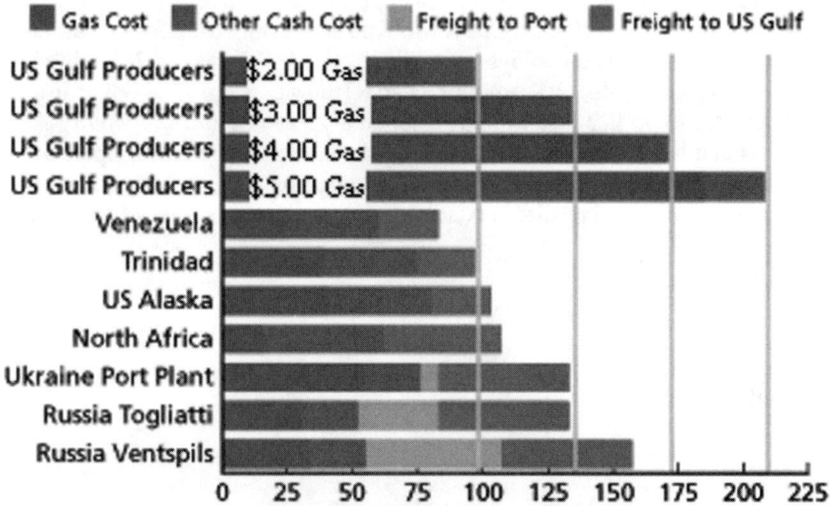

Figure 6.20. Ammonia production costs in different countries [153]
(Reproduced by permission of Fertecon)

6.8. MATERIALS OF CONSTRUCTION

If the synthesis gas contains traces of carbon oxides, ammonium carbamate will form upon mixing with the ammonia in the recirculating gas from the synthesis loop. The carbamate will clog and/or corrode downstream equipment. To avoid this condition, consider controlling the carbon oxides level in fresh makeup gas at less than 5 ppm[88].

The compressor interstage coolers are usually constructed of carbon steel and use water as the cooling medium. Low-velocity areas in the water passage are prone to scaling and microbial corrosion. These factors may lead to tube failures. Improved design of exchangers can eliminate low-velocity areas while a proper cooling water treatment program can control microbial corrosion[88].

Startup heaters – electrical or direct-fired types – are used to heat synthesis gas for the converter during startup. Some of the problems encountered with these heaters are hydrogen-induced tracks, overheating and flame impingement, thinning at bends, and furnace explosions. Normally, SS321 is used for startup heater coils and the downstream pipeline[88].

Many ammonia synthesis converters operate at 150 to 200 bar and around 515°C. Under these conditions, nitriding and hydrogen embrittlement can occur. The pressure shell is a multi-layer or multi-wall carbon steel vessel. The internal catalyst baskets, contained in the shell, are made of SS 321 material[88].

Nitriding of pipes and catalyst-support grids are usually encountered in ammonia plants. The nitriding effect is more pronounced in low-alloy steels above 450°C. Austenitic steels with a high-nickel content offer considerably more resistance. Alloys of the Cr-Ni-Mo type are usually used for the ammonia converter internals. These alloys contain 12% to 25% Cr, 5% to 25% Ni, along with Molybdenum (Mo), Vanadium (V) and Tungsten (W). The boiler usually contains SA 213 or T22 material. T22 contains 2.25% Cr and 1% Mo or related materials that have been improved. Due to the high pressure and temperature in the reactor, atomic hydrogen is absorbed by diffusion. The subsequent reaction with carbon in the steel results in blistering and cracking from decarburization[88].

7

USES OF AMMONIA

Anhydrous ammonia is produced in about 80 countries. Worldwide (in 2001) about 86% was used for nitrogen fertilizer production, including about 4% that was directly applied to the fields[153]. In the United States the distribution of ammonia use differs slightly from the worldwide uses. In the USA only 80% of the ammonia is used to make fertilizers and of the ammonia that goes into fertilizers, 20% is used as a direct application fertilizer. Chemical intermediates (such as acrylonitrile and caprolactam) account for 19% of the ammonia use while the remaining 1% is used in pulp and paper, metals and refrigeration applications[37]. In 2001 the United States consumed 22% of all ammonia that was used in non-fertilizer uses. The USA also consumed 80% of all ammonia that was used as a direct application fertilizer[153]. A summary of worldwide uses of ammonia in 2001 is shown in Figure 7.1.

7.1. CHEMICAL PRODUCTION AND OTHER USES

Ammonia solutions are used to clean, bleach and deodorize; to etch aluminum; to saponify (i.e., hydrolyze) oils and fats; and in the manufacture of chemicals. Some of the specific uses of ammonia are listed below.

7.1.1. Acrylonitrile

Acrylonitrile is used in the production of acrylic fibers and various resins. Acrylonitrile is produced by the catalytic oxidation of propylene and ammonia. About 0.48 tonnes of ammonia are needed to produce one tonne of acrylonitrile[57].

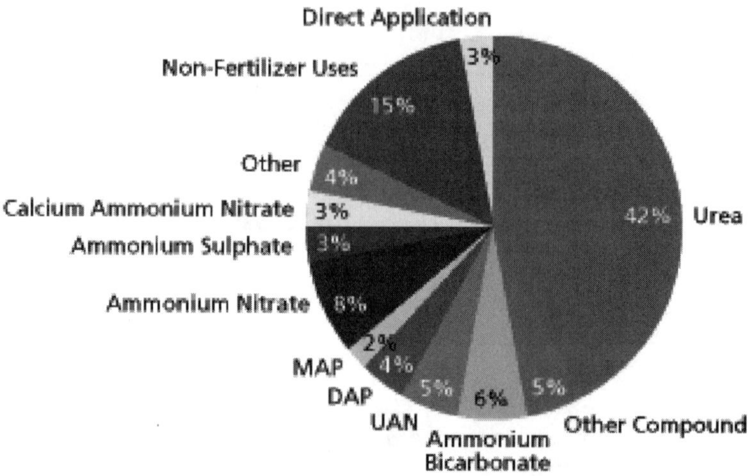

Figure 7.1. Worldwide Uses of Ammonia in 2003
(Reproduced by Permission of Fertecon)

7.1.2. Caprolactam

Caprolactam is used principally in the production of nylon 6 fibers. In the production of caprolactam, ammonia is used to maintain a reaction pH of 7 in one step, and in another step ammonia is used to neutralize sulfuric acid. Both of these steps result in the production of a low-grade ammonium sulfate fertilizer. In the classical process 5 kg of $(NH_4)_2SO_4$ are produced per kilogram of caprolactam. In 1992 about 33% of the worldwide caprolactam capacity was based on DSM's caprolactam process that only produces 1.8 kg of $(NH_4)_2SO_4$ per kilogram of caprolactam[46].

7.1.3. Miscellaneous Amines and Nitriles

A small but significant amount of ammonia is also consumed in the production of various amines and nitriles[57].

7.1.4. Metallurgy

In this metallurgy applications ammonia is cracked to produce a "dissociated ammonia" that consists of 75% hydrogen and 25% nitrogen. Dissociated ammonia is used as a reducing atmosphere in heat-treating several metals, as a source of nitrogen in the nitriding and carbo-nitriding processes for case-hardening of steels, for the reduction of metal oxides, for sintering, for brazing, for glass-to-metal seals and for metallizing. Metal-treating customers purchase metallurgical grade (99.99% NH_3) or refrigeration grade (99.98% NH_3)[57].

7.1.5. Refrigeration

Anhydrous ammonia is used as a refrigerant in industrial, closed-circuit refrigeration systems. Ammonia's high latent heat, low vapor density, chemical stability and low iron corrosivity promote its use. It is primarily used in large installations. Ammonia is used in large industrial systems (rather than halocarbon refrigerants) because ammonia absorbs about six times as much heat per unit weight and energy costs of operation are substantially lower[57].

7.1.6. Pulp and Paper

In the production of ammonia-based sulfite pulp, the amount of ammonia used per ton of pulp varies from mill to mill. The industry average is about 125 pounds (56.7 kilograms) of ammonia per ton of sulfite pulp. This industry also uses ammonia for stack gas scrubbing and wastewater treatment[57].

7.1.7. Phosphate Ore Flotation

A small amount of ammonia is used as a modifying reagent in the froth flotation of phosphate ores.

7.1.8. Household Cleaners

Household ammonia, liquid window cleaners, liquid all-purpose cleaners and various other household cleaning products constitute this small market for ammonia. Ammonia is also used in the manufacture of ammonium salts of alcohol, ether and sulfates. These compounds are widely used as surfactants in detergents that are in liquid dishwashing formulations[57].

7.1.9. Air Pollution Control

A number of processes use ammonia for scrubbing sulfur oxides (SOx) and/or nitrogen oxides (NOx) from industrial and electrical power plant stack gases. Sulfur dioxide reacts with ammonia solutions to form a sulfite/bisulfite liquor. At sulfite pulp mills, this liquor is recirculated for use in the cooking step. At sulfuric acid plants, this liquor can be oxidized or acidulated with sulfuric acid to make ammonium sulfate. A regenerative ammonia-based scrubbing system is used to scrub SOx from tail gases at some Claus sulfur plants. A catalytic process that uses ammonia for reducing NOx in stack gases was installed at the coal gasification plant at Beulah, North Dakota[57].

7.1.10. Uranium Concentrate Production

Ammonia is used to precipitate the final uranium concentrate salt (yellowcake) prior to drying or calcination. It is generally economically preferred over sodium hydroxide or hydrogen peroxide[57].

7.1.11. Water Purification

Ammonia is combined with chlorine to purify some municipal and industrial water supplies. They act as delayed sterilizing agents and permit higher residual chlorine concentrations than does chlorine alone – without producing a chlorine taste and odor[57].

7.1.12. Waste Treatment

Small amounts of ammonia are added when needed as the nitrogen source for the bacteria in industrial and municipal biological waste treatment systems. Other industrial plants use minor amounts of ammonia to neutralize acid in plant wastes[57].

7.1.13. Duplicating Processes

Ammonia is used as a process chemical in photographic copy machines[57].

7.1.14. Fabric Treatment

Permanent press, 100% cotton fabrics are produced via a licensed process involving mercerization in a bath of liquid anhydrous ammonia. This Sanfor-Set process was commercialized in 1975[57].

7.1.15. Semiconductor Industry

The semiconductor industry is an area where the use of ultra-high purity ammonia (99.99995% to 99.99999%) is growing in 2002. This ammonia is referred to as Grade 6.5 and Grade 7.0 – in reference to the number of nines in the purity. At least three companies announced the startup or expansion of new high purity ammonia plants in 2002. Praxair has a facility in Kingman, Arizona; BOC Edwards started up a 200 tonne per year plant in Taiwan that will service the Asian region; and Air Products started up a new plant in Hometown, Pennsylvania in addition to their existing plants in Morrisville, Pennsylvania and Catoosa, Oklahoma.

High purity ammonia is used in gallium nitride (GaN) manufacturing processes to provide high brightness blue and white LED's (Light Emitting Diodes), in high performance optoelectronics (such as liquid crystal displays – LCD's and flat panel displays – FPD's), and in high power electronic devices (such as lasers and laser diodes). Extremely low moisture levels in the ammonia are a key factor in the production of high quality LED's and FPD's.

7.1.16. Miscellaneous

Relatively small amounts of ammonia are used in resin beds for regeneration of ion exchange resins at plants manufacturing corn sweeteners. Minor amounts of ammonia are also consumed:

- As corrosion inhibitors at petroleum refineries and natural gas plants,
- In the rubber industry for the stabilization of natural and synthetic latex to prevent coagulation during transportation and storage,
- By the pharmaceutical industry as a pH adjuster in solvent extraction reactions,
- In the food and beverage industry as a source of nitrogen for the growth of yeast and microorganisms,
- For pH control in yeast production,
- As a curing agent in the production of leather , and
- As a slime and mold preventive in tanning liquor[57].

8

GENERAL AMMONIA INFORMATION

8.1. AMMONIA PROPERTIES

Ammonia is a colorless, flammable, toxic, alkaline gas. Its water solution is alkaline because much of the dissolved ammonia reacts with water (H_2O) to form ammonium hydroxide (NH_4OH), which is a weak base.

Ammonia dissolves alkali metals, barium, calcium and strontium and forms an unstable blue solution. This solution contains the metal ion and free electrons that slowly decompose, release hydrogen and form the metal amide. Compared to water, liquid ammonia is less likely to release protons (H^+ ions), is more likely to take up protons (to form NH_4^+ ions) and is a stronger reducing agent[219].

Ammonia takes part in many chemical reactions. Ammonia reacts with strong acids to form stable ammonium salts[219]:

- With hydrogen chloride, it forms ammonium chloride,
- With nitric acid, it forms ammonium nitrate, and
- With sulfuric acid, it forms ammonium sulfate.

Ammonium salts of weak acids are readily decomposed into the acid and ammonia. Ammonium carbonate [$(NH_3)_2CO_3$ H_2O] is a colorless-to-white crystalline solid commonly known as smelling salts. In a water solution it is sometimes called "aromatic spirits of ammonia". Ammonia also reacts with Lewis acids (electron acceptors) such as sulfur dioxide or sulfur trioxide or boron trifluoride[219].

Another kind of reaction, commonly called ammonolysis, occurs when one or more of the hydrogen atoms in the ammonia molecule is replaced by some other atom or radical. For example, chlorine gas (Cl_2) reacts directly with ammonia to form monochloramine (NH_2Cl) and hydrogen chloride (HCl)[219].

Ammonia also takes part in oxidation and reduction reactions. It burns in oxygen to form nitrogen gas (N_2) and water. In the presence of a catalyst (e.g., platinum) it is oxidized in air to form water and nitric oxide (NO). It reduces hot-metal oxides to the metal (e.g., cupric oxide to copper metal)[219].

Some of the physical properties of ammonia are listed in Table 8.1 and Table 8.2.

Table 8.1. Physical Properties of Ammonia

Property	Value
Molecular Weight	17.03
Boiling Point (°C)	-33.35 at 760 mm Hg
Freezing Point (°C)	-77.7
Decomposition Temperature (°C / °F)	450 to 500 / 842 to 932
Color	Colorless
Critical Temperature (°C)	133.0
Critical Pressure (kPa / psi)	1,425 / 1,657
Specific Heat (J/kg °K)	
0 °C	2,097.2
100 °C	2,226.2
200 °C	2,105.6
Heat of Formation of Gas, ΔH_f (kJ /kg mol, Btu/lb-mol)	
0 °K	-39,222 (-16,862)
298 °K	-46,222 (-19,872)
Solubility in Water (weight %)	
0 °C	42.8
20 °C	33.1
25 °C	31.8
40 °C	23.4
60 °C	14.1
Specific Gravity of Anhydrous Ammonia	
-40.0 °C	0.690
-33.4 °C	0.682
0.0 °C	0.639
40.0 °C	0.580
Vapor Pressure	116.6 psig at 21 °C (70 °F) 7,500 mm Hg at 25 °C
Vapor Density (Air = 1.0)	0.6 at 0 °C
Flammable Limits in Air (% by Volume)	
Lower Explosion Limit (LEL)	15
Upper Explosion Limit (UEL)	28
Autoignition Temperature (°C / °F)	651 / 1,204
NFPA Ratings for Storage Vessels	
Health	3 out of 4
Flammability	1 out of 4
Reactivity	0 out of 4

Table 8.2. Densities of Aqueous Ammonia at 15 °C

Ammonia (Weight %)	Density (g/liter)
8	0.970
16	0.947
32	0.889
50	0.832
75	0.733
100	0.618

8.2. AMMONIA TOXICITY

Ammonia is a strong local irritant, and the primary target organ is the pulmonary system. Ammonia or ammonium hydroxide can penetrate the cornea rapidly and lead to damage of the iris, cataract and glaucoma. Oral ingestion of aqueous ammonia can corrode the mucous membranes of oral cavity and constrictions of the esophagus may result[74].

When ammonia enters the body, it is converted to urea and excreted by the kidneys. The capacity for detoxification via urea is sufficient to eliminate the ammonium ion when ammonia is inhaled in non-irritating concentrations. Repeated inhalation can cause a higher tolerance because the mucous membranes become increasingly resistant. Ammonia is not considered to be carcinogenic nor is it mutagenic. The effects of different ammonia concentrations are summarized in Table 8.3[74].

Table 8.3. Health Effects of Different Ammonia Concentrations

Ammonia Concentration (ppm)	Health Effect[a]
5	Threshold Detection Limit
50	Easily Perceived
50 to 72	No Significant Impairment to Respiration
100	Irritation to Nose and Throat Burning Sensation in eyes
200	Headache and Nausea
> 300	Irritation in Respiratory Tract Difficulty in Breathing and Eye Irritation
250 to 500	Rapid Heart Beating
> 455	Respiratory and Eye Irritation Corneal Opacities
700	Immediate Onset of Burning Sensation in the Eyes
1,000	Immediate Coughing

[a] The TLV has been set at 25 ppm (17 mg/m^3) as an 8-hour, time-weighted average (TWA) with a short-term exposure limit (STEL) of 35 ppm (24 mg/m^3).

8.3. AMMONIA QUALITY

Most industrial uses of ammonia require a higher-purity level of product than agricultural uses. Typical ammonia specifications are shown in Table 8.4[74].

Table 8.4. Ammonia Quality Specifications

		Commercial Grade		Refrigeration Grade	
		USA	Germany	USA	Germany
Purity	Wt%, minimum	99.5	99.5	99.98	99.98[a]
Water	Wt%, maximum	0.5	0.2	0.015	0.02
Inerts[b]	mL/g, maximum	c	c	0.1	0.08
Oil	ppm by weight	5.0	5.0	3.0	c
Free of H_2S, Pyridine and Naphthenes					

[a] Allowable boiling point change on vaporization of 5% to 97% of the test sample, 0.9 °C.

[b] The non-condensabl e gases dissolved in ammonia are H_2, N_2, CH_4, and Ar. Their amounts depend on the methods of synthesis and storage. The inerts amount to about 50 mL/kg for atmospheric storage.

[c] Not Specified

Commercial grade is produced by ammonia synthesis while Refrigeration grade is normally made from industrial raw ammonia by distillation. Refrigeration grade is also used in metallurgy[74]. A Research grade that contains 99.999% ammonia is available as well as high purity grades for semiconductors that have a purity as high at 99.99999% (Grade 7.0). Commercial grade is sufficient for most nonagricultural uses. For ammonia shipped or pipelined in the United States, water content must be at least 0.2 weight % to inhibit stress corrosion cracking of the carbon steel[57].

Various concentrations and purities of aqueous ammonia are on the market. The typical concentration is 25% to 30% ammonia and the iron content is less than 10 ppm. If the ammonia content is above 25%, it must be shipped in pressure vessels because of its elevated vapor pressure. For more stringent purity requirements for aqueous ammonia, the containers should be made of seawater-resistant aluminum (magnesium alloyed) or austenitic steels[74].

Ammonia is readily detectable in air in the range of a few parts per million by its characteristic odor and alkaline reaction. Specific indicators, such as Nessler's reagent (HgI_2 in KOH), can detect ammonia in a concentration of 1 ppm. For the quantitative determination of ammonia in air, synthesis gas and aqueous solutions, these methods can be used[74]: Acidimetry and Volumetric Analysis By Absorption, Gas Chromatography, Infrared Absorption, Thermal Conductivity Measurement, Electrical Conductivity Measurement, Measurement of Heat of Neutralization, and Density Measurement (for aqueous ammonia).

8.4. DISTRIBUTION AND STORAGE

Ammonia production requires storage facilities to smooth over fluctuations in production, usage and shipments. Ammonia is stored in bulk in large capacity containers installed above or below ground. It is distributed to point of use almost exclusively as a liquid. Gaseous ammonia sometimes is used within plants[74].

Three methods have been used for storing liquid ammonia and the determining factors for the type of storage are temperature and the flow of the ammonia into and out of storage[74]:

- Pressure storage at ambient temperature in spherical or cylindrical pressure vessels having capacities up to about 1500 tonnes.
- Atmospheric storage at –33°C in insulated cylindrical tanks for amounts to about 50,000 tonnes per vessel.
- Reduced pressure storage at about 0°C. Usually in insulated, spherical pressure vessels for quantities up to about 2,500 tonnes per sphere.

Ammonia is delivered in small containers, tank trucks, tank cars, barges and via pipeline. The most common small containers are 1) cylindrical steel bottles and pressurized flasks that contain about 20 to 200 kg and 2) polyethylene canisters and metals casks. The steel bottles are used in laboratories and small refrigeration systems while the canisters are used for 25% aqueous ammonia[74].

Trucks have ammonia capacities up to 100 m^3 while jumbo rail cars hold up to 150 m^3. The different designs are[74]:

- Pressure vessels for anhydrous ammonia with a maximum operating pressure of 2.5 MPa (25 bar / 37.5 psig),
- Atmospheric vessels for 25% aqueous ammonia, and
- Vessels designed for elevated pressure for high-concentration aqueous ammonia (up to 1.6 MPa or 16 bar).

Truck transportation rates for anhydrous ammonia are relatively high because of liability insurance rates and the need to use specialized equipment. Distribution is further complicated by attempts to restrict the movement of ammonia in some jurisdictions. In 1990, the U.S. Department of Transportation classified anhydrous ammonia as a non-flammable gas and required shipments to be marked with the words “inhalation hazard.” International shipments are required to carry the “inhalation hazard” and “poison gas” markings[57].

Distribution in rail cars and trucks primarily serves smaller processing operations and wholesale merchants. However, rail transport of liquid ammonia also supplements marine and pipeline shipments[74].

Liquid ammonia shipments by barge constitute a larger volume than by road or rail. Barge shipments supply users without their own production and to meet needs caused by operating breakdowns. Refrigerated ammonia is the preferred method for marine shipments. Barges and ships with pressure vessels have capacities of about 400 to 2,000 tonnes. On the other hand refrigerated ammonia tankers have capacities up to 50,000 tonnes[74].

Several ammonia pipeline systems appeared in the United States in the 1960's. Their construction resulted from the building of large plants near the sources of natural gas and from developing a large market for liquid anhydrous ammonia in direct fertilization. Pipelines are the lowest cost method to move ammonia over the long distances between the producers along the Gulf Coast and the consumers in the upper Midwest. Some pipelines are up to 3,000 kilometers long and up to 10" in diameter[74].

An integrated pipeline system has not been developed in Europe. Only a few pipelines with lengths of less than 50 kilometers have been installed. However in the former Soviet Union, a 2,424-km-long pipeline delivers ammonia from Grigorowski/Odessa to Togliatti/Wolga[74].

8.5. MATERIALS OF CONSTRUCTION

Carbon steels such as BS1515, BS15110213 and ASTM A516 are the conventional materials used in the construction of ammonia storage units. For refrigerated (atmospheric) storage, low-temperature carbon steel is used[88].

One of the major problems encountered in the storage and transport of anhydrous liquid ammonia is the stress-corrosion-cracking (SCC) of carbon steel equipment. Cracks most often occur at the weld joints, where the leftover stress is at a maximum. The leftover stress is that which remains even after heat treatment. The hardness of the material and the presence of impurities and oxygenates in ammonia aggravate SCC[88].

Some of the steps that can be taken to help minimize the impact of SCC in ammonia storage are: Complete stress relief, operation without air contact and the addition of small amounts of water (0.2%) as an inhibitor. Low-temperature carbon steels have considerably more resistance to SCC than normal carbon steels. This makes them the preferred material of construction for large atmospheric liquid ammonia storage tanks that operate at –33°C[88].

Cavitation corrosion occurs in pumps that have flow conditions that allow bubble formation on the surface of impellers. These bubbles, upon formation , break with enough force to rupture the protective film of the stainless steel. Plants can prevent this by designing a system that avoids bubble formation (i.e., provide sufficient Net Positive Suction Head – NPSH – for the pump), by polishing rotating parts to remove bubble formation sites and by using alloys with greater corrosion resistance and strength[88].

8.6. AMMONIA PRICE

In the United States, two ammonia pipelines deliver ammonia to the Mississippi delta and to the Texas and Oklahoma panhandle region. Direct sales by basic producers are generally at wholesale prices. Basic producers also sell to distributors and jobbers for resale to end users buying smaller quantities. Distributors of ammonia to industrial end users have their own storage facilities, packaging plants and distribution networks. They are able to supply end users that buy in less-than-truckload quantities (a minimum of 1,000 to 2,000 pounds or 450 to 910 kilograms) or in cylinders or bottles[57].

Another important aspect of the ammonia business is seasonality. The major end use for anhydrous ammonia (excluding production of downstream nitrogen products) is as a direct application fertilizer. The application season, particularly in Canada and the northern United States, is limited and significant storage capability is needed. As a result, it is normal to expect a plant to produce a downstream nitrogen product (such as urea) that is 1) easier to store and handle and 2) does not have such a limited application season. Industrial markets are normally nonseasonal[57].

A variety of contractual arrangements between producing and marketing companies are possible. Long-term fixed-price contracts are not considered workable due to the volatility in the market and the uncertainty regarding future natural gas prices. The normal long-term contract simply guarantees continuity of supply and calls for prices to be negotiated every three to six months[57]. Historical ammonia prices in the United States and in Russia are shown in Figure 8.1.

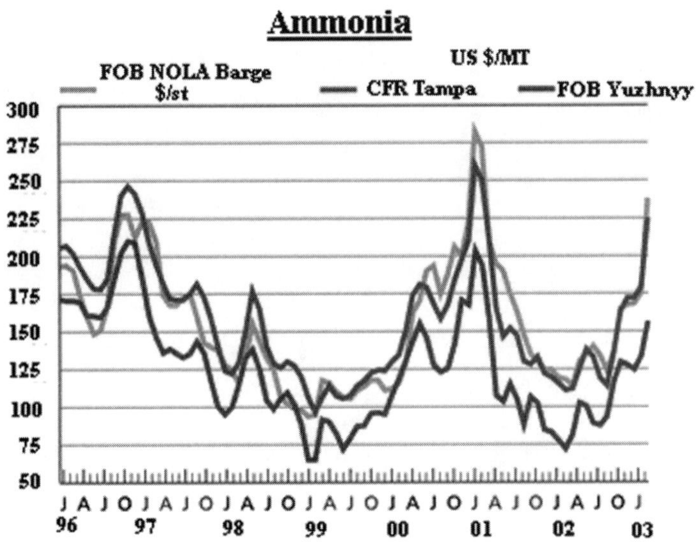

Figure 8.1. Historical ammonia prices: 1996 to 2003.
(Reproduced by permission of Fertecon)

As shown in Figure 8.2, international ammonia prices fluctuate on a short-term basis. Over the longer term, they follow a cyclical pattern, moving up and down around a relatively stable base.

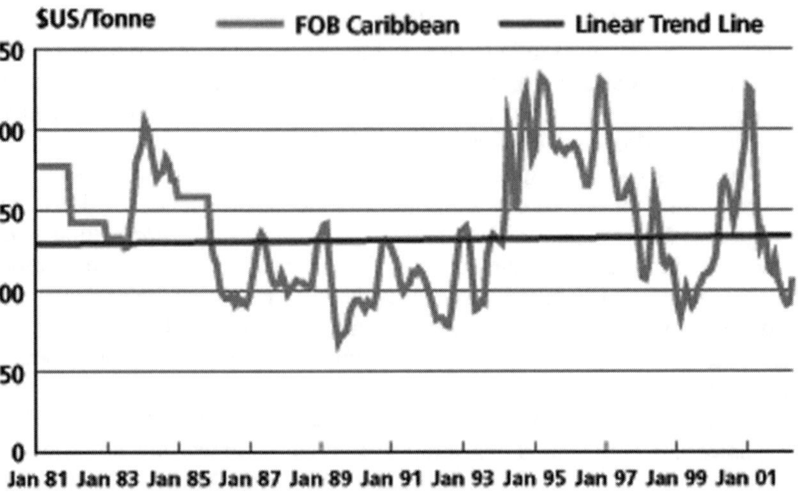

Figure 8.2. International ammonia prices: 1981 to 2001.
(Reproduced by permission of Fertecon)

9

NITRIC ACID

Nitric acid is a strongly acidic, corrosive liquid that is produced commercially by the oxidation of ammonia and subsequent reaction of the oxidation products with water. Pure nitric acid is colorless. The characteristic yellow-brown color generally associated with concentrated solutions is due to dissolved nitrogen dioxide[91].

Nitric acid is an active compound and its salts, the nitrates, are found in all fertile soils. The alchemists obtained nitric acid by heating alum and copper sulfate with nitrate in a retort. Owing to its powerful corrosive action, they named it aqua fortis or "strong water."[7] It is also known as azotic acid, hydrogen nitrate and nitryl hydroxide.

Being a powerful oxidizing agent (electron acceptor), nitric acid reacts violently with many organic materials and the reactions may be explosive. As a general rule, oxidizing reactions occur primarily with the concentrated acid and favor the formation of nitrogen dioxide (NO_2). The acidic properties tend to dominate the dilute acid, and this results in the preferential formation of nitrogen oxide (NO)[53].

Nitric acid reacts with all metals except the precious metal series and certain alloys. Although chromium, iron and aluminum readily dissolve in dilute nitric acid, the concentrated acid forms a metal oxide layer that protects (passivates) the metal from further oxidation[53].

9.1. PHYSICAL PROPERTIES

Pure anhydrous nitric acid (100%) is a colorless liquid that solidifies at a temperature of -41.6°C to form white crystals. It boils at 84.1°C. When it boils in light, a partial decomposition occurs with the formation of NO_2 via the reaction shown in Eq. (9.1):

$$HNO_3 \leftrightarrow H_2O + 2\,NO_2 + \tfrac{1}{2}\,O_2 \qquad (9.1)$$

This means that anhydrous nitric acid should be stored below 0°C to avoid decomposition. The nitrogen dioxide remains dissolved in the nitric acid and creates a yellow color at room temperature and a red color at higher temperatures. While the pure acid tends to give off white fumes when exposed to air, acid with dissolved nitrogen dioxide gives off reddish-brown vapors which leads to the common name "red fuming acid"[53].

Nitric acid is miscible with water and distillation gives an azeotrope with a concentration of 68.4% HNO_3 and a boiling temperature of 121.9°C at atmospheric temperature. Two solid hydrates are known – the monohydrate ($HNO_3 \cdot H_2O$) and the trihydrate ($HNO_3 \cdot 3H_2O$)[53].

Nitrogen oxides are soluble in nitric acid, and this property influences all the physical characteristics that depend on the concentration of the oxides. This mainly includes the vapor pressure above the liquid and the boiling temperature as well as the color. "Red Fuming Nitric Acid" is a term that, according to "Sax's Dangerous Properties of Industrial Materials" applies to nitric acid that contains more than 17% NO_2 and has a density of more than 1.48 g/cm^3 [53]. White fuming nitric acid usually contains 90% to 99% by weight HNO_3, from 0% to 2% by weight dissolved NO_2 and up to 10% by weight water.

The TLV for nitric acid has been set at 2 ppm as an 8 hour time-weighted average (TWA) with a short-term exposure limit (STEL) of 4 ppm.

Some properties of nitric acid are given in Table 9.1, Table 9.2 and Table 9.3.

Table 9.1. Physical Properties of Nitric Acid Solutions[92, 93, 94]

wt. % HNO_3	Density (at 20° C) g/cm^3	Freezing Point °C	Boiling Point, °C	Specific Heat (at 20°C) J/(g•K)^a	Partial Pressure (at 20°C) Pa^b HNO_3	H_20
0.0	0.99823	0	100.0	4.19		2333
10.0	1.0543	- 7	101.2	3.73		2266
20.0	1.1150	- 17	103.4	3.39		2026
30.0	1.1800	- 36	107.0	3.18		1760
40.0	1.2463	- 30	112.0	3.01		1440
50.0	1.3100	-20	116.4	2.85	27	1053
60.0	1.3667	- 22	120.4	2.64	120	653
70.0	1.4134	-41	121.6	2.43	387	347
80.0	1.4521	-39	116.6	2.22	1400	120
90.0	1.4826	- 60	102.0	1.97	3600	27
100.0	1.5129	- 42	86.0	1.76	6000	0

^{a} To convert J/ (g•°K) to cal/(g•°C), divide by 4.184.
^{b} To convert Pa to atm, divide by 1.013 x 10^5.

Table 9.2. Viscosity and Thermal Conductivity of Nitric Acid Solutions[92, 93, 94]

wt. % HNO_3	Viscosity (@ 20°C) mPa• s (cP)	Thermal Conductivity (@ 20°C) W/(m•K)[a]
0.0	1.0	0.61
10.0	1.1	0.57
20.0	1.2	0.54
30.0	1.4	0.50
40.0	1.6	0.47
50.0	1.9	0.43
60.0	2.0	0.40
70.0	2.0	0.36
80.0	1.9	0.35
90.0	1.4	0.31
100.0	0.9	0.28

[a] To convert W (m•°K) to Btu (h•ft•°F), divide by 1.7307.

Table 9.3. Thermodynamic Properties of Nitric Acid and Its Hydrates[95]

	HNO_3	$HNO_3{\bullet}H_2O$	$HNO_3{\bullet}\ 3H_2O$
Nitric acid, wt. %	100.0	77.77	53.83
Freezing point, C	-41.59	-37.62	-18.47
Heat of formation (at 25°C). k J/mol[a]	-173.35	-472.07	-888.45
Free energy of formation (at 25°C). kJ/mol[a]	-79.97	-329.29	-810.99
Entropy (at 25 C), kJ/(mol•°K)[a]	155.71	217.00	347.17
Heat of fusion, kJ/mol[a]	10.48	17.52	29.12
Heat of vaporization (at 20°C), kJ/mol[a]	39.48		

[a] To convert Joules to calories, divide by 4.184.

9.2. PROCESSES

The oldest nitric acid production method involved the reaction of $NaNO_3$ with sulfuric acid in heated cast iron retorts. The evolved nitric acid vapors were condensed and collected in stoneware vessels.

The Birdeland-Eyde or Arc process made nitric acid by 1) passing air through an electric arc, 2) forming nitric oxide, 3) oxidizing NO with air and 4) absorbing the resulting oxides of nitrogen in water. The reactions are shown in Eqs. (9.2) through (9.5):

$$N_2 + O_2 \rightarrow 2\,NO \qquad (9.2)$$

$$2\,NO + O_2 \rightarrow 2\,NO_2 \quad (9.3)$$

$$2\,NO_2 + H_2O \rightarrow HNO_2 + HNO_3 \quad (9.4)$$

$$3\,HNO_2 \rightarrow HNO_3 + 2\,NO + H_2O \quad (9.5)$$

The arc is spread into a disc by an electromagnet. The process operated in Norway from 1905 to 1930. It used hydroelectric power, but it was made obsolete by the ammonia oxidation process. This process was first used in the United States in 1917[1].

The Ostwald process is the basis for the modern family of processes that make nitric acid by the catalytic oxidation of ammonia. Wilhelm Ostwald, a German physical chemist, discovered it in 1900. The process was used by Germany during World War I to make explosives after the Allied blockade cut off the regular German supply of nitrites from Chile and other places[96].

Some of the named processes that have been developed based on the Ostwald process are listed below[1]:

- The CNA (**C**oncentrated **N**itric **A**cid) process is a general name for processes that make nitric acid more concentrated than the 70% made in conventional processes.
- The CONIA process makes nitric acid simultaneously at two different concentrations.
- The DSN (**D**irect **S**trong **N**itric) process is a general name for processes that concentrate nitric acid from the 50% to 70% range up to the 98% range by adding dinitrogen tetroxide.
- The HOKO process also makes nitric acid simultaneously at two different concentrations.
- The Hycon process makes nitric acid at a range of concentrations. It was developed in 1968 and patented (U.S. Patent 3,542,510) by the Chemical Construction Corporation.
- The SABAR (**S**trong **A**cid **B**y **A**zeotropic **R**ectification) process makes nitric acid by the atmospheric oxidation of ammonia. The nitrous gases from the oxidation are absorbed in azeotropic nitric acid in the presence of oxygen under pressure. The process was developed by Davy McKee who built plants based on this technology from 1974 to 1986.

Some of the principal variations between these processes are different operating temperatures, different operating pressures and different concentrations of the nitric acid product. Other differences include catalysts and spent catalyst recovery systems. In the 1990's, new air emission control requirements necessitated the modification of some plants and prompted the re-evaluation of processes for new plants. This resulted in a trend toward plants with relatively higher capital costs but lower emissions. In general, the ammonia requirement is about 0.29 units per unit of 100% HNO_3[91].

9.2.1. Chemistry

In the Ostwald process ammonia is catalytically oxidized in a heterogeneous reaction. The key steps in the process are: 1) Oxidation of ammonia to nitric oxide (NO), 2) Oxidation of NO to nitrogen dioxide (NO_2) and 3) Absorption of NO_2 in water to produce nitric acid. These steps are summarized in the following sections[91, 97].

9.2.1.1. Oxidation Reactions

Ammonia reacts with air on platinum/rhodium alloy catalysts in the oxidation section of nitric acid plants. Nitric oxide and water are formed in this step according to Eq. (9.6).

$$4\ NH_3 + 5\ O_2 \rightarrow 4\ NO + 6\ H_2O \quad \Delta H_{298} = -54\ \text{kcal/mol}\ (-226\ \text{kJ/mol}) \qquad (9.6)$$

Simultaneously N_2O, N_2 and H_2O are formed in side reactions according to Eqs. (9.7) and (9.8).

$$4\ NH_3 + 3\ O_2 \rightarrow 2\ N_2 + 6\ H_2O \quad \Delta H_{298} = -75.8\ \text{kcal/mol}\ (-317\ \text{kJ/mol}) \qquad (9.7)$$

$$4\ NH_3 + 4\ O_2 \rightarrow 2\ N_2O + 6\ H_2O \qquad (9.8)$$

The yield of nitric oxide in Eq. (9.7) depends on pressure and temperature as indicated in Table 9.4[97]. The formation of nitrogen by direct combustion of ammonia [see Eq. (9.7)] usually represents the most significant yield loss. Small amounts of nitrous oxide (N_2O) are also formed, but much lower reaction temperatures (around 500 °C) are required for N_2O to become a significant yield loss[104].

Table 9.4. Effect of Reaction Pressure and Temperature on Nitric Oxide Yield

Pressure (bar)	Temperature (°C)	Nitric Oxide Yield (%)
Below 1.7	810 to 850	97
1.7 to 6.5	850 to 900	96
Above 6.5	900 to 940	95

The space velocity in the ammonia oxidation reactor can also affect yield and should be high to avoid decomposition of ammonia in the feed gas. This decomposition can occur on the reactor walls according to Eq. (9.7). Another potential yield loss is shown in Eq. (9.9).

$$4\ NH_3 + 3\ NO_2 \rightarrow {}^7/_2\ N_2 + 6\ H_2O \tag{9.9}$$

This reaction is minimized by keeping the catalyst bed shallow and the bed velocity high enough to avoid back mixing. The effect of reactor residence time on nitric oxide yield is shown in Table 9.5[98].

Table 9.5. Effect of Reactor Residence Time on Nitric Oxide Yield

Reactor Residence Time, Seconds	Nitric Oxide Yield, Percent NH_3 Oxidized
0.280	82.1
0.110	85.7
0.061	90.2
0.023	91.8

At typical catalyst temperatures of 800°C to 940°C, nitric oxide (NO) is thermodynamically unstable and slowly decomposes into nitrogen and oxygen. Decomposition losses are minimized by avoiding excessive catalyst contact time and by rapidly cooling the gases as they exit the converter. To achieve ammonia conversions of 93% to 98% the catalyst contact time must be as short as 0.0010 to 0.0001 seconds[104].

The enthalpy of the hot reaction gases is used to produce steam and/or to preheat the waste gas (tail gas). The heated waste gas is discharged to the atmosphere through a gas turbine for energy recovery. The combustion gas (after this heat transfer for energy recovery) has a temperature of 100°C to 200°C. It is then further cooled with water. The water produced in Eqs. (9.6) to (9.8) is then condensed in a cooler-condenser and transferred to the absorption column[97].

NO is oxidized to NO_2 as the combustion gases are cooled [see Eq. (9.3)]. The other possible side reactions are shown in Eq. (9.10) and Eq. (9.11).

$$2\ NO + O_2 \leftrightarrow 2\ NO_2 \qquad \Delta H_{298} = -13.6\ \text{kcal/mol}\ (-57\ \text{kJ/mol}) \tag{9.3}$$

$$4\ NO \rightarrow 2\ N_2O + O_2 \tag{9.10}$$

$$2\ NO_2 \leftrightarrow N_2O_4 \qquad \Delta H_{298} = -6.8\ \text{kcal/mol}\ (-28.6\ \text{kJ/mol of}\ NO_2) \tag{9.11}$$

For the reaction in Eq. (9.3), secondary air is added to the gas mixture obtained from ammonia oxidation to increase the oxygen content to such a level that the waste gas leaving the plant has an oxygen content between 2% and 4%

by volume. Low temperatures strongly favor the formation of NO_2. Below 150°C, equilibrium is almost totally in favor of NO_2 formation. This is a slow reaction, but the rate constant for NO_2 formation rapidly increases with reductions in temperature[104].

The rate of reaction, and therefore the time needed to achieve a given extent of oxidation, is proportional to pressure cubed [see Eq. (9.12)]. This is of great significance in plant design and economics. The volume or size of equipment needed to oxidize the nitric oxide is inversely proportional to pressure cubed[104]:

$$d(P_{NO})/dt = -k\,(P_{NO})^2(P_{O2}) \qquad (9.12)$$

Nitrogen dioxide rapidly forms an equilibrium mixture with its dimer, dinitrogen tetroxide or nitrogen peroxide [see Eq. (9.11)]. The formation of the tetroxide is favored by low temperature and high pressure[104].

9.2.1.2. Absorption Reaction

The reaction in Eq. (9.13) is exothermic and continuous cooling is required within the absorber[104].

$$3\,NO_2 + H_2O \leftrightarrow NO + 2\,HNO_3 \quad \Delta H_{298} = -11.0 \text{ kcal/mol } (-46 \text{ kJ/mol of } NO_2) \qquad (9.13)$$

Equilibrium for this reaction is a function of temperature with low temperature favoring the formation of nitric acid. The mechanisms and rate-controlling steps for acid formation change with concentration and the extent of oxidation to nitrogen peroxide. At high concentrations of nitrogen peroxide (>40 wt %), N_2O_4 is the primary route to acid formation. At lower concentrations of nitrogen peroxide, a combination of mechanisms involving NO_2, N_2O_3, and HNO_2 become more important[104].

The nitric acid produced in the absorber contains dissolved nitrogen oxides and is then bleached by the secondary air.

9.2.1.3. Overall Reaction

The overall ammonia oxidation reaction that results in nitric acid production is shown in Eq. (9.14).

$$NH_3 + 2\,O_2 \rightarrow HNO_3 + H_2O \quad \Delta H = -98.7 \text{ kcal/mol} \qquad (9.14)$$

9.2.2. Catalysts

The kinetics of the ammonia oxidation reaction are limited by the mass transfer of the chemical species (NH_3, O_2, NO and H_2O) to and from the vicinity of the catalyst surface. If the catalyst area available to reactants is higher than the minimum area permitting the reaction, the mass transfer limits the kinetics.

If the available area is too small, the kinetics are limited by the catalytic surface. In a reactor, the concentration of ammonia decreases progressively. Therefore to convert the last percent of ammonia more and more catalytic area is needed[237].

To achieve maximum ammonia conversion, one catalyst supplier has developed a catalyst pack that consists of several types of gauze[237]:

- At the top of the pack, the gauzes consist of warp knitted gauzes with large voidage and large diameter wire. Such gauzes maintain high ammonia selectivity and minimize mass transfer. This should limit rhodium enrichment and help maintain mechanical strength of the catalyst pack.
- In the middle of the pack, the gauzes consist of warp knitted gauzes with lower voidage and possibly smaller diameter wire.
- At the bottom of the pack, special Bispin® gauzes with very high efficiency convert the remaining ammonia. (A Bispin® gauze is a woven structure that contains a series of straight wires with another wire wound around the straight wire.)
- These catalyst packs are generally designed for a nitrogen loading of less than 36 tons nitrogen per meter^2 per day.

The catalyst typically consists of several woven or knitted wire gauze sheets with a composition of about 90% platinum, 5% to 10% rhodium and up to 5% palladium. The primary functions of rhodium are to add mechanical strength to the gauze and to improve yield. The palladium is present to replace a portion of the more expensive rhodium[91, 104]. Figure 9.1 shows that conversion efficiency does not improve if more than 5% rhodium is present in the catalyst[144].

Catalyst loading for most plants is reported to be from 1.5 to 1.9 kg per tonne / hour of nitric acid production and is probably lowest for high pressure operations. In general, the wire diameter of the gauze varies inversely with pressure. The number of layers of gauze also varies from four at atmospheric pressure up to 24 to 45 layers for high pressure operations[104].

The performance of a catalyst pack varies with its age. Changes in efficiency are affected by structural and compositional changes to the catalyst. A new gauze has smooth rounded wires of catalyst. After a short period of operation, conversion efficiency increases as the surface roughens and metal migrates to the surface to form small metal growths that increase surface area of the catalyst. As these growths increase, the catalyst structure is weakened and catalyst is lost. The catalyst surface becomes enriched with inactive rhodium oxide (Rh_2O_3), and ammonia conversion efficiency begins a gradual decline[104].

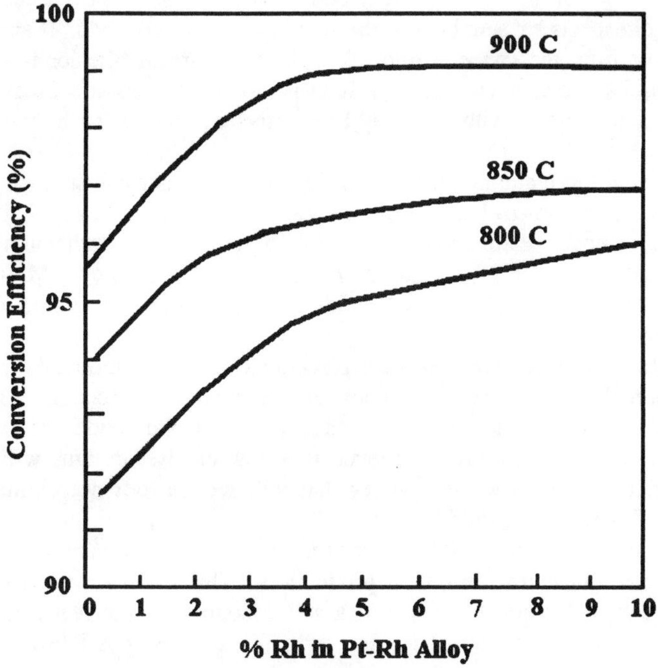

Figure 9.1. Effect of rhodium content on catalyst efficiency in ammonia oxidation.

The life of the catalyst depends on several variables. High-pressure oxidation operations (which also operate at higher temperatures, per Table 9.4) usually require the catalyst to be regenerated more frequently. The presence of trace amounts of iron, calcium, MoS_2 (lubricants) and phosphorus in the ammonia feed has been shown to have a deleterious effect on catalyst life. Deposits of iron oxides tend to catalyze the conversion of ammonia and oxygen to nitrogen and water – rather than to nitric oxide. Air pollution or other contamination of the air that is used in Eq. (9.6) can also poison the catalyst. These effects, as well as poor ammonia-air mixing and poor gas distribution across the catalyst, may reduce the yield by up to 10%[91, 97]. High pressure operations have the shortest operating time of 2 to 3 months. This increases to 4 to 6 months for medium pressure plants and as much as 8 to 12 months for low pressure combustion. The advantages of low pressure operation are offset by the capital cost for larger process equipment[104].

When the platinum catalyst no longer provides the desired reaction rates or product composition (due to low yield), new sheets of catalyst may be added to the old sheets. Or the old sheets may be replaced by a set of new sheets. In this case the old sheets are sent back to the supplier to be refurbished. If any catalyst that is lost from the gauze is trapped in the downstream filter or is recovered during equipment cleaning, it is also returned to be reprocessed. Catalyst packs that are contaminated with iron may be reactivated by pickling in hydrochloric acid[91].

Some of the materials that have been examined as catalysts include: Pure Platinum, Platinum-Iridium Alloys, Various Compositions of Platinum-Rhodium Alloys, Platinum-Palladium Alloys, Platinum-Ruthenium Alloys, Platinum-Rhenium Alloys, Platinum-Tungsten Alloys, Fe_2O_3-Mn_2O_3-Bi_2O_3 (Braun Oxide), CoO-Bi_2O_3, CoO with Al_2O_3, Thorium, Cerium, Zinc and Cadmium.

In the 1990's research on the replacement of the noble metals (platinum, rhodium, palladium) in the catalyst has centered on oxides of cobalt with smaller amounts of co-catalysts such as lanthanum metal and oxides of scandium, yttrium and lanthanum. It is believed that new catalyst designs will provide more-efficient, lighter-weight gauzes that will reduce (but not eliminate) the amount of platinum required [91].

In 2000, U.S. Patent 6,073,467 was issued for a three-dimensional, knitted, noble metal gauze for nitric acid production. This catalyst design claims to provide elevated product yields and a stable course for the reaction. It also claims to extend catalyst service life with the smallest possible noble metal losses.

9.2.2.1. Precious Metal / Catalyst Recovery

During the life of the catalyst, it becomes polycrystalline in appearance. Some weight loss due to catalyst erosion and vaporization (as platinum oxides) also occurs. This is more pronounced at the higher converter temperatures that are employed in the higher-pressure processes. In general the catalyst pack is more likely to lose platinum than it is to lose palladium or rhodium. The rate of metal loss is a strong function of temperature. As much as a tenfold increase has been reported for a temperature change from 820 to 920°C. Gross catalyst losses are around 0.05 to 0.10 grams/tonne HNO_3 for atmospheric to medium pressure plants. These losses increase to 0.20 to 0.40 grams/tonne HNO_3 for high pressure processes[104].

The precious metals combine with other impurities in the process stream to create deposits that reduce the performance of the heat exchangers. This causes a reduction in capacity, a decrease in steam production, or the overheating of equipment[237].

Several filter systems that recover as much as 50% of the lost catalyst have been investigated. These include filters of glass wool, asbestos fibers, Raschig rings and marble chips. Shortcomings of these recovery systems have been overcome by "getter" systems that use palladium alloy gauzes and recover 60% to 80% of the lost catalyst. The "getter" gauze is placed as close to the platinum catalyst pack as possible so that the catalyst components can be recovered while they are in the vapor form. Although pure palladium is the most effective “getter”, it becomes too brittle and therefore a palladium alloy is used. Additional catalyst is recovered during periodic cleaning of the ammonia oxidation reactor/converter and during cleaning of downstream heat exchangers[91, 97].

The platinum recovery on palladium gauzes depends on the transfer of platinum oxide vapor from the gas stream to the surface of the palladium gauzes. The speed of transfer depends on the design conditions of the nitric acid plant and also on the structure of the palladium gauzes[237].

The MTL® (Mass Transfer Limited) gauzes use Pd/Ni (95/5) or Pd/Cu (95/5) alloys. The base metals in the alloys are not stable at the nitric acid plant operating conditions. They are rapidly oxidized and vaporized in the gas stream. This rapid oxidation makes the wire swell by a factor of 2 after ten days of operation. The larger surface area results in a large improvement in the capture of platinum oxide vapor[237].

At temperatures lower than 800°C, the palladium oxide is stable and progressively recovers the palladium wire. This makes it inactive for further platinum recovery. For this reason MTL® gauzes are not used in atmospheric nitric acid plants. The speed of diffusion of platinum into the core of the palladium wire decreases over time due to a lower differential of concentration of platinum on the surface and in the core of the wire[237].

9.2.3. Operating Pressure

All nitric acid plants are based on the same basic chemical operations: 1) Oxidation of ammonia with air to give nitric oxide, 2) Oxidation of the nitric oxide to nitrogen dioxide and 3) Absorption in water to give a solution of nitric acid. The efficiency of the first step is favored by low pressure whereas that of the second step is favored by high pressure. These considerations, combined with economic reasons give rise to two types of nitric acid plants – single pressure and dual pressure[97].

The high mono-pressure process has been favored in the United States because of 1) its lower capital cost and 2) the traditionally lower energy and ammonia prices found in the United States. In Europe, where allowable capital payback periods and energy costs have traditionally been higher, the dual-pressure process evolved[104].

In the single pressure plant, the oxidation and absorption steps take place at essentially the same pressure. In dual pressure plants, absorption takes place at a higher pressure than the oxidation stage. The oxidation and absorption steps can be classified as[97]: Low Pressure (Below 1.7 bar), Medium Pressure (Between 1.7 and 6.5 bar) and High Pressure (Between 6.5 and 13 bar).

The main unit operations in nitric acid plants are the same for all types of operating pressures. These steps are[97]: 1) Ammonia Evaporation, 2) Ammonia Filtration, 3) Air Filtration, 4) Air Compression, 5) Air/Ammonia Mixing, 6)

Catalytic Reaction of Ammonia and Air, 7) Energy Recovery by Steam Generation and/or Gas Re-Heating, 8) Gas Cooling, 9) Dual Pressure Only ➔ NOx Compression, 10) Absorption with Nitric Acid Production, and 11) Tail Gas Energy Recovery.

Most new plants built in the 1990's have a dual-pressure design. Ammonia oxidation is operated at a medium pressure while absorption is operated at a high pressure. High pressure systems offer the advantage of lower capital costs (as much as 30% lower than some lower-pressure processes) and have been preferred in the United States for this reason. Relative disadvantages are higher catalyst loss and lower ammonia conversion. Lower pressure oxidation designs have the best efficiencies in ammonia conversion and catalyst conservation. These plants have been preferred in Europe and other areas where ammonia has been relatively expensive and where allowable payout times have been longer[91].

Typical raw material and utility consumption figures for both single and Dual-Pressure processes are given in Table 9.6[99].

Table 9.6. Raw material and utility consumptions in different nitric acid plants

	H&H	M&M	M&H
Operating Pressure (Bar-a)	11	9	4.5 / 12
Typical Run Time (Days)	70	120	210
Ammonia Usage[1] (ton/ton)	0.291	0.285	0.279
Platinum Usage[2] (mg/ton)	50	35	30
Power Usage[3] (kW-hr/ton)	1	2	2
HP Steam Export[4] (ton/ton)	0.51	0.60	0.54
LP Steam Import[5] (ton/ton)	0.27	0.21	-0.06
Cooling Water[6] (m^3/ton)	130	120	110

(Reproduced by permission of Monsanto Enviro-Chem Systems, Inc.)

[1] Includes catalytic NOx abatement to 200 ppmv and average ammonia conversion efficiency over the entire gauze run length.

[2] Described as net of platinum recovery systems.

[3] Add 25 kW for HVAC, lighting and instrumentation.

[4] Steam conditions are 42 kg/cm^2 g and 400°C.

[5] Steam conditions are saturated at 3.5 kg/cm^2 g.

[6] At 11°C temperature rise and 2.5 kg/cm^2 pressure allowance.

9.2.3.1. Single Pressure Plant

The Single Pressure design (see Figure 9.2) is generally used in smaller plants or in larger plants where minimization of capital is a critical design consideration. Ammonia oxidation and absorption of NO_2 occur at the same relative pressure. This reduces the complexity and capital cost compared to the dual pressure plant. However efficiency may be lower in both the front and back ends of the plant. Single pressure plants can be designed to operate 1) at low to medium pressure to favor the ammonia oxidation reaction, 2) at high pressure to favor the absorption reactions or 3) at any operating pressure in between[99].

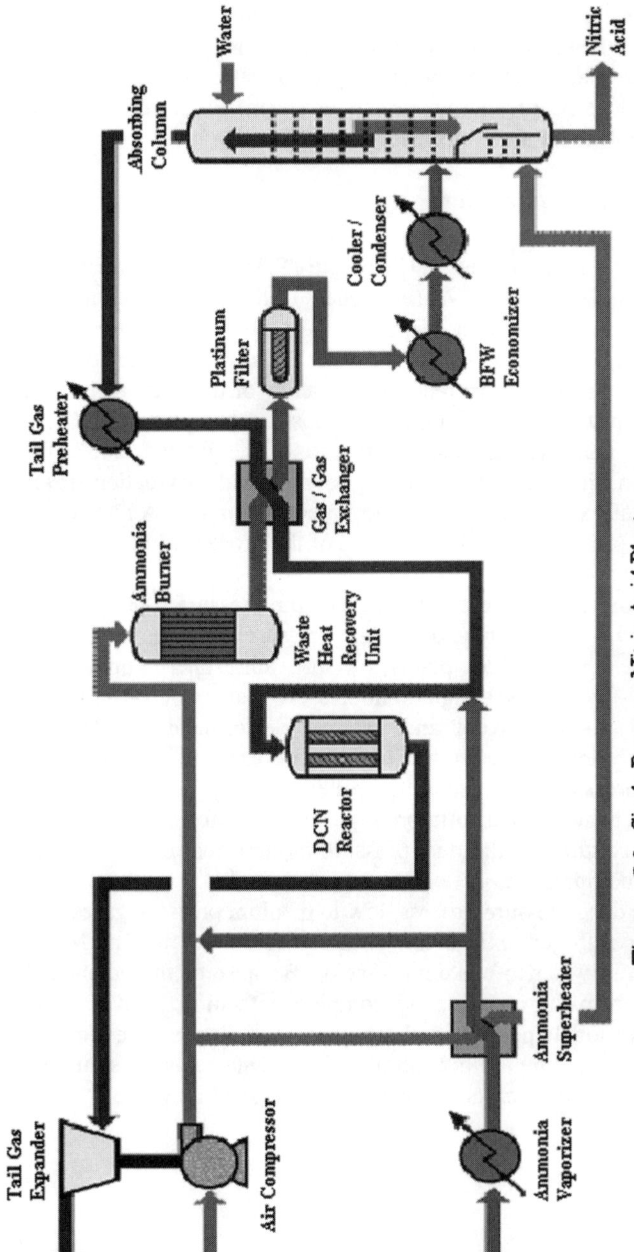

Figure 9.2. Single Pressure Nitric Acid Plant.
(Reproduced by Permission of Monsanto Enviro-Chem Systems, Inc.)

Single pressure processes are either medium pressure (0.3 to 0.6 MPa or 3 to 6 atm), or high pressure (0.7 to 1.2 MPa or 7 to 12 atm). Higher catalyst temperatures and higher operating pressures accommodate a more efficient recovery of process energy, either as steam or as reheated tail gas, which provides power for air compression[104].

9.2.3.2. Dual Pressure Plant

The Dual Pressure design (See Figure 9.3) is generally used in larger plants, or in mid-size plants where higher utility/raw material costs dictate a minimization of operating expense. Ammonia oxidation occurs at low or medium pressure. This decreases the gas density and the nitrogen loading on the gauze. The result is an increase in efficiency of the ammonia oxidation reaction and minimization of platinum catalyst loss. Absorption of NO_2 occurs at high pressure to maximize the partial pressure of the gas reactants. The result is an increase in the rate of the nitrogen monoxide oxidation reaction and the solubility of oxygen and NO_2 in the aqueous solution. A nitrous gas compressor boosts the pressure from the front end of the process (ammonia oxidation) to the back end (absorption)[91, 99].

Dual-pressure processes have a medium pressure (0.3 to 0.6 MPa) front end for ammonia oxidation and a high pressure (1.1 to 1.5 MPa) tail end for absorption. Some older plants still use atmospheric pressure for ammonia oxidation. Compared to high monopressure plants, the lower oxidation pressure improves ammonia yield and catalyst performance. Platinum losses are significantly lower and production runs are extended by a longer catalyst life. Reduced pressure also results in weaker nitric acid condensate from the cooler condenser, which helps to improve absorber performance. Due to the split in operating conditions, the dual-pressure process requires a specialized stainless steel NO_x compressor.

In the dual-pressure process, low to medium pressure gases (0.3 to 0.6 MPa) containing nitrogen oxides are compressed to 1.1 to 1.5 MPa for efficient absorption in water to make nitric acid. Because of the potential for unreacted ammonia from the converter to combine with nitric acid to form ammonium nitrate, additional operating and safety precautions are necessary to deal with the accumulation of ammonium nitrate solids. One reported solution is a periodic injection of steam to parts of the compressor (see Figure 9.3)[104].

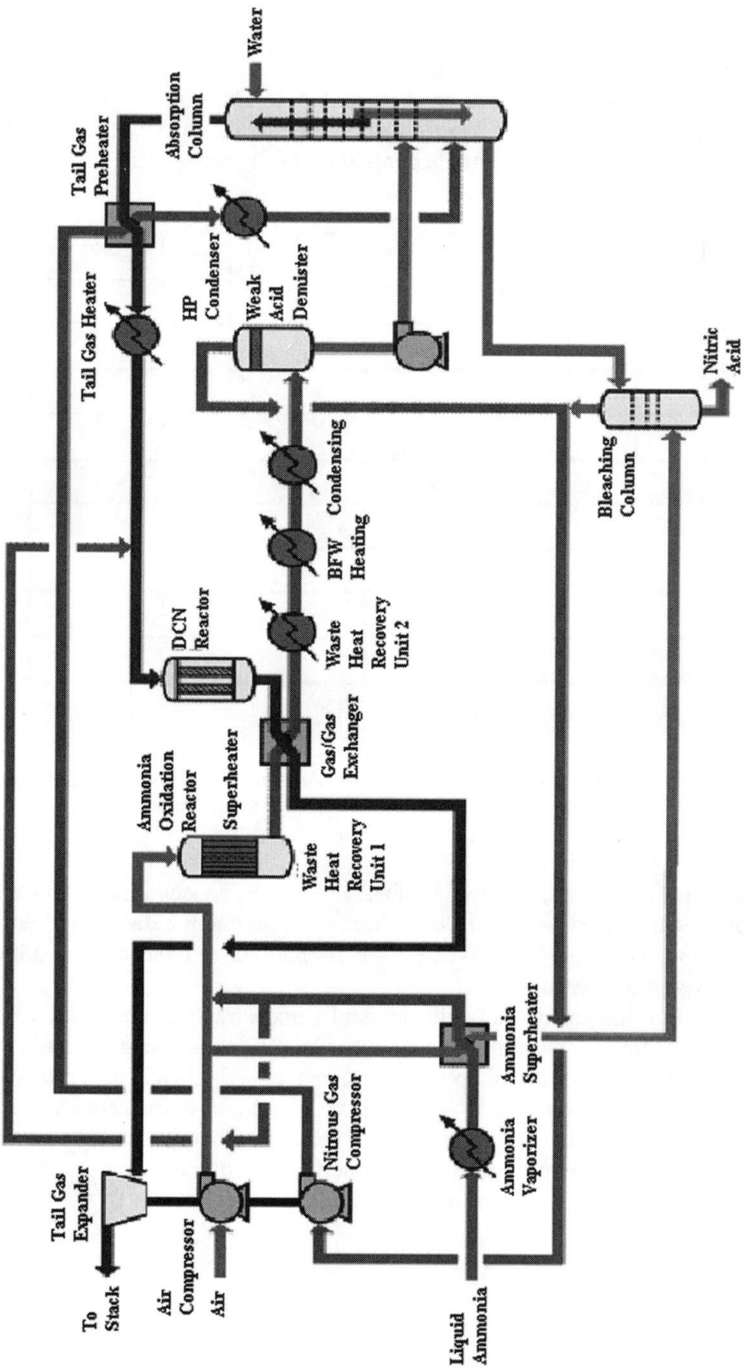

Figure 9.3. Dual Pressure Nitric Acid Plant.
(Reproduced by Permission of Monsanto Enviro-Chem Systems, Inc.)

9.2.4. Concentration of Nitric Acid

The vapor-liquid equilibrium for the nitric acid / water system at atmospheric pressure is shown in Figure 9.4. This figure shows that a concentration of 68.4 weight % nitric acid is the maximum (i.e., the azeotropic point) that can be obtained by simple distillation of the weak acid[220].

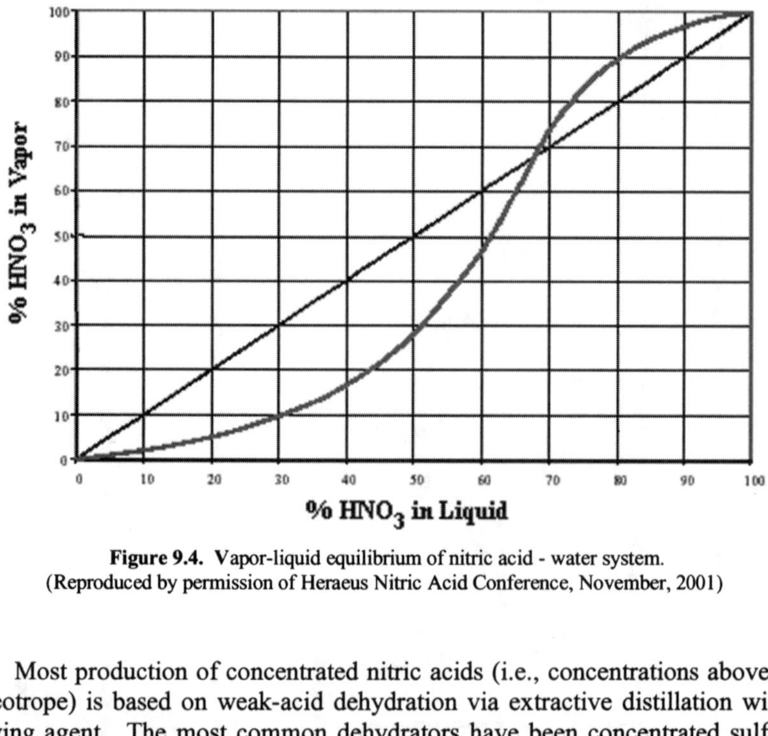

Figure 9.4. Vapor-liquid equilibrium of nitric acid - water system.
(Reproduced by permission of Heraeus Nitric Acid Conference, November, 2001)

Most production of concentrated nitric acids (i.e., concentrations above the azeotrope) is based on weak-acid dehydration via extractive distillation with a drying agent. The most common dehydrators have been concentrated sulfuric acid and magnesium nitrite[91].

Figure 9.5 illustrates the sulfuric acid concentrations that lead to the production of higher concentrations of nitric acid. At a concentration of 67 weight % H_2SO_4, the azeotropic point has vanished, and 99 weight % nitric acid can be distilled. The nitric acid is the lighter phase and is extracted as vapor. These vapors are condensed overhead and a portion of the nitric acid is returned to the distillation column as reflux. The sulfuric acid and water go with the bottom liquid phase and are concentrated for reuse in the process[104, 220].

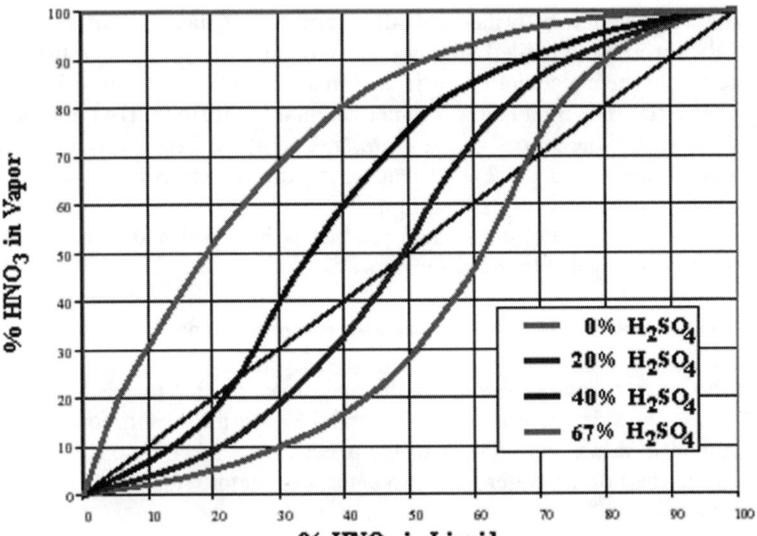

Figure 9.5. Vapor-liquid equilibrium of nitric acid - water system in the presence of sulfuric acid. (Reproduced by permission of Heraeus Nitric Acid Conference, November, 2001)

An example of the dehydration process is the Kvaerner Chemetics process that uses concentrated sulphuric acid to eliminate the azeotrope and enable colorless, strong nitric acid up to 99% to be distilled from the mixed acids. The residual acid leaving the bottom of the processing tower is approximately 70% sulphuric acid[101].

Alternative methods of integrated concentration have been developed that do not require separate dehydrating agents. These processes differ from the direct process in that weak acid reacts with concentrated nitrogen dioxide to produce an acid that is sufficiently superazeotropic that distillation into concentrated acid is economically feasible. The weaker azeotropic acid may be recycled for concentration or used as it is[91].

In the magnesium nitrate process, an aqueous solution containing 72 wt.% magnesium nitrate is fed to the middle of a distillation column and a 60% nitric acid solution is fed to the same tray. Nitric acid (95% to 98%) is taken overhead while a 55% magnesium nitrate solution is removed from the bottoms. The bottoms stream is concentrated in an evaporator and returned to the column. The magnesium nitrate feed is restricted to 72% because at higher concentrations, it is quite viscous and has high melting points. The magnesium nitrate is a less effective dehydrating agent than sulfuric acid. Therefore, processes that use magnesium nitrate use rectification to increase the concentration of nitric acid up to 99 wt. %[104].

Uhde developed another process for the production of highly concentrated nitric acid (98% to 99%) that uses oxygen but no dehydrating agents. The NO from ammonia oxidation is converted to NO_2 by highly concentrated HNO_3. The NO_2 is cooled with brine, and after removal of the residual water, it is physically absorbed in chilled, highly-concentrated nitric acid. Then the NO_2 is separated from the concentrated nitric acid in a distillation column and liquefied by refrigeration. The liquid NO_2 is dimerized in a $N_2O_4/HNO_3/H_2O$ mixture that reacts with supplemental oxygen at a pressure of about 720 psi to form highly concentrated nitric acid. The acid that leaves the reactor contains about 20% dissolved N_2O_4, which is separated in a distillation column and returned to the reactor. The concentrated nitric acid product is bleached and sent to storage, and a portion is recycled to the absorber.

9.2.5. Water Balance in Dual Pressure Azeotropic Plants

In the reactor water is formed in the main reaction that makes NO [see Eq. (9.6)] and in the side reactions (Eqs. 9.7 and 9.8). In the Absorption step water is consumed to dilute the product to the azeotropic concentration (~68 weight %). A major part of the water may also enter the reactor with the air that is used to feed the reactor[220].

Overall there is a surplus of one mole of water per mole of ammonia that is fed to the plant - if the burner is 100% efficient. If the burner efficiency is 95%, 5% of the ammonia feed is converted to N_2 or N_2O according to Eq. (9.7) or Eq. (9.8) with no extra water. Therefore a maximum nitric acid concentration of 76.6% can be produced[220].

The Absorption Column (see Figure 9.3) requires a minimum process water feed to ensure safe and stable operation, and the plant must also keep the NOx content in the tailgas within the allowable limits. If the oxygen content in the tailgas is assumed to be 2.5 volume %, then the amount of intake air can be determined. And the maximum water content of the air is calculated to be 27 grams per kg of dry air[220].

The water balance in the nitric acid process consists of reaction water, dilution water and water carried with the process streams (see Figure 9.6)[220].

Figure 9.7 shows that the admissible combinations of ambient air temperature and humidity for the production of azeotropic acid lie below the line. If the air conditions lie above the line, a cooler must be installed to remove water from the air before it enters the Ammonia Oxidation reactor[220].

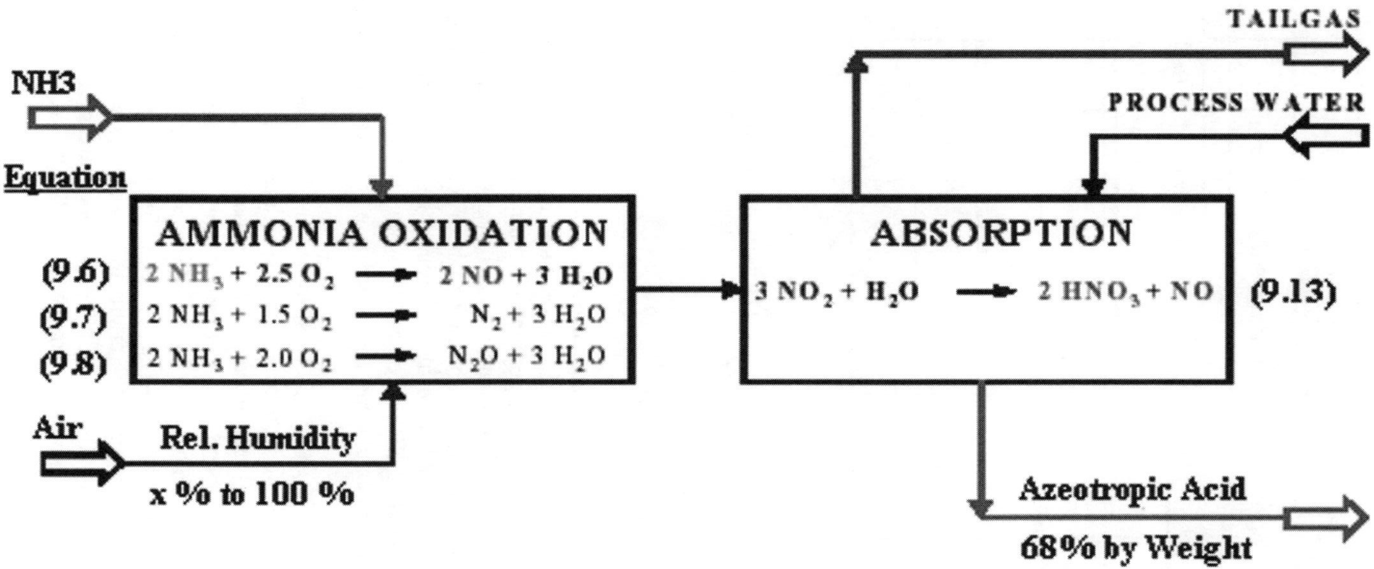

Figure 9.6. Water balance for an azeotropic nitric acid plant.
(Reproduced by permission of Uhde GmbH and copyright by Uhde)

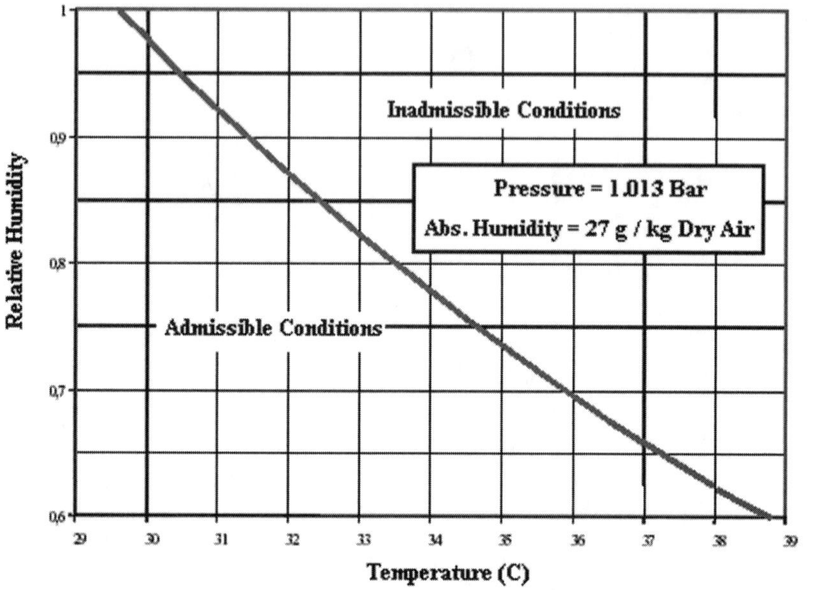

Figure 9.7. Theoretical limit curve for intake air humidity.
(Reproduced by permission of Uhde GmbH and copyright by Uhde)

9.2.6. Direct Strong Acid Processes

Direct strong nitric (DSN) processes use many of the production steps that are used for weak acid production. However they do not need pure oxygen, sulfuric acid or magnesium nitrate to produce concentrated nitric acid[100, 104].

In the DSN processes ammonia and air react over a platinum catalyst, and the process gas stream is cooled to the point where weak acid condensate is formed. Tail gases are also reheated by hot process gases and provide energy recovery in a gas expander. DSN processes differ from weak acid plants in that additional or different process steps are needed to achieve superazeotropic strengths of nitric acid[100, 104].

In one process the NO_2 is partially absorbed in the water generated by oxidation of ammonia to produce condensate as dilute nitric acid. Nitrogen dioxide (NO_2) is also partially absorbed in the process water in the absorber to produce more dilute nitric acid. At the same time, the remaining NO_2 is absorbed in 80 to 90 wt. % nitric acid in the NO_2 absorber to form red fuming nitric acid. Nitrogen peroxide (N_2O_4) is separated from the red fuming nitric acid by air and reacts with water in the dilute nitric acid. As a result, the acid concentration reaches 80 to 90 wt.% HNO_3, and this stream is distilled to produce concentrated nitric acid (98 wt. % HNO_3)[100]. The "weak" or dilute acid

(50% to 60% HNO_3) can be used directly for the manufacture of ammonium nitrate and other products or it can be further concentrated by various means[91]. This process also uses SN-1 stainless steel to avoid corrosive problems that may be found in conventional concentrated nitric acid processes[100].

The key to producing strong acid is obtaining high partial pressures of nitrogen peroxide. In the HOKO process, N_2O_4 is absorbed in chilled 98 wt. % acid, distilled off and condensed. The N_2O_4 is mixed with oxygen and weak nitric acid in an autoclave at 5 MPa to produce strong acid[104].

As mentioned earlier in this chapter, the SABAR process absorbs nitrous gases in azeotropic nitric acid in the presence of oxygen and then makes strong acid by azeotropic rectification. Some of the key reactions in this process are illustrated by Eq. (9.15) and Eq. (9.16).

$$2\,NO_2 \leftrightarrow N_2O_4 \tag{9.15}$$

$$4\,NO_2 + 2\,H_2O + 4\,O_2 \rightarrow 4\,HNO_3 \tag{9.16}$$

Nitrogen peroxide in the process gas is chemically absorbed into azeotropic or greater strength acid. Air stripping and rectification of the resulting acid stream produces strong acid of 98 to 99 wt. % strength[104].

DSN plants have found little application in the United States, but several have been built in Europe. Most production of concentrated acids (i.e., concentrations above the azeotrope) is based on weak-acid dehydration via distillation with a drying agent. The most common dehydrators have been concentrated sulfuric acid and magnesium nitrite. However future concentrated nitric acid plants may avoid these processes because of the perceived difficulty in recycling or disposing of spent dehydrating agents[91, 104].

9.3. STABILIZERS

Over a period of time, concentrated nitric acids tend to decompose according to Eq. (9.17):

$$2\,HNO_3 \rightarrow 4\,NO_2 + H_2O + O_2 \tag{9.17}$$

As a result, pressure will build up in storage vessels. Also nitric acid is very corrosive. Therefore, some stabilizers and/or corrosion inhibitors are used.

Corrosion of aluminum by red fuming nitric acid is reduced by adding 4% by weight of hydrogen fluoride. Decomposition of concentrated acid is reduced by such substances as quaternary ammonium compounds, organic sulfones, inorganic persulfates and organic sulfonium compounds[102].

9.4. POLLUTION ABATEMENT

The primary pollution problem in nitric acid plants is the abatement of NOx in tail gases. Three options exist to reduce these emissions to acceptable levels: 1) Capture the NOx and convert it to additional nitric acid, 2) Capture the NOx and convert it to nitrate-nitrite sales, or 3) Render the NOx harmless by converting it to non-polluting compounds. The processes that have been developed to reduce emissions at existing and new plants can be classified into four general categories: Absorption, Adsorption, Selective Catalytic Reduction (SCR) and Non-Selective Catalytic Reduction[91].

The main environmental factor that affects nitric acid process selection is the concentration of acid forming oxides of nitrogen (NOx) in the tail gas emission. In the United States, gaseous emissions from newly constructed nitric acid plants are limited to 1.5 kilograms NOx per tonne of nitric acid produced (100% basis) with a maximum opacity of 10%. This discharge limit corresponds to about 200 to 230 ppmv of nitrogen oxides in vented tail gas, whereas concentrations after absorption may contain as much as 2,000 to 3,000 ppmv of nitrogen oxides. In Western Europe, NOx emissions are limited by EU regulations to 200 ppm[91, 104].

In an economic comparison of abatement systems, a 1991 EPA study indicates that extended absorption to be the most cost-effective method for NOx removal. Selective Reduction matches its performance only in small-capacity plants of about 200 to 250 tonnes per day. Nonselective abatement systems were indicated to be the least cost-effective method of abatement. The results of any comparison depend on the cost of capital versus variable operating costs. A low capital cost for SCR is offset by the ammonia required to remove the NOx. Higher tail gas NOx concentrations make this method less attractive. The investment for extended absorption is partially recovered by increased yield of nitric acid product[104].

See Chapter 4 for additional comments about NOx abatement.

9.4.1. Absorption Abatement

Absorption Abatement or extended absorption refers to modifications that involve the addition of increased absorption capacity or optimization of the existing absorption system to oxidize and react the nitrogen oxides with water to form acid.. Tail gases are passed through an absorber containing either water or an aqueous solution of ammonia, urea or sodium hydroxide. When water is used as the absorbent, the resultant weak acid is recycled. This increases nitric acid yields by 1% to 3%. When other absorbents are used, the recovered NOx is typically consumed in the production of nitrogen solutions for fertilizer use. If sodium hydroxide is the absorbent, pure sodium nitrite and sodium nitrate may be recovered[91, 104].

Bubble cap trays hold their liquid seal on shutdown and are therefore preferred for minimizing NOx abatement problems during start-up and shutdown. Low partial pressures of nitrogen oxides lead to low rates of oxidation and a need for large amounts of gas-phase holdup. Extended absorption requires relatively few trays but large oxidation volumes, and often employs refrigeration to promote the oxidation process and minimize column size. This method of abatement is most effective for high pressure absorption in which abatement to less than 200 ppmv NOx can be achieved in a single column. In general, medium pressure plants use two absorption columns to achieve concentrations of 500 ppmv NOx. Cold tail gases are reheated by heat exchange with hot process gas to increase power recovery in the expander[104].

The gas at the outlet of the absorber may vary within the following limits during stable operation:

- NOx 100 to 3,500 ppmv
- N_2O 300 to 3,500 ppmv
- O_2 15 to 4% by volume
- H_2O 0.3% to 2% by volume
- Flow 3,100 to 3,400 Nm^3/tonne of HNO_3

Start-up and shutdown periods will normally increase the NOx content of the tail gas at the stack during the few hours needed for the process to stabilize or for NOx to be cleared from the plant. During the ammonia oxidation some nitrous oxide (NO_2) is formed. The amount depends on combustion conditions (pressure, temperature), catalyst composition and age, and design of the oxidation reactor[97].

The minimum emission levels achieved in 2001 without added pollution abatement are:

- For Medium Pressure Absorption: 1,000 to 2,000 ppmv
- For High Pressure Absorption: 100 to 200 ppmv

The absorption is influenced by other process parameters such as the cooling water temperature, absorber design technology and operating pressure. Whereas a dual-pressure plant or a high-pressure, single-pressure plant (with high pressure absorption) may give acceptable emission levels, the medium-pressure absorption plant must be followed by an abatement system[97].

Chemical absorption has also been developed for NOx removal. One such method uses weak acid (25 to 30 wt. %) to remove nitrogen oxides from the tail gas to form nitrous acid (HNO_2). Steam stripping regenerates the clean acid for recycle. The removed NOx is compressed and fed to the plant's main absorption column[104].

9.4.2. Adsorption Abatement

Adsorption Abatement uses acid-resistant molecular sieves to absorb the NOx from the tail gas. The adsorbent is periodically regenerated and the NOx recovered, converted to NO_2 and recycled to the plant for recovery as nitric acid. The tail gas may also be mixed with a reducing agent (usually natural gas or ammonia) and passed through a catalyst to reduce the NOx to elemental nitrogen. When natural gas is used, conditions must be carefully controlled and facilities installed to control the heat released by natural gas that reacts with both the NOx and the oxygen[91].

9.4.3. Selective Catalytic Reduction

Selective Catalytic Reduction (SCR) is normally used in new nitric acid plants. In this process ammonia reacts with nitric oxide and nitrogen dioxide but only to a lesser extent with oxygen to selectively reduce the NOx compounds to N_2. The reactions are shown below[97, 104]:

$$6\,NO + 4\,NH_3 \rightarrow 5\,N_2 + 6\,H_2O \tag{9.18}$$

$$6\,NO_2 + 8\,NH_3 \rightarrow 7\,N_2 + 12\,H_2O \tag{9.19}$$

$$3\,O_2 + 4\,NH_3 \rightarrow 2\,N_2 + 6\,H_2O \tag{9.20}$$

Potential catalysts include vanadium pentoxide, platinum, iron/chromium oxides and zeolites. The catalysts are usually shaped like honeycombs or flat parallel plates. The operating temperature is usually between 210°C and 410°C, and operating pressure has only a minor effect on overall efficiency. The ammonia leakage through the catalyst beds depends on the catalyst efficiency. Tail gas from the absorber is preheated to a minimum temperature for good operation of the SCR catalyst and to improve expander energy recovery . The reactant gas is mixed with the preheated tail gas and the mixture is fed to the reactor containing the catalyst bed[97, 104].

For effective abatement, a slight excess of ammonia is used and this leaves 5 to 20 ppmv of ammonia in the treated tail gas. Higher concentrations of ammonia present a potential safety hazard. The biggest concern is that ammonia at low temperatures can form ammonium nitrate and nitrite. These compounds might accumulate in the downstream equipment and pose an explosion hazard upon being heated[104].

The advantages of the SCR system are[97]: ammonia is readily available in a nitric acid plant; a low NOx content can be achieved; the increase in tail gas temperature is negligible; and no oxygen is consumed.

The disadvantages of the SCR system are[97]: the tail gas temperature after the expander must be kept high enough to avoid any ammonium salt deposits; a small ammonia slip exists and must be treated; and the system can only start when temperatures are high enough.

The capital cost of an integrated SCR unit for a new 1,000 tonne/day plant is estimated to be ~1.5% of the total capital cost of the nitric acid plant. This cost includes the cost of the SCR catalyst but excludes spare parts. The capital cost of an end-of-pipe SCR unit for an existing 1,000 tonne/day plant is estimated to be 3% to 6% of the total capital cost of the nitric acid plant. But this is very dependent on the type of nitric acid process. The SCR will increase operating costs by ~1.1% when NOx in the tail gas is reduced from 1,000 ppmv to 200 ppmv. The maintenance cost of the SCR unit is typically 2.5% of the capital cost[97].

9.4.4. Non-Selective Catalytic Reduction

Nonselective abatement uses a catalyst and fuel (usually a gaseous hydrocarbon) to reduce nitrogen oxides to nitrogen and, in the process, combust any remaining free oxygen in the tail gas. This process consumes significantly more fuel than a selective reduction system, but the energy from burning the fuel is mostly retrieved as power in the expander[104].

Platinum, palladium and rhodium (either in pellet form or as a honeycomb) are typically used as catalysts in these systems. The minimum temperature for inlet tail gas depends on the fuel and its ignition temperature. Hydrogen, which has the lowest ignition temperature, requires about 150 to 200°C[104].

A large temperature rise due to combustion (about 130°C for every volume percent of oxygen) places a limit on the total oxygen content in the absorber tail gas. Free oxygen tends to be consumed preferentially and fuel must be fed in a slight stoichiometric excess to oxygen. Unreacted hydrocarbons are discharged to the atmosphere along with small amounts of carbon monoxide and hydrogen cyanide. Relatively few plants use this method of abatement[104].

9.4.5. Absorption Plus Selective Catalytic Reduction

Monsanto Enviro-Chem offers NOx abatement technology that is licensed from Rhodia of France. It includes a High Efficiency Absorption (HEA) section for extended absorption and a catalytic reduction section (SCR) for Catalytic Destruction of NOx (i.e., the DCN technology)[99].

The HEA technology is based on the enhanced oxidation of the nitrous acid (HNO_2) intermediate in the liquid phase. Conventional absorption technology is based upon the partial oxidation of the HNO_2 intermediate, which evolves nitrogen monoxide (NO) gas. This must be reoxidized to nitrogen dioxide (NO_2) in a relatively slow gas phase reaction[99].

With Rhodia's high activity catalyst in the DCN process, efficient catalytic reduction of NOx to nitrogen and water is possible at temperatures as low as 180°C without measurable ammonia in the tail gas. This catalyst has operated for over 10 years in some installations[99].

The use of the HEA Column plus the DCN Reactor in series (see Figure 9.8) allows the optimization of the operating conditions and the operating cost. The HEA Column removes the majority of the NOx and stabilizes the NOx content in the tail gas. The DCN reactor destroys the remaining NOx as required to meet the emission limits. A comparison of the operating conditions for the Monsanto technology is given in Table 9.7[99].

Table 9.7. Operating Conditions in NOx Abatement Systems

		DCN[a] Only	HEA Plus DCN	HEA[b] Only
NOx Gas to HEA	ppmv	N/A	3,500	3,500
NOx Gas to DCN	ppmv	3,500	700	N/A
NOx Gas to Turbine	ppmv	200	200	200
Nitric Acid Recovered	kg/ton – HNO_3	N/A	27	31
Ammonia Consumed	kg/ton – HNO_3	8.5	1.3	N/A
Gas Temperature Rise	°C	43	6.5	0
Pressure Drop	kg/cm^2	0.07	0.2	0.35
Capital Cost	Ratio	1.0	2.5	3.5

(Reproduced by permission of Monsanto Enviro-Chem Systems, Inc.)

[a] DCN: Destruction by Catalysis of NOx

[b] HEA: High Efficiency Absorption

9.4.6. Nitrous Oxide Removal

As discussed above the preferred technology for NOx removal in nitric acid plants is selective catalytic reduction (SCR) using ammonia as a reductant and in many cases vanadium-pentoxide-type catalysts. Unfortunately, this process does not remove of N_2O (nitrous oxide)[221].

Several different approaches are being examined to reduce N_2O emissions. Primary measures are aimed at preventing N_2O from being formed. Secondary and Tertiary abatement measures have focused on two locations for N_2O removal: 1) directly downstream of the ammonia oxidation catalyst and 2) in the tail gases (see Figure 9.2 or Figure 9.3)[222].

9.4.6.1. Prevent Nitrous Oxide Formation in the Reactor

One of the Primary measures that started in the early 1990's was the use of knitted gauzes that gave higher ammonia conversion and reduced N_2O formation. In the late 1990's catalyst packs were introduced in which the geometry and density of the gauzes throughout the pack are tailored to local reaction conditions. These catalysts give increased ammonia conversion as well as longer campaign lives[222].

Cobalt oxide (Co_3O_4) catalysts are being used in some nitric acid plants as an alternative to platinum-rhodium (Pt-Rh). They generate less N_2O, cost less and have a longer campaign life than Pt-Rh gauzes. (A paper in 2000 reported a conversion rate of ammonia to nitrous oxide as low as 0.5% over cobalt oxide catalyst)[222].

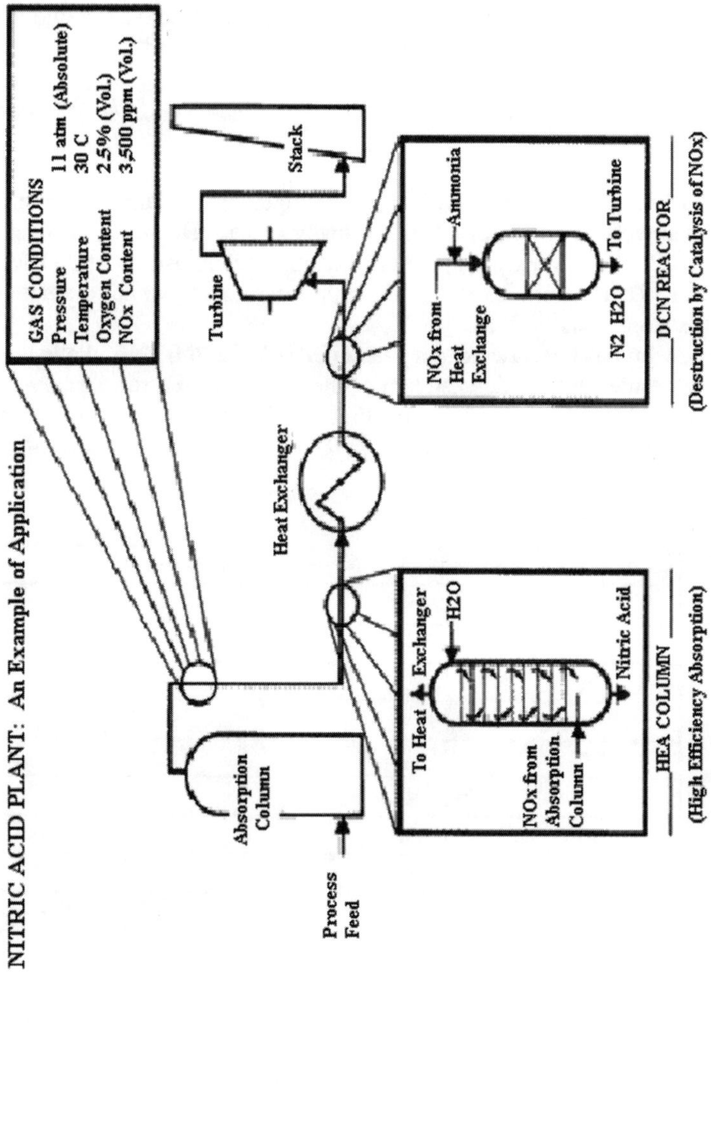

Figure 9.8. NOx Abatement System with DCN (Destruction by Catalysis of NOx). (Reproduced by Permission of Monsanto Enviro-Chem Systems, Inc.)

However cobalt oxide does have some drawbacks. Lower ammonia conversion efficiencies have been reported - as low as 88% to 92% in a high pressure plant compared with a typical value between 94% and 95% for Pt-Rh gauzes. The optimum operating temperature is 70 to 80°C lower than for Pt-Rh gauzes, and this could result in difficulties with the steam balance in a revamped plant. Cobalt oxide catalysts also suffer from reversible deactivation due to the reduction of Co_3O_4 to CoO in the upper parts of the catalyst bed[222].

9.4.6.2. Nitrous Oxide Removal Between the Reactor and the Absorption Column Inlet

Catalytic decomposition of N_2O at high temperature would offer a simple and cost effective way to remove N_2O for many plants. However, catalyst life may be short due to the high temperature and the decomposition of NO would be catalyzed. This would cause production losses and would dramatically increase operating costs[221].

Norsk Hydro has been awarded a patent (EP 0 359 286 B1) for the inclusion of a suitable residence time downstream of the ammonia oxidation catalyst that allows most of the N_2O to decompose. Figure 9.9 shows that an empty space below the gauze of 3.5 meters would allow 90% of the N_2O to decompose[222, 294].

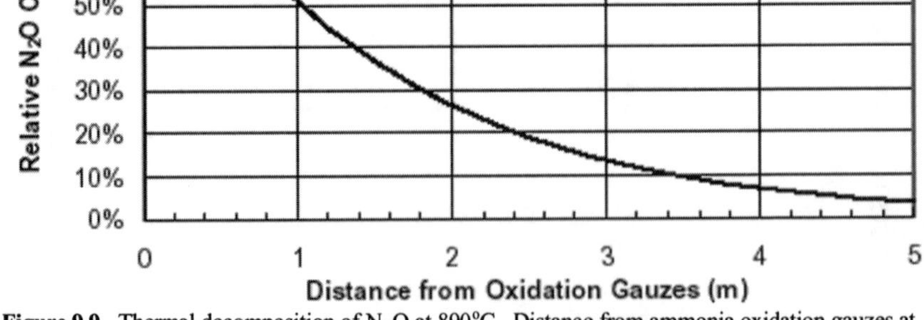

Figure 9.9. Thermal decomposition of N_2O at 890°C. Distance from ammonia oxidation gauzes at 2.2 m/s gas velocity.

(Produced from formula in Norsk Hydro patent)

Several proposals have been made to install a catalyst in the space beneath the ammonia oxidation gauzes in the converter. In 1999 BASF was testing a Cu-Zn-Al spinel catalyst in a production plant. They reported an 80% reduction in N_2O (from 1000 ppm down to 200 ppm) in an atmospheric combustion plant and a 70% reduction (down to 350 ppm) in the 5.5 bar combustion section of their dual pressure plant in Antwerp[222].

A patent was issued to DuPont (US 5,478,549) in 1995 for a zirconium oxide catalyst that can be installed below the gauzes and catchment system. The German company, L & C Steinmüller, submitted a patent application (DE 198 05 202 A1) in 1999 for a wide range of N_2O abatement catalyst that can be installed below the gauzes[222].

9.4.6.3. Nitrous Oxide Removal Upstream of Expander

A safer option may be to destroy the nitrous oxide after the absorption section, in the tail gases of the nitric acid plant. The process conditions are less severe, and NOx decomposition is not a problem[221].

The Energy Research Centre of the Netherlands (ECN) developed a catalytic system for N_2O removal from the tail gases of a nitric acid plant. The technology can be used both upstream of the expander (where the temperature is 250 - 500°C and pressure is 4 to 12 bara) and downstream the expander (atmospheric pressure, temperature 100 to 200°C)[221].

The simplest reaction for the destruction of N_2O is shown in Eq. (9.21). Unfortunately many catalysts for this reaction are sensitive to water. And nitric acid plant tail gases contain 0.5 to 1.0% water. This means relatively high temperatures (> 400°C) are needed to make Direct Decomposition of N_2O possible[221].

$$2\,N_2O \rightarrow 2\,N_2 + O_2 \qquad (9.21)$$

Destruction of N_2O can be carried out at lower temperatures by adding a reductant. In this case an iron-containing zeolite catalyst is used for the selective catalytic reduction of N_2O using hydrocarbons as a reductant. The catalyst did not deactivate in a 2000-hour test under demanding conditions (450°C, 6% H_2O). Hydrocarbons such as propane (or LPG) and methane (widely available as natural gas) can be used as the reducing agent[221].

Equation (9.22) shows that destruction of one mole of the greenhouse gas N_2O results in the release of three moles of CO_2. However N_2O contributes 310 times more to global warming than CO_2. Therefore the process has a theoretical efficiency of more than 99%[221].

$$N_2O + C_3H_8 + 4.5\,O_2 \rightarrow N_2 + 3\,CO_2 + 4\,H_2O \qquad (9.22)$$

Figure 9.10 shows how N_2O destruction varies with tail gas temperature and with the fuel that is used[221].

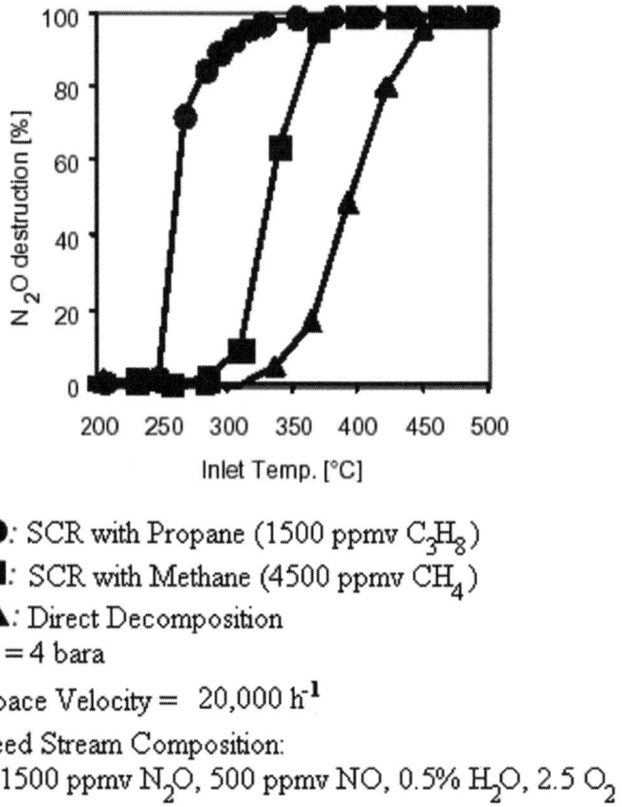

Figure 9.10. Nitrous oxide destruction at different tail gas temperatures. (Reproduced by permission of Energy Research Centre of the Netherlands)

The SCR of N_2O with hydrocarbons is an exothermic reaction and the adiabatic temperature rise is 30 to 100°C, depending on the concentration of added hydrocarbon. In an existing nitric acid plant, the expander is usually designed to work at a very well-defined temperature, so the heat of the N_2O removal reaction has to be removed. A tail gas cooler can be used for this purpose. Direct decomposition of N_2O produces only 3 to 5°C of heat so cooling of the gases before they enter the expander is normally not necessary with this process[221].

9.4.6.4. Nitrous Oxide Removal Downstream of Expander

At atmospheric pressure and temperatures of 100 to 200°C, both direct decomposition and SCR are not possible unless the tail gases are preheated. Figure 9.11 shows how N_2O can be removed downstream of the expander. After startup by an additional burner, the heat generated by the selective catalyst reduction (SCR) of N_2O maintains the off gases at the required temperature. Addition of 500 ppmv of propane (C_3H_8 / N_2O = 1:3) is sufficient. At atmospheric pressure and a space velocity of 20,000 h^{-1}, 80% N_2O destruction efficiency is reached at 350°C[221].

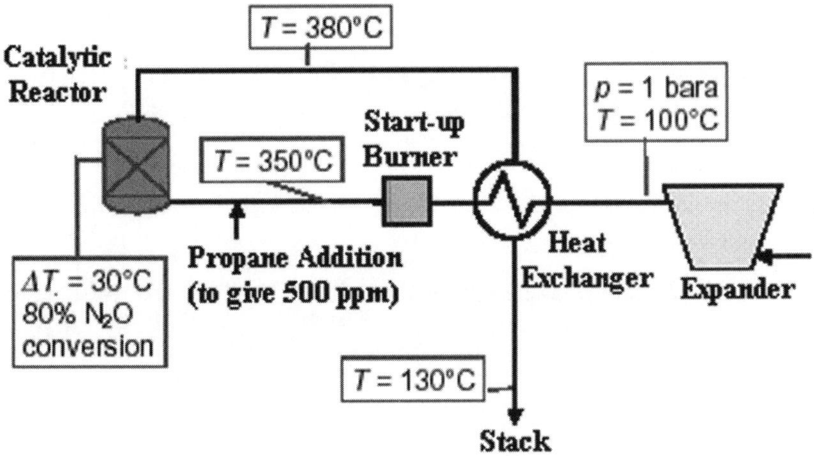

Figure 9.11. Nitrous oxide removal facility downstream of expander. (Reproduced by permission of Energy Research Centre of the Netherlands)

9.4.6.5. Combined NOx and Nitrous Oxide Removal

The selective catalyst reduction of both NOx and N_2O can take place when a second metal-exchanged, zeolite catalyst is combined with the catalyst that removes N_2O. Propane is used as the reductant. Destruction efficiency of both NOx and N_2O is higher than 80% at a tail gas temperature above 280°C, a pressure of 4 bara and a space velocity of 15,000 h^{-1}. At atmospheric pressure, conversions of NOx and N_2O are somewhat lower. A space velocity of 15,000 h^{-1} and a temperature of 350°C results in 80% destruction of both NOx and N_2O[221].

9.4.6.6. Economic Evaluation

Table 9.8 describes the plant that was the basis for the economic evaluation while Table 9.9 compares the investment and costs[221].

As shown in Table 9.9, Direct Catalyst Decomposition is an economic option for N_2O reduction in an existing nitric acid plant. The cost per tonne of CO_2 equivalents removed range from 0.4 to 1.0 EUR (based on 2001 costs) - depending on the temperature and the space velocity (i.e., catalyst loading)[221].

Table 9.8. Typical Nitric Acid Plant Characteristics

Characteristics	Value
HNO_3 Capacity	600,000 tonnes / year
N_2O emissions	5,000 tonnes / year
CO_2 Equivalents	1,500,000 tonnes / year
Tail Gas Flow	200,000 Nm^3/hr
Pressure Upstream of Expander	10 bara
Temperature Upstream of Expander	350°C
Pressure Downstream of Expander	1 bara
Temperature Downstream of Expander	100°C

(Reproduced by permission of Energy Research Centre of the Netherlands)

Table 9.9. Economic Evaluation of N_2O Abatement Systems[e]

		Direct[a] Decom-position	SCR[b] Upstream of Expander	SCR[c] Downstream of Expander
Investment Costs	10^6 EUR	2.5	4.5	5.1
Exploitation Costs				
Capital Cost	10^3 EUR/yr	341	622	724
Fixed Operating Costs	10^3 EUR/yr	126	265	294
Energy[d]	10^3 EUR/yr	-	2,297	603
Recovery Losses	10^3 EUR/yr	90	453	272
Revenue from Steam	10^3 EUR/yr	-	526	-
Annual Costs	10^3 EUR/yr	557	3,111	1,893
N_2O Removal Efficiency	%	70	90	90
N_2O Reduction	tonnes/yr	3,248	4,183	4,183
Cost per tonne N_2O	EUR/tonne N_2O	171	704	416
CO_2 Equivalents	10^3 tonnes/yr	1,006	1,282	1,285
Cost per tonne CO_2	EUR/tonne CO_2	0.5	2.3	1.5

(Reproduced by permission of Energy Research Centre of the Netherlands)

[a] Direct Decomposition Upstream of Expander: Tail Gas Temperature = 450 °C; Space Velocity = 45,000 h-1; N_2O conversion = 70%; Retrofit Factor = 50%

[b] SCR of N_2O Upstream of Expander Using 1900 ppmv Propane: Tail Gas Temperature = 350 °C; Space Velocity = 20,000 h-1; N_2O conversion = 90%; Retrofit Factor = 50%

[c] End-of-Pipe SCR of N_2O Using 500 ppmv Propane: Tail Gas Temperature = 100 °C; Space Velocity = 20,000 h-1; N_2O conversion = 90%; No Retrofit Factor

[d] Includes the costs of LPG as reducing agent at 0.35 EUR/kg, which includes transportation costs.

[e] All cost and investment estimates are based on 2001 costs.

The SCR technology can be used in any nitric acid plant. The costs per tonne of CO_2 equivalents removed range from 1.3 to 3.0 EUR. The reductant (natural gas or LPG) is the largest cost item in these processes and makes up 20% to 60% of the total costs[221].

9.5. MATERIALS OF CONSTRUCTION

9.5.1. Materials for Weak Acid Processes

Stainless steels have excellent corrosion resistance to weak nitric acid and are the primary materials of construction for weak acid processes. Low carbon stainless steels are preferred because of their resistance to corrosion at weld points[104].

Catalyst baskets and gauze supports (where the temperature may reach 900°C) must be resistant to oxidation, nitriding and distortion from high temperatures. Typical materials of construction are high strength alloys made of iron-nickel-chromium, nickel-chromium and nickel-chromium-tungsten-molybdenum[104].

High chromium (20% to 27%) stainless steel is used in the cooler condenser and the tail gas preheater. Compared to low carbon stainless steel, it has better corrosion resistance to nitric acid at elevated temperatures so the extra cost is justified by the longer life[104].

Zirconium is used in nitric acid service for cooler condensers, tail gas preheaters and reboilers. It rivals tantalum in its corrosion resistance to nitric acid at all concentrations up to the boiling point. Its resistance extends up to 230°C and 65 wt %. However it is susceptible to stress corrosion cracking, which can be prevented by avoiding high, sustained tensile stresses[104].

Duplex stainless steels (4% nickel, 23% chrome) may offer cost advantages in absorption columns. These materials provide the ductability of austenitic stainless and the stress-corrosion cracking resistance of ferritic stainless steel[104].

In Europe nitric acid below 70% concentration should be shipped in railcars and tank trucks made of the materials listed in Table 9.10[97].

Table 9.10. Materials of Construction for Railcars and Tank Trucks in Europe[97]

	Suitable Materials of Construction		
V2A	DIN/EN 1.4306	=	UNS S 30403 (AISI 304L)
	DIN/EN 1.4541	=	UNS S 32100 (AISI 321)
	DIN/EN 1.4550	=	UNS S 34700 (AISI 347)
	DIN/EN 1.4465	=	UNS S 31050
Antinite	DIN/EN 1.4361	=	UNS S 30601
	(Recommended for Concentrations Above 85%)		

Warning: Materials DIN/EN 1.4301 or 1.44001 [UNS S 30400 or S 31600 (AISI 304 or 316)] are not suitable for nitric acid tanks and should be avoided.

Note: All hoses must be made of polyethylene, polypropylene or PTFE.

Pumps can be used for both loading and unloading nitric acid. Standard chemical pumps may be used for nitric acid concentrations up to 70% if they are made of suitable materials (for example, DIN/EN 1.4408 = UNS I 92900 (ASTM A 296 Grade C78M). Magnetically coupled, side-channel pumps made from Antinite (DIN/EN 1.4361 = UNS S 30601) or compressed-air diaphragm pumps made of PTFE/TFM composite are recommended for highly concentrated nitric acids above 70%[97].

9.5.2. Materials for Strong Acid Processes

Materials of construction for strong acid (98 to 99 wt %) are aluminum, tantalum, borosilicate glass, glass-lined steel, high silica cast iron and high-silica stainless steels. Some stainless steels can be used for storage and shipment of up to 95 wt % nitric acid (see Table 9.10). At higher concentrations, corrosion rates for stainless steel become excessive and aluminum is the commonly accepted material of construction. The use of aluminum alloys is restricted to high acid concentrations of 80 to 100 wt % and temperatures less than 38°C[104].

High silica stainless steels (4 to 7 wt %) provide good corrosion resistance to concentrated nitric acid, but not to weaker acid strengths. These steels are commonly used in direct strong acid processes. High-silica cast iron also displays good corrosion resistance to strong acid and is used mostly in columns and pump casings[104].

Titanium is resistant to nitric acid from 65 to 90 wt % and in dilute acid below 10 wt %. It is subject to stress-corrosion cracking above 90 wt %, and it is not used in red fuming acid service because of the potential for a pyrophoric reaction[104].

Tantalum exhibits good corrosion resistance to nitric acid over a wide range of concentrations and temperatures. It is used in applications where the nitric acid is close to or above its normal boiling point[104].

Glass offers good resistance to strong acid at high temperatures. However, it is subject to thermal shock and a gradual loss in integrity as materials such as iron and silica are leached out into the acid. Nonmetallic materials such as PTFE, PVDC, PVDF and furan are mainly restricted to weak acid service at ambient to moderate temperatures[104].

9.6. PRODUCTION

It is estimated that 65% of the worldwide nitric acid production is used to make ammonium nitrate (AN) for use in fertilizers. Other AN uses (mainly explosives) account for about 15% of the market. The balance is consumed in a variety of non-AN industrial applications. World nitric acid production in 1999 was estimated to be 53 million tonnes per year. The vast bulk of the tonnage is used captively, and the merchant market probably involves only about 10% of the total. International trade has little impact on the nitric acid balance. The average industry operating rate is usually in the 80% to 85% range. The rate dropped slightly lower in the early 1990's and recovered partially as a result of capacity shutdowns[91].

World demand for nitric acid will continue to be largely dependent upon demand for solid ammonium nitrate fertilizer and nitrogen fertilizer solutions that incorporate ammonium nitrate. Since the early 1980's, urea has been displacing ammonia nitrate as a fertilizer. The resulting reduction in demand for nitric acid has been partially offset by the increased use of ammonium nitrate in explosives and by the growth in production of polyurethane foams and nylon-6,6. World nitric acid production declined by about 5% between 1987 and 1999, but it is projected to increase marginally by 2005. The declines occurred mainly in the former Soviet Union, Eastern Europe and Western Europe. Production increases occurred mainly in the United States and Africa[91, 104]. Nitric Acid production in the United States is listed in Table 9.11.

Table 9.11. Nitric Acid Production

Year	Thousand Metric Tons
1990	7,196
1991	7,191
1992	7,381
1993	7,488
1994	7,905
1995	8,020
1996	8,351
1997	8,557
1998	8,423
1999	8,115
2000	7,981

Nitric acid is sold commercially in various concentrations that are expressed in degrees Baumé (Be), which is an alternative specific gravity scale. Table 9.12 lists the standard merchant grades. Various grades of concentrated or fuming nitric acid (95% or more HNO_3 / NO_2) are available. Concentrated acid fumes strongly and decomposes in light or at elevated temperatures[104].

Table 9.12. Nitric Acid Grades and Their Content

	Grade (degrees Baumé)			
	36 Degrees	38 Degrees	40 Degrees	42 Degrees
Specific Gravity	1.330	1.335	1.381	1.408
Percent HNO_3	52.3%	56.5%	61.4%	67.2%

9.7. STORAGE AND DISTRIBUTION

Nitric acid is normally stored in flat-bottomed, roofed tanks that are made from low-carbon, austenitic stainless steel. Most concentrations of nitric acid are available in tank cars and by truck. Stainless steel is necessary for concentrations up to 80% to 85%. Stronger solutions are less corrosive and may be stored in aluminum. Approximately 90% of nitric acid production is consumed on site to make ammonium nitrate (AN) fertilizers and some industrial explosives. Thus, the merchant market for nitric acid accounts for only 10% of the total. The production of AN fertilizers and most industrial explosives do not require acid concentrations in excess of the azeotropic composition of 68.8%[91].

In the United States the Department of Transportation (DOT) defines three categories of nitric acid: 1) nonfuming, more than 70 wt % acid; 2) nonfuming, less than 70 wt % acid; and 3) red fuming nitric acid. All must be labeled "corrosive". Red fuming acid must also be labeled "oxidizer" and "poison."

Nitric acid is subject to a certain degree of self-decomposition, depending on the concentration and temperature. The general rule is: the higher the concentration or the temperature, the faster the decomposition rate. During decomposition nitrogen dioxide and dinitrogen tetroxide are formed in addition to water and oxygen, and they are absorbed in the nitric acid. This leads to an orange-brown color and an increase in the dinitrogen tetroxide concentration. Nitric acid must be transported from the manufacturer to the user within the shortest possible time, particularly in the summer months[97].

Close scheduling is recommended for the transport of nitric acid that has a concentration above 99%. For truck shipments the maximum recommended transport time is three (3) days while rail shipments have a recommended transport time of five (5) days[97].

Tank trucks and rail cars that contain nitric acid below 70% concentration must be made of special steel that is resistant to nitric acid. Suitable materials include[97]: 304L Stainless Steel (DIN/EN 1.4306 = UNS S 30403), 321 Stainless Steel (DIN/EN 1.4541 = UNS S 32100), 347 Stainless Steel (DIN/EN 1.4550 = UNS S 34700), (DIN/EN 1.4465 = UNS S 31050), and Antinite (DIN/EN 1.4361 = UNS S 30601) – Recommended for concentrations above 85%.

Materials that are not recommended for nitric acid service are 304 stainless steel (DIN/EN 1.4301 = UNS S 30400) and 316 stainless steel (DIN/EN 1.4401 = UNS S 31600). All hoses for nitric acid loading and unloading should be made of polyethylene (PE), PTFE or polypropylene (PP)[97].

9.8. USES OF NITRIC ACID

Some of the uses for nitric acid are listed below.

- Ammonium Nitrate (AN) production for fertilizers and explosives.
- Adipic Acid: Nitric Acid oxidizes cyclohexanone-cyclohexanol mixtures (so-called KA oil) to adipic acid, nylon-6,6, other resins and plasticizers .
- Nitrobenzene: Nitrobenzene is made by the direct nitration of benzene with nitric/sulfuric acid mixtures primarily for aniline production. And aniline is a raw material for the manufacture of methylene diphenyl di-isocyanate (MDI) that is used to make rigid foams.
- Toluene Diisocyanate (TDI): Toluene diisocyanates (2,4- and 2,6 isomers) are produced from toluene diamine derived from dinitrotoluene, which is produced by the nitration of toluene with nitric/sulfuric acid mixtures. TDI is used to make flexible polyurethane foams, elastomers and coatings.
- Potassium Nitrate production for tobacco fertilizer and industrial markets.
- Sodium Nitrate production for fertilizers, explosives and as a heat transfer medium.
- Other Metal Nitrates are made by the reaction of nitric acid with the metal or its salt.
- Steel Industry:
  - Nitric acid is used by the steel industry to remove surface oxides (pickling) of stainless steels, to brighten and clean surfaces after salt-bath descaling and to prepare stainless steel surfaces for corrosion resistance (passivation).
  - Nitric acid pickling is generally restricted to low-carbon stainless steels of the 200 (nickel-chrome-manganese), 300 (nickel-chrome) and 400 (chrome) series.
  - The amount of acid used per ton of stainless steel varies widely, depending on the surface area per unit weight, the temperature of the treated surface, the pretreatment conditions (annealing and rolling temperatures) and the grade of the stainless steel. Industry estimates vary widely, ranging from about 5 to 50 pounds of nitric acid consumed per tonne of stainless steel produced.
  - Nitric acid may also be used to treat surfaces of nickel and chrome alloys, in metal etching and in treatment of refractory metals such as zirconium.
  - An estimated 25 to 30 thousand tonnes of nitric acid are used in these metal treatment markets.

- Electronics Industry: Electronics-grade nitric acid (70%) is used by semiconductor and printed circuit board manufacturers. Nitric acid cleans contaminants from the wafer's surface during several steps in semiconductor manufacture. Acetic and hydrofluoric acids may be blended with nitric acid to etch away metal. Nitric acid is also used for cleaning in the manufacture of printed circuit boards.
- Nitrochlorobenzene: Nitrochlorobenzene isomers are made by nitration of monochlorobenzene. Some of chemicals that can be made from these isomers are phosalone insecticides, antioxidants, rubber chemicals, AZO intermediates for dyes and diamine compounds used in specialty textiles and fibers. An estimated 30,000 tonnes of nitric acid are consumed annually to make nitrochlorobenzenes.
- Cellulose Nitrate: Cellulose nitrate (nitrocellulose, nitrocotton or pyrolyxin) is made by the action of nitric/sulfuric acid mixtures on cellulose. Theoretically, three nitro groups per glucose unit are required for complete nitration. The degree of nitration varies and is a function of the desired end use of the cellulose nitrate. Explosive uses require a high degree of substitution. Surface coating formulations may use nitrocellulose with a lower degree of nitration. An estimated 25,000 tonnes per year of nitric acid (100% basis) are consumed for industrial nitrocellulose (75% for surface coating uses and 25% for printing inks). Less than 10,000 tonnes per year of explosive nitrocellulose are consumed in the United States.
- Several nitrate paraffins (e.g., nitromethane, nitroethane and nitropropane) are used as solvents for cellulosic resins and for vinyl and epoxy surface coatings. Nitromethane is also used as a chlorinated solvent stabilizer. It is estimated that the annual consumption of nitric acid for nitroparaffin production is in the 30,000 tonne per year range.
- Non-military Explosives: Nitric acid is consumed in the production of several explosive chemicals and materials although there is no commercial production of trinitrotoluene (TNT). Nitric acid consumption for these explosives is estimated to be about 10,000 thousand tonnes per year.
- Nitric acid is used to digest crude uranium concentrates (U_3O_8, yellowcake) in some uranium purification processes. About 1,000 tonnes of nitric acid may be used to process yellowcake in the United States.
- Mixed Fertilizers: An estimated 40,000 tonnes of nitric acid is used to produce mixed fertilizers other than those based on ammonium nitrate.
- Concentrated Nitric Acid: An estimated 295,000 tonnes per year of concentrated nitric are consumed in the United States.

10

AMMONIUM NITRATE

Ammonium nitrate (AN) was first prepared in the sixteenth century. Its early industrial development was primarily for use in explosives. However, about 1940 its use as a fertilizer developed rapidly. Ammonium nitrate is produced mainly by the reaction of gaseous ammonia with aqueous nitric acid:

$$HNO_3\,(l) + NH_3\,(g) \rightarrow NH_4NO_3\,(aq) \quad \Delta H = +26 \text{ kcal/g-mol} \qquad (10.1)$$

The resulting AN solution may be handled in various ways[103]. It can be stored as a solution, used in down-stream plants or sold as a solution; it can be formed into solid AN by prilling or granulation, or it can be mixed with a solid filler. The most common filler is calcium carbonate in the form of ground limestone, dolomite or by-product calcium carbonate. This product is known as "Calcium Ammonium Nitrate" (CAN) and can be prilled or granulated. Granular products that contain AN and either ammonium or calcium sulphate are also manufactured.

10.1. PROCESSES

As shown in Figure 10.1[295], the typical AN production process has three main unit operations: Neutralization, Evaporation and Solidification (prilling and granulation)[103].

Some precautions that have been suggested in AN manufacture are[107]:

1. operation should be conducted below 155°C,
2. when the pH is less than 4.0 for a prolonged period of time, type 304L stainless steel should be used; and
3. when the pH is over 4.0 for solids, over 6.0 for solutions and the temperature is less than 115°C, aluminum may be used.

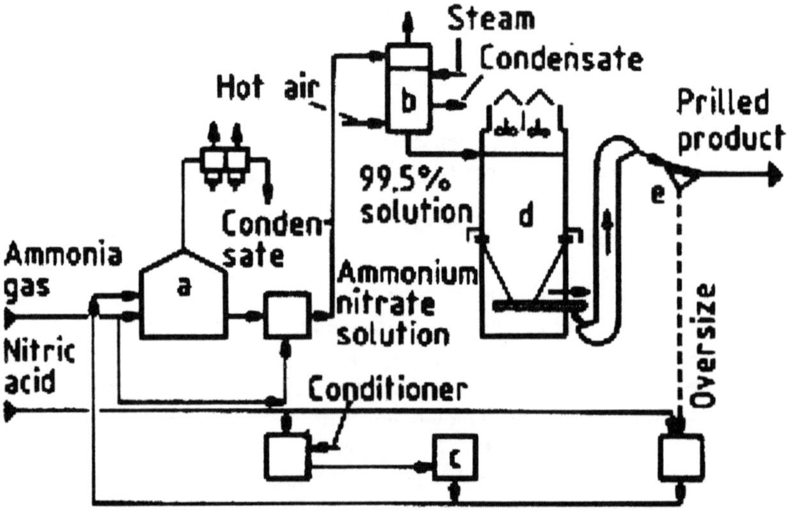

Figure 10.1. Ammonium nitrate process sketch. (Reproduced by permission of Wiley-VHCH)

10.1.1. Neutralization

The exothermic neutralization of nitric acid with ammonia gas produces AN solution and steam. The nitric acid is commonly preheated using equipment of suitable corrosion resistance. Preheating is usually employed if the available concentration of nitric acid is in the lower end of the 50% to 70% range. Steam or hot condensate from the AN process is used to preheat the nitric acid[103].

Neutralization can be performed in a single stage or in two stages. A two-stage neutralizer operates with a low pH in the first stage (acidic conditions) and a neutral pH in the second stage. In most neutralizers the pressure, temperature and concentration are linked by the boiling point characteristics of AN solutions with only two of these variables being independent. Ammonia gas may contain small quantities of inerts such as hydrogen, nitrogen and methane. These inerts will be vented from the neutralizer system at a point that depends upon the design of the particular process[103].

Neutralizers may be free-boiling vessels, circulating systems or pipe reactors. The water in the nitric acid is evaporated by the heat of reaction (550 to 620 BTU/lb of AN, depending on the acid strength). At least ten different types and designs of neutralizers are used in Europe. The factors that influence the choice of neutralizer are[103]:

- A two-stage neutralizer produces most of the boil-off steam in the first stage and most of the ammonia emissions in the second stage. This reduces the total ammonia emissions.
- A single-stage neutralizer is inherently simpler and cheaper.
- Neutralization at an elevated pressure will produce steam at a higher temperature (and AN at a higher concentration).
- The control of the neutralizer is a critical parameter. The pH and the temperature must both be strictly controlled to limit the losses from the neutralizer and to avoid a safety incident due to high temperature excursions.
- At the operating temperature of the neutralizer, impurity control is of great importance because a safety incident will also be a significant environmental incident. Some manufacturers do not recycle AN screenings to the neutralizer for this reason. Recycling is especially to be avoided if an organic anti-caking additive contaminates the screenings. It should be noted that an acidic solution of AN is more unstable than an alkaline solution.
- The components must be mixed quickly and thoroughly in the reactor to avoid local overheating, losses of nitrogen and decomposition of the ammonium nitrate[295].

The heart of a forced circulation process is the neutralization loop (see Figure 10.2). It consists of a neutralizer, the vapor separator and the circulation pump. Due to its low investment costs, vacuum neutralization is chosen in most cases. For higher energy efficiency, pressure neutralization units are more applicable. The heat of reaction can be used for either concentration of the AN solution or for clean steam production[226].

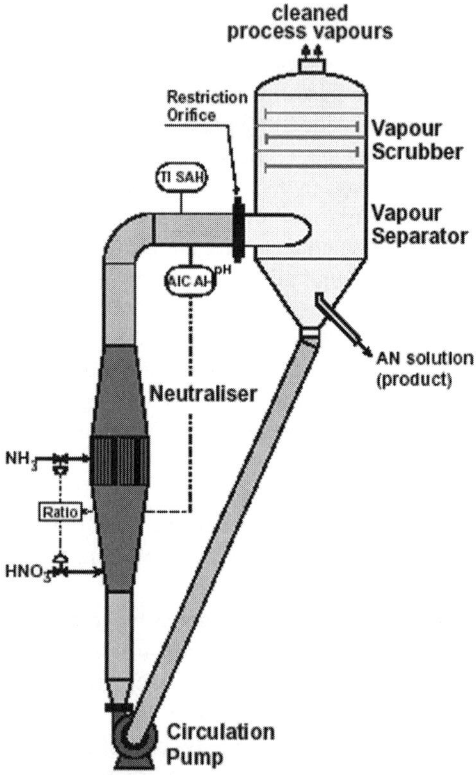

Figure 10.2. Ammonium nitrate neutralization loop.
(Reproduced by permission of Uhde GmbH and copyright by Uhde)

The main advantages of this process are[226]:

- No boiling occurs during the chemical reaction because a constant pressure is maintained in the reactor. This prevents ammonia slip and avoids any hazard related to bubbles in the AN solutions.
- The neutralization loop has a well-defined mixing profile.
- The neutralization loop has a well-defined temperature profile.
- The acid feed design avoids corrosion problems and safety risks.
- The ammonia sparger design provides a smooth reaction and good mixing of the reactants.
- The vapor scrubbing units can produce a liquid effluent that contains as little as 15 ppm N.

10.1.1.1. Carnit Process

The Carnit process for production of concentrated AN solutions requires no external heat supply. The reaction of ammonia and nitric acid occurs in a recycle flow loop where the pressure is higher than the vapor pressure of the solution. The recycle solution, which is slightly ammoniacal, supplies heat for the final concentration and for production of export steam. The free ammonia in the production off-take is neutralized before pressure reduction and subsequent concentration steps[227].

A specific feature of the Carnit process is the self-regulating combination of the adiabatic flash. This occurs upon pressure reduction with the first isobaric concentration step where the flashed vapors supply the heat of evaporation. Efficient pH control achieves a liquid effluent with less than 50 ppm of AN[227].

Energy self-sufficiency can be maintained even with 45 wt% acid. When 97.5% solution is produced with 60 wt% nitric acid feed, 128 kg/tonne AN of clean, 6 bar steam is produced[227].

The Carnit process can be licensed from Kemira SA (Belgium).

10.1.1.2. UCB Process

In the UCB process, a heat exchanger in the pressure reactor uses a part of the heat of reaction to make steam (see Figure 10.3)[295].

In this process ammonia and 52% to 63% nitric acid are preheated and sprayed into the sump of the reactor. The reactor pressure is about 4.5 bar; the temperature is 170 to 180°C and the pH is kept between 3 and 5 by controlling the ratio of reactants. This pH range reduces the amount of nitrogen that is lost in the process steam. A 75% to 80% NH_4NO_3 solution leaves the reactor and is concentrated to 95% in a falling film evaporator[295].

The heat of reaction generates 1) process steam in the reactor and 2) 5.5 bar pure steam. The process steam is used to preheat boiler feed water and nitric acid as well as operate the falling film evaporator. The pure steam is fed to the plant steam header[295].

10.1.1.3. Stamicarbon Process

The Stamicarbon process is another process that works under pressure. The neutralizer is a loop reactor that opens into a separator. The reaction solution is circulated without the use of a pump (see Figure 10.4)[295].

In this process 60% nitric acid, preheated ammonia and a small quanitty of sulfuric acid are introduced at the lower end of the reaction loop. The reactor operates at 4 bar and 178°C. The initial ammonium nitrate solution has a concentration of 78%[295].

The steam that leaves the top of the separator is used to concentrate the NH_4NO_3 solution to 95% in a vacuum evaporator. Excess steam is condensed, and ammonia is recovered from the condensate so it can be returned to the

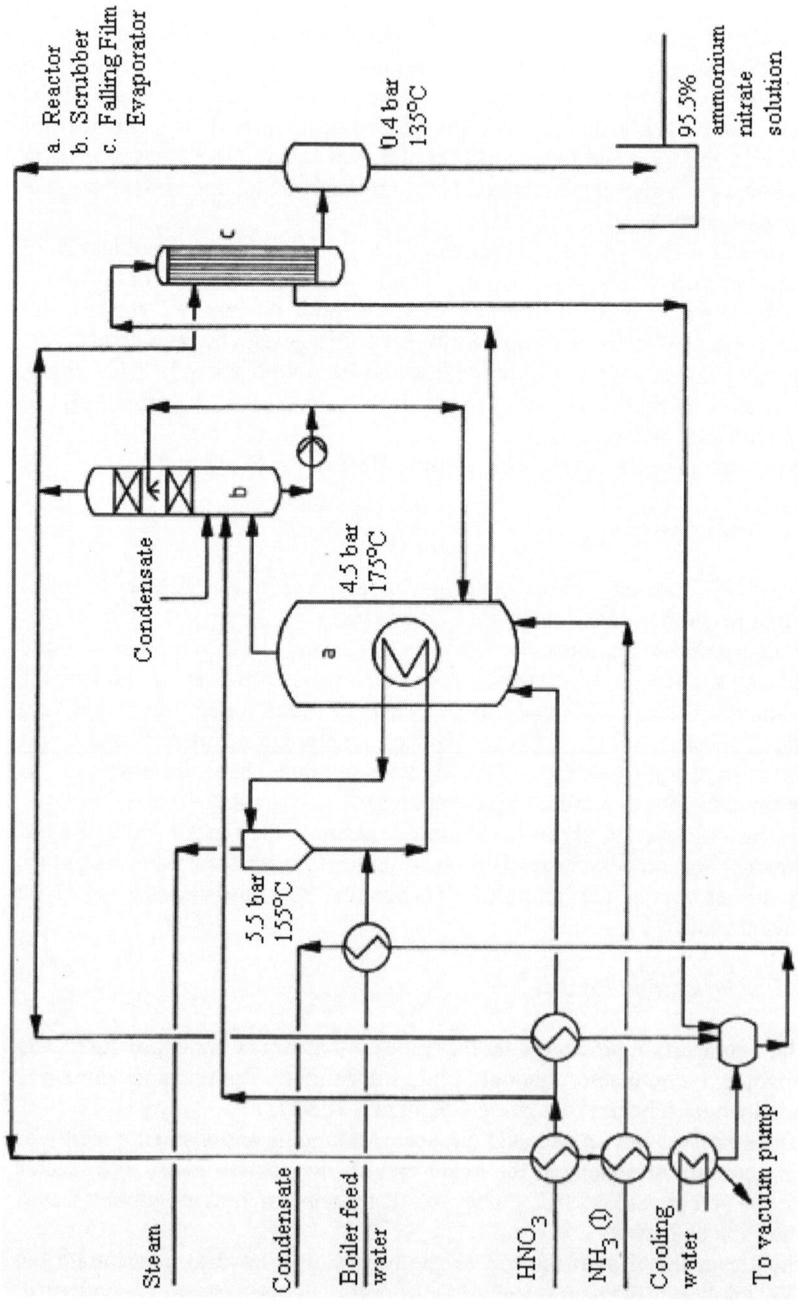

Figure 10.3. UCB ammonium nitrate process. **(Reproduced by permission of Wiley-VCH)**

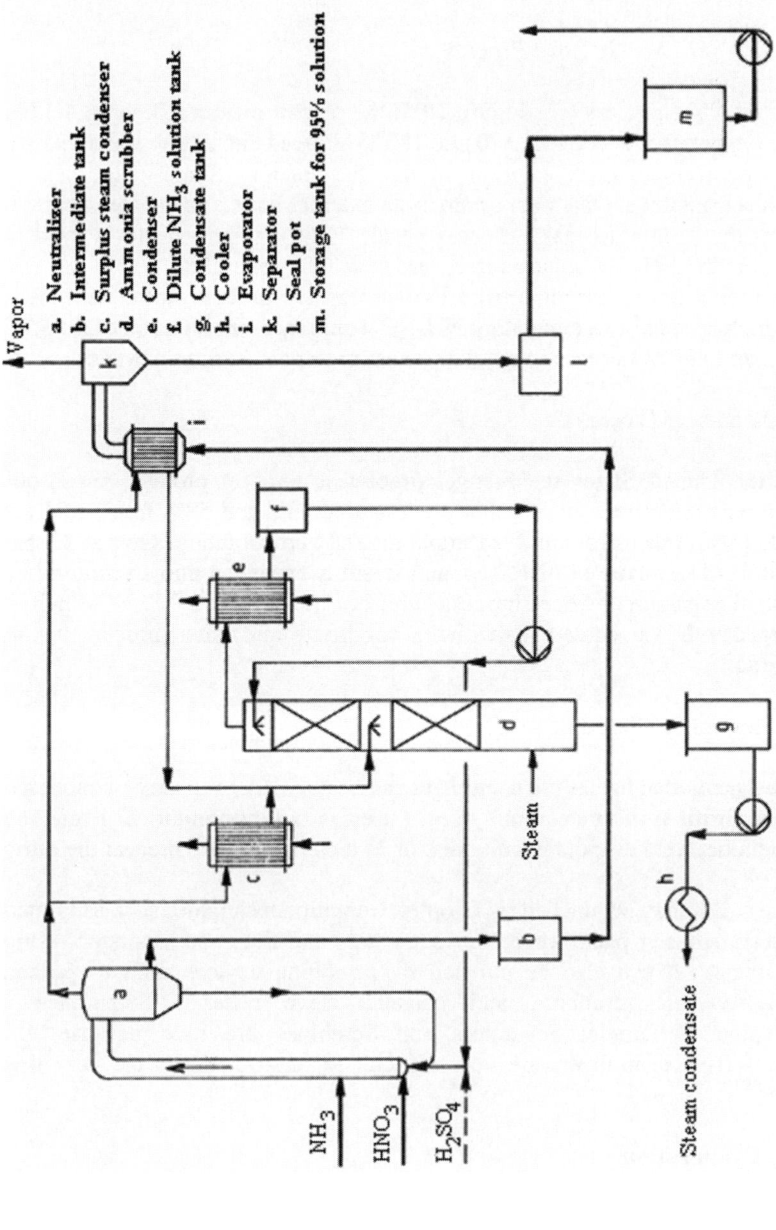

Figure 10.4. The Stamicarbon ammonium nitrate process. (Reproduced by permission of Wiley-VCH)

reactor. In a second evaporator, the concentration can be increased to a range of 98 to 99.5% by using fresh steam[295].

10.1.1.4. NSM / Norsk Hydro Process

In the NSM process (see Figure 10.5) the reactor pressure is about 4.5 bar and the temperature is between 170 and 180°C. Forced circulation and a thermal siphon effect circulate the solution through the reactor. Some of the heat of reaction is used to generate pure steam in an external boiler; and some vaporizes water in the reactor to make process steam. This process steam is used to concentrate the NH_4NO_3 solution from less than 80% up to 95%[295].

Ammonia losses are minimized by washing the process steam with nitric acid, which is added to a circulating NH_4NO_3 solution. Concentration of the 95% solution up to 99.5% is carried out with steam in a special vacuum evaporator[295].

10.1.1.5. Stengel Process

In the United States the Stengel process is used to produce anhydrous ammonium nitrate directly. In this process ammonia and 58% nitric acid are preheated with fresh steam and fed into a packed vertical tube reactor at 3.5 bar and 240°C. The mixture of NH_4NO_3 and steam is expanded into a vacuum in a centrifugal separator. After stripping with hot air, a 99.8% NH_4NO_3 melt is discharged onto a cooled steal belt, solidified and then broken up or granulated[295].

10.1.2. Steam Purification

The steam that leaves the neutralizer can be purified, or it can be condensed and then purified. The steam may be 1) used in the evaporator; 2) it may be used to preheat and evaporate ammonia; or 3) it can be used to preheat the nitric acid[103].

The steam may be purified by Droplet Separation techniques such as knitted wire mesh demister pads, wave plate separators and fiber pad separators using PTFE fibers. It can also be purified by scrubbing devices such as packed columns, venturi scrubbers, and irrigated sieve plates. Sometimes a combination of Droplet Separators and Scrubbers are used because AN emissions from neutralizers are very difficult to remove due to the very fine particles[103].

10.1.3. Evaporation

The evaporator is normally needed to remove the majority of the water that is present in the AN solution. It must produce a solution with the required concentration at a temperature that avoids crystallization. The acceptable water content depends on the process that is used to make the finished product. It is

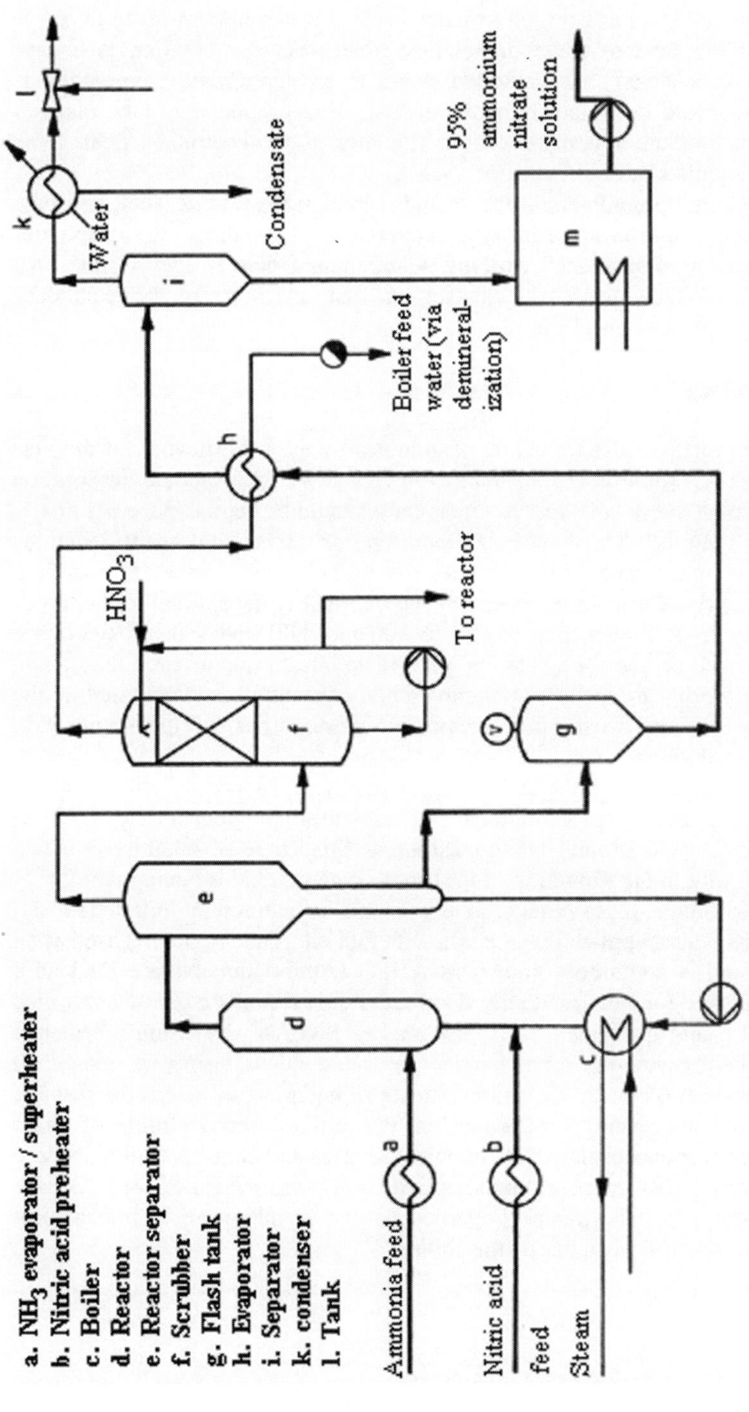

Figure 10.5. NSM / Norsk Hydro ammonium nitrate process. (Reproduced by permission of Wiley-VCH)

normally below 1% for a prilled product while a water content of up to 8% is needed for the feed to some granulation processes. Evaporation is always performed with steam, and saturated steam at an appropriate temperature is preferred to avoid decomposition of the AN. Evaporation may take place at atmospheric pressure or under vacuum. The latter allows neutralizer steam to be reused but requires more investment[103].

Evaporators in commercial use include circulatory systems, shell and tube heat exchangers and falling film-type evaporators. The falling film evaporator has the advantages of a small working volume and a short residence time. All evaporators produce steam that must be purified, and some of the techniques mentioned above can also be applied to this steam[103].

10.1.4. Prilling

Prilling refers to the formation of granules by the solidification of droplets of AN. The AN solution is concentrated to 96% to 99% by vacuum evaporation to form a "melt". For safe operation, the melt should be kept above a pH of 4.5. The melt may be mixed with an additive or nucleating agent (such as diammonium phosphate) that stabilizes the prills against temperature cycling through the crystal transition phases. Then the melt is transported to the top of the prilling tower (which may be 50 to 100 feet tall) and sprayed through a system of fixed orifices or a rotating, perforated bucket into a rising air stream. High density prills are made by pumping a hot, concentrated (99%+) melt to the prilling tower. Low density prills, which are more porous, are made with 95% to 96% AN solutions[240].

The droplets crystallize and condense into hard, spherical "prills" that are dried, cooled, and sized for shipment. If calcium ammonium nitrate (CAN or nitro chalk) is made, ground calcium carbonate (limestone or dolomite) is added to the melt prior to the formation of the droplets when CAN is being made[103, 105].

Sasol in South Africa produces a porous, prilled ammonium nitrate (PPAN) that finds its widest application in a mix with fuel oil. This mixture is used as an explosive and is commonly known as ANFO (Ammonium Nitrate Fuel Oil). Standard PPAN contains randomly distributed closed pores of an uncontrolled variable size and quantity. Sasol also makes EXPAN® by using a patented process where polymeric microspheres are entrained uniformly in individual prills. Surfactants are added prior to the prilling process to ensure that the microspheres are evenly distributed in the prill. The addition of these microspheres (or encapsulated gas bubbles) reduces and controls prill density to desired levels. This improves the sensitivity and performance of the explosive while retaining the desirable properties of the standard prills (mechanical strength, oil absorption and free-flowability)[106].

10.1.5. Granulation

Granulation is a more general term and refers to techniques using processes such as agglomeration, accretion or crushing to make a granular fertilizer. In contrast to the prilling technique, granulation requires a more complicated plant and a variety of equipment is used rotating pans and drums, fluidized beds and other equipment. Granular products can be made in a wider choice of particle size than prills[103].

In the late 1990's the fertilizer industry was starting to prefer granular product versus prills or crystals. This preference is mainly due to the spreading of bulk blending as a way to customize the fertilizer mix that is applied in what is called precision agriculture. This blending requires hard granules with low dust levels so they can withstand the handling imposed on the product. Fluid Drum Granulation (FDG) is one way to obtain such products[228].

Some granulation processes can use AN containing up to 8% water, but this water must still be removed in the process. Granulators used in AN/CAN plants include rotary pans and fluidized beds. The calcium carbonate will normally be added in the process before the granulator if CAN is to be produced. The AN is added in the granulator as a spray of hot concentrated solution. No further drying of the granules will normally be required. The granules are screened and the fines and crushed oversize returned to the granulator[103].

Typical CAN granulators include drums and pugmills. The calcium carbonate may be mixed with the AN solution before granulation or in the granulator itself. Granules from this process will normally require drying in a fluidized bed or rotary dryer. It may not be necessary to add any additional heat when drying CAN as the granules can have sufficient heat to provide the necessary driving force. Such a process is known as an autothermal process. The granules are screened after the dryer[103].

10.1.5.1. Pipe Reactor Plus Fluidized Drum Granulator

In 1990 Incitec (Australia) installed a process that combined these feature: 1) direct production of high strength ammonium nitrate melt using a pipe reactor with 2) granulation and drying using a fluidized drum granulator. The advantages of this process are lower capital cost, lower effluent discharges and limited operator surveillance[229].

The AZF pipe reactor process from La Grande Paroisse was combined with the fluidized drum granulator process from Kaltenbach-Thüring S.A. The design melt concentration from the plant is 92% with 94% being achievable. The steam from the process is used to preheat the nitric acid feed and to vaporize the ammonia feed[229].

The fluidized drum granulation process had not been used before to produce granular AN. However, it had been used to fatten urea prills. Extensive pilot plant trials were carried out to define the parameters for granulating AN. The process uses a standard granulation circuit except that both granulation and drying are achieved in one small drum[229].

10.1.6. Air Abatement Equipment

Emissions from the prilling and granulation sections of AN and CAN plants can be treated by a range of abatement equipment. Particulate material from some granulation plants is relatively coarse in particle size, whereas the prill tower emissions contain very fine particles. Candle filters are normally required for a prill tower emission. They can abate particulate emissions down to 15 mg/m^3 of air. For coarser material dry devices such as bag filters or dry cyclones are used. Particulate emissions can be of higher concentration, perhaps up to 30 to 50 mg/m^3, but the recovered material is a solid that can more readily be recycled to the process.

10.2. PRODUCTION

Production of fertilizer grade AN is concentrated mainly in Europe and North America. Apparent demand in the United States was 18% below the 1998 level of 10 billion tons. This was mainly due to weakness in the fertilizer sector. US production of urea-ammonium nitrate (UAN) solutions also dropped significantly (from 3.8 million tons in 1998 to 2.9 million tons in 2001) from 1999 to 2001. In 2001 US ammonium nitrate capacity remains well in excess of domestic requirements as the industry operated at less than 70% of nameplate US capacity. Table 10.1 summarizes US supply and demand[240].

Table 10.1. Ammonium Nitrate Supply and Demand[240]

	Thousands of Tons						
	1995	1997	1998	1999	2000	2001	2002
Capacity	9,662	11,054	11,008	11,008	10,444	10,444	10,444
Production	8,489	8,604	9,079	7,630	7,498	7,101	
Imports		794	781	837	1,030	901	1,050
Exports		99	52	60	31	24	21
Inv. Change	-81	+27	na	na	na	+40	
Demand	9,265	9,306	9,856	8,629	8,375	8,090	8,250

(Reproduced by permission of Mannsville Chemical Products Corp.)

Demand fluctuations are difficult to predict because of the variety of factors that influence fertilizer demand and farm economics in any given year. Demand from the explosives industry has been more stable. This is because demand from the coal mining industry accounts for two-thirds of total US explosives consumption. And coal production has grown from 1.06 billion tons per year in

1996 to 1.12 billion tons in 2001. However AN demand for fertilizers is likely to trend up very slowly. AN demand for explosives is expected to grow modestly[240].

International trade amounts to about 4 million tonnes (on a nitrogen equivalent basis) per year. The trade is divided equally between concentrated ammonium nitrate (33.5% to 33.5% nitrogen) and calcium ammonium nitrate (26% to 28% nitrogen). The Former Soviet Union is the largest net exporter with 1.3 million tonnes exported between 1996 and 1998 compared to zero prior to 1990. In 2000, Russia entered into an agreement with the US that limits Russia's exports of AN. Duties may be imposed on imports of UAN from Russia, the Ukraine and Belarus once an investigation is complete. And in July 2001 the International Trade Administration set an anti-dumping duty of 156.29% ad valorem on imports of AN from the Ukraine[36, 240].

As a large volume commodity chemical, AN prices fluctuate with changes in the supply/demand balance and with raw material costs. The list price has been $150 per ton since 1995 and is expected to stay at that level through 2004. Between 1995 and 2001 the average sales price has varied from a low of $128 per ton in 2000 to a high of $143 per ton in 2001[240].

10.3. STORAGE AND DISTRIBUTION

Solid AN in packages must be stored in a general warehouse that have been approved for AN duty. Bulk AN and CAN must be protected from moisture as both products are hygroscopic. Large bulk warehouses may be air-conditioned due to local climatic conditions and because of the anti-caking additive that is used to protect the product. Some additives can also reduce the water uptake rate. The anti-caking agents may be internal to the finished particle or applied as a coating to the outside. Some of these additives and agents include kieselguhr, limestone and clay[103, 240].

AN may explode under certain conditions such as 1) contamination and confinement with high temperature and high pressure that is sufficient to cause decomposition into the higher oxides of nitrogen or 2) confinement with high energy shock. The United States Environmental Protection Agency (EPA) classifies AN that contains more than 0.2% combustible substances as an explosive[240].

Some of the physical and chemical properties of ammonium nitrate are given in Table 10.2[103].

Table 10.2. Ammonium Nitrate Physical and Chemical Properties[103]

Property		
Appearance		White or Off-White Granules or Prills
Odor		Odorless
pH (10 g/100 ml)		> 4.5
Melting Point		160°C to 170°C (depending on moisture content)
Decomposition Point		> 210°C
Molecular Weight		80.05
Solubility in Water	0°C	118.3 g / 100 g
	20°C	190.0 g / 100 g
	80°C	576.0 g / 100 g
Bulk Density		830 to 1,100 kg/m^3
Nitrogen Content		35% (100% Ammonium Nitrate)

AN solution may be stored prior to use in down-stream plants or prior to sale. It must be stored at a temperature above the crystallizing temperature of the solution. Tanks may be lagged and/or heated. The solution may be circulated through a heat exchanger or heated with a steam coil. Gaseous ammonia is normally added in small quantities to maintain the solution at the correct pH because AN solutions lose ammonia during storage[103]. Steps should also be taken to avoid contamination with decomposition catalysts such as chlorides and organic materials.

Ammonium nitrate decomposes in two ways. Controlled decomposition with careful heating is the commercial process for producing nitrous oxide:

$$NH_4NO_3 \xrightarrow{200^\circ C - 260^\circ C} N_2O + H_2O \quad \Delta H = -6.7 \text{ kcal/g-mol} \tag{10.2}$$

Explosive decomposition by heating or by explosive shock is the basis for its use as an explosive:

$$2\,NH_4NO_3 \rightarrow 2\,N_2 + 4\,H_2O + O_2 \quad \Delta H = -28.2 \text{ kcal/g-mol} \tag{10.3}$$

The oxygen that is released can be used to more than triple the explosive effects by mixing an organic such as fuel oil with the AN. This mixture, known as ANFO, is much cheaper and much safer to handle than dynamite:

$$3\,NH_4NO_3 + (CH_2)_n \rightarrow 3\,N_2 + 7\,H_2O + CO_2 \quad \Delta H = -102.5 \text{ kcal/g-mol} \tag{10.4}$$

10.4. USES

Commercial grades of AN include solutions that contain about 83% AN and solid products. Both grades are used for direct fertilizer applications, or they are blended with other materials to form mixed solid and liquid fertilizers. High density, prilled AN is usually used for direct applications while other solid AN is mixed with phosphate and potash for multi-nutrient fertilizers. Solid AN has lost market share to urea because it has a lower nitrogen content (33-35% for AN vs. 45-46% for urea), plus it costs more per ton. In addition to direct application, AN solutions are mixed with urea to make nitrogen solution fertilizers[240].

Ammonium nitrate / fuel oil (ANFO) explosives are a mixture of 94% AN and 6% fuel oil. They are low-cost explosives that are used for coal and metal mining, quarrying and construction. Low density prills are used because they are porous and absorb the fuel oil. ANFO is safe to handle and easy to make[240].

AN is also an ingredient in slurry type explosives, which are mixtures of AN, dynamite (TNT), water and aluminum. They are available as gels or with the consistency of thick soup. They can be used in open pit and underground mines as well as water-filled bore holes. Amatol explosives - mixtures of AN and TNT - have low sensitivity, and can be cast, pressed or granulated - depending on the ratios of the ingredients[240].

Thiokol (Brigham City, UT) has patented a co-crystalized form of potassium dinitramide and AN. This material is being developed as an oxidizer for space launch motors and other uses[240].

The AN end uses are summarized in Table 10.3[240].

Table 10.3. Ammonium Nitrate End Uses

End Use	Percent
Liquid Fertilizers	44
Solid Fertilizers	32
Explosives	21
Miscellaneous	3

(Reproduced by permission of Mannsville Chemical Products Corp.)

11

UREA

Urea, which is also known as carbamide and carbonyldiamide, was discovered in urine by Rouelle in 1773. It was first synthesized from ammonia and cyanic acid by Woehler in 1828[108]. In the early 1900's urea was produced on an industrial scale by the hydration of cyanamide, which was obtained from calcium cyanamide[109]:

$$CaCN_2 + H_2O + CO_2 \rightarrow CaCO_3 + CNNH_3 \quad (11.1)$$

$$CNNH_3 + H_2O \rightarrow CO(NH_2)_2 \quad (11.2)$$

After development of the ammonia process by Haber and Bosch in 1913, the production of urea from ammonia and CO_2 (both of which are formed in ammonia synthesis) developed rapidly. In 2001 urea is prepared on an industrial scale exclusively by this method which uses the Basaroff reactions[109].

$$\underset{\text{Ammonia}}{2\,NH_3} + \underset{\text{Carbon Dioxide}}{CO_2} \leftrightarrow \underset{\text{Ammonium Carbamate}}{NH_2COONH_4} \quad (11.3)$$

$$\underset{\text{Ammonium Carbamate}}{NH_2COONH_4} \leftrightarrow \underset{\text{Urea}}{CO(NH_2)_2} + \underset{\text{Water}}{H_2O} \quad (11.4)$$

The reaction in Eq. (11.3) is fast and exothermic and essentially goes to completion under the high pressure reaction conditions that are used industrially. The reaction in Eq. (11.4) is slower and is endothermic. It does not go to completion. The conversion (on a CO_2) basis is usually 50% to 80%. The conversion increases with increasing temperature and NH_3/CO_2 ratio. It decreases with increasing H_2O/CO_2 ratio[110].

When urea melt is subjected to heat, particularly during concentrating operations, some biuret is formed, depending on temperature and residence time.

$$\underset{\text{Urea}}{2\,CO(NH_2)_2} \quad \leftrightarrow \quad \underset{\text{Biuret}}{NH_2CONHCONH_2} \quad + \quad \underset{\text{Ammonia}}{NH_3} \qquad (11.5)$$

Biuret is undesirable for some industrial applications and in fertilizers used as foliar sprays. Prilled urea normally contains more biuret (0.8% - 1.5%) than granular urea because prilled urea goes through an extra evaporation step.

Urea is an odorless, white solid that contains 46.65% nitrogen, the most of any ordinarily solid fertilizer. Urea is not flammable, but it will melt and decompose in a fire to give off ammonia. When it is dissolved in water, it hydrolyzes very slowly to ammonium carbamate and eventually decomposes to ammonia and carbon dioxide. Urea dust is not regarded as hazardous. However it is recommended that exposure should be limited to < 10 mg/m^3 (8 hr TWA) for inhalable dust and < 5 mg/m^3 (8 hour TWA) for respirable dust. Some of the other properties of urea are given in Table 11.1[108, 110].

Table 11.1. Physical Properties of Urea

Chemical Formula	NH_2CONH_2
Molecular Weight	60.06
Freezing/Melting Point	132.6°C
Boiling Point	Decomposes
Density, d_4^{20}	1.3230 g/cm^3
Heat of Solution in Water	+ 251 J/g (60 cal/g)
Bulk Density	0.74 g/cm^3
Specific Heat (J/(kg °K)	
0°C	1.439
50°C	1.661
100°C	1.887
150°C	2.109

Properties of Saturated Aqueous Solutions of Urea

Temperature (°C)	Solubility in Water (g/100 g Solution)	Density (g/cm^3)	Viscosity (mPa-s =cP)	Water Vapor Pressure (kPa)
0	41.0	1.120	2.63	0.53
20	51.6	1.147	1.96	1.73
40	62.2	1.167	1.72	5.33
60	72.2	1.184	1.72	12.00
80	80.6	1.198	1.93	21.33
100	88.3	1.210	2.35	29.33
120	95.5	1.221	2.93	18.00
130	99.2	1.226	3.25	0.93

11.1. UREA PROCESSES

In 2001 most new urea plants were licensed by Snamprogetti (SNAM), Stamicarbon (STAC) or Toyo. SNAM utilizes thermal stripping while STAC and Toyo use CO_2 Stripping. At the end of 1996, about 70 SNAM plants, 125 STAC and 7 Toyo plants had been built. STAC will design plants for over 3,000 tonnes per day; SNAM designs plants for about 2,800 tonnes per day; and Toyo designs plants for about 2,300 tonnes per day as single train units[108].

Since urea is made from ammonia and carbon dioxide (both of which are produced in an ammonia plant), all urea plants are located adjacent to or in close proximity to an ammonia plant. Figure 11.1 shows how ammonia and urea plants can be connected[111].

The Snamprogetti (SNAM) process is shown in Figure 11.2[108]. In this process NH_3 and CO_2 are converted to urea via ammonium carbamate at a pressure of 150 to 160 bar and a temperature of 180°C to 190°C. An NH_3-to-CO_2 molar ratio of 3.5 is used in the reactor to give a CO_2 conversion of 65%. The reactor effluent enters a falling-film stripper where a large part of the unconverted carbamate is decomposed by the stripping action of the excess NH_3. Because of the high stripping temperatures (200-210°C), bimetallic tubes (made of zirconium and 25-22-2 stainless steel) are needed to avoid corrosion in the tubes and erosion near the bottom of the stripper[110].

Residual carbamate and CO_2 are recovered downstream of the stripper in two successive stages. The first stage operates at 17 bar, and the second stage operates at 3.5 bar. Ammonia and CO_2 vapors from the stripper top are mixed with the recovered carbamate solution from the High Pressure (HP) / Low Pressure (LP) sections, condensed in the HP carbamate condenser and fed to the reactor. The urea solution that leaves the LP decomposition stage is concentrated in the evaporation section to make a urea melt[110].

The synthesis section of the Stamicarbon (STAC) process is shown in Figure 11.3. An updated version of this process, Urea 2000plus, is offered by DSM (Stamicarbon's parent company)[108, 110]. In this process NH_3 and CO_2 are converted to urea via ammonium carbamate at a pressure of 140 bar and a temperature of 180°C to 185°C. An NH_3-to-CO_2 molar ratio of 2.95 is used in the reactor to give a CO_2 conversion of 60%, and an NH_3 conversion of 41%. The reactor is lined with 316L stainless steel. It contains sieve trays to provide good contact between the gas and liquid phases and to prevent back-mixing[108, 110].

The reactor effluent, which contains unconverted NH_3 and CO_2, is stripped in a falling-film type shell and tube exchanger at reactor pressure. The stripping agent, CO_2, flows upward, which is countercurrent to the effluent stream. The stripper tubes are make of 25-22-2 stainless steel, which has lasted almost 30 years in some plants. The stripped-off NH_3 and CO_2 are then partially condensed and recycled to the reactor. The heat from this condensation is used to make 4.5 bar steam. Some of this steam can be used in downstream sections of the plant while some is sent to the turbine on the CO_2 compressor[108, 110].

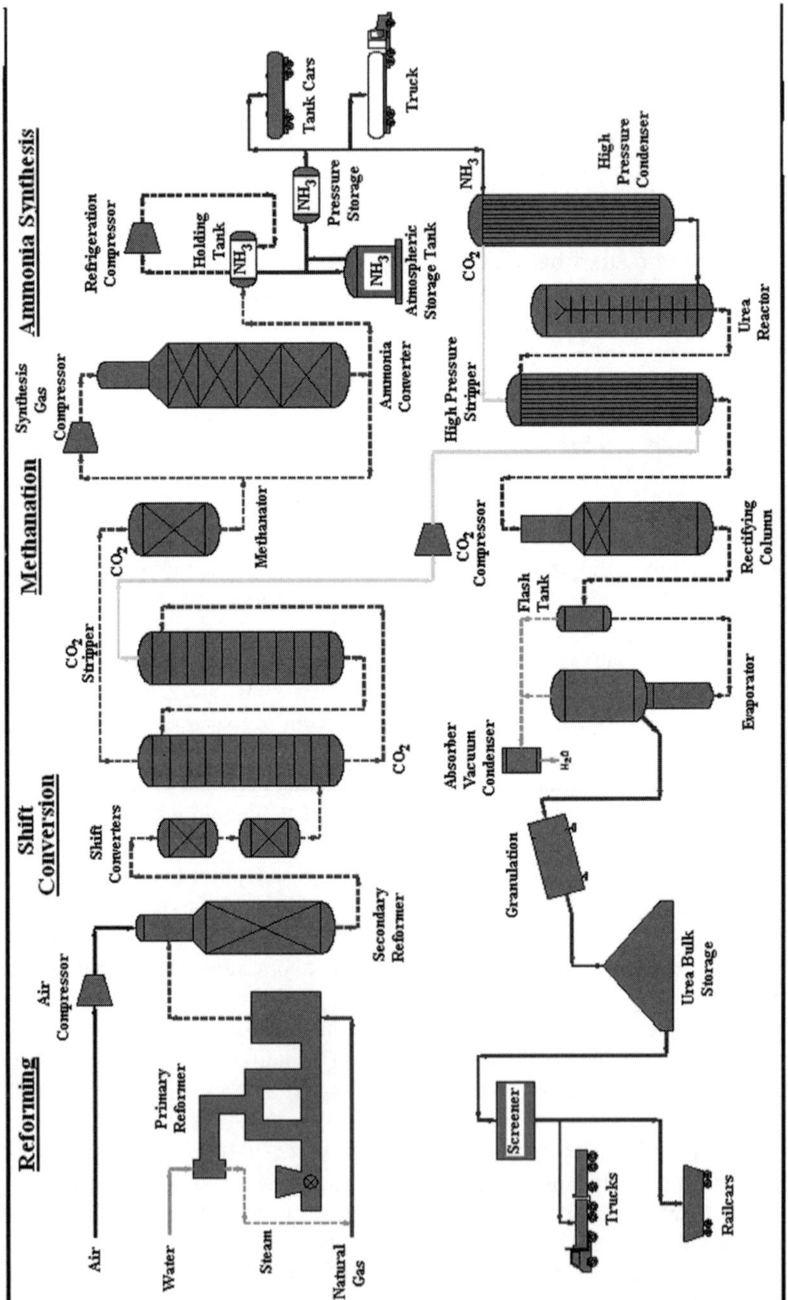

Figure 11.1. Agrium's Carseland nitrogen operations: ammonia and urea production. (Reproduced by permission of Agrium Inc.)

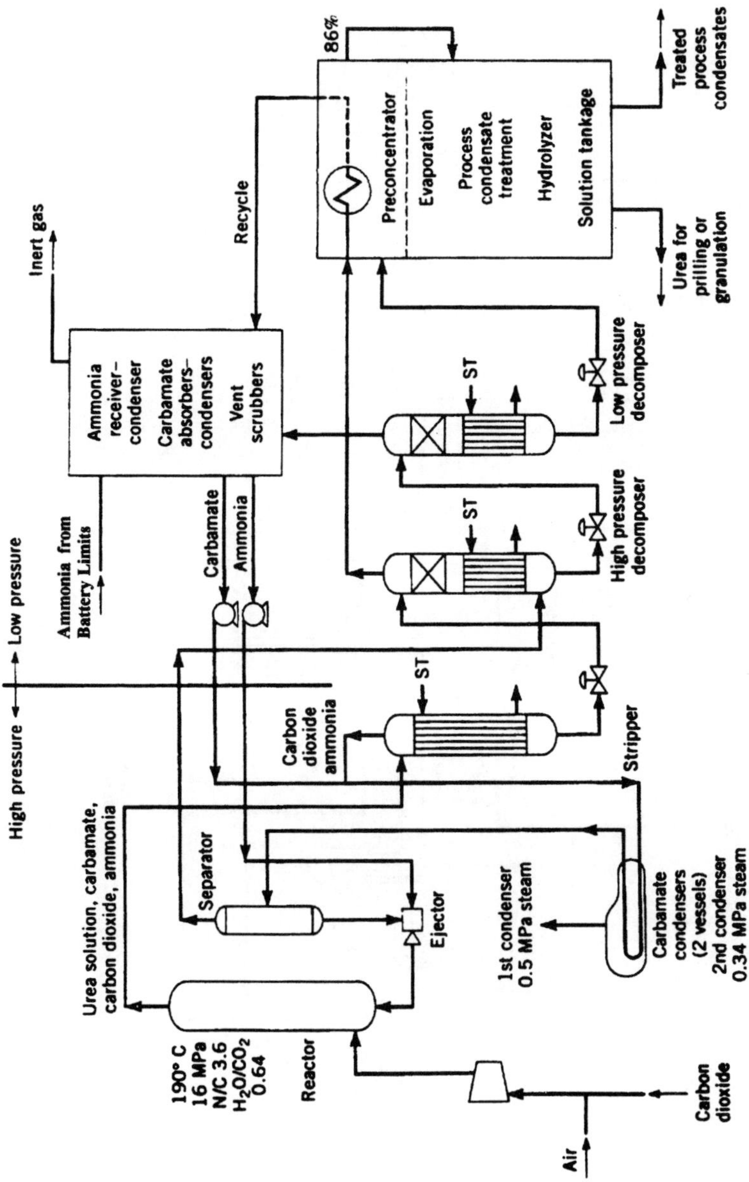

Figure 11.2. Snamprogretti thermal stripping urea process. (This material is used by permission of John Wiley & Sons, Inc.)

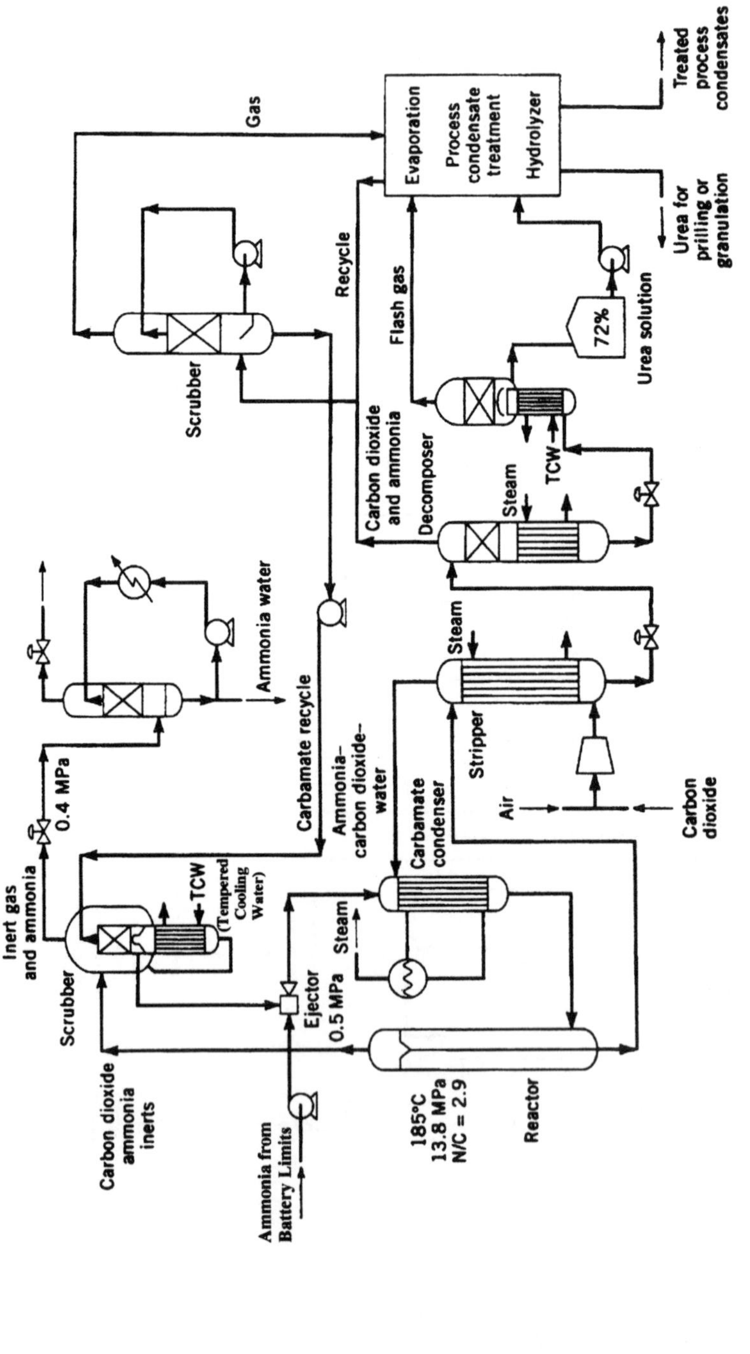

Figure 11.3. Stamicarbon CO_2 stripping process.
(This material is used by permission of John Wiley & Sons, Inc.)

The NH_3 and CO_2 in the stripper effluent are vaporized in a 4.0 bar decomposition stage and subsequently condensed to form a carbamate solution. This solution is recycled to the reactor bottom along with the uncondensed NH_3 and CO_2 where the carbamate is converted into urea. The urea solution that leaves the bottom of the reactor is then evaporated to 99.8% for prilling or plus 95% for granulation[108, 110].

The synthesis section of the Toyo Engineering-ACES (**A**dvanced **P**rocess for **C**ost and **E**nergy **S**aving) process is shown in Figure 11.4. In this process the synthesis section operates at 175 bar with an NH_3-to-CO_2 molar ratio of 4.0 and a temperature of 185 to 190°C. Liquid ammonia is fed directly into the reactor while gaseous CO_2 is sent from the centrifugal CO_2 compressor at the bottom of the stripper[108, 110].

The reactor effluent is stripped at essentially reactor pressure using CO_2 as the stripping agent. The overhead gas mixture from the stripper is fed to two carbamate condensers in parallel. Here the gases are condensed and recycled under gravity to the reactor along with absorbent solutions from the HP scrubber and absorber. The heat generated in the first carbamate condenser is used to generate 5.0 bar steam, and the heat from the second condenser is used to heat the solution leaving the stripper bottom after pressure reduction[108, 110].

The ACES stripper uses a ferrite-austenite stainless steel as do the carbamate condensers. The reactor and scrubber are constructed with 316L urea-grade stainless steel[110].

The inerts in the synthesis section are purged to the scrubber from the reactor top for recovery and recycle of NH_3 and CO_2. Unreacted NH_3 and CO_2 from several sources are eventually recycled to the reactor via HP and LP absorbers[108, 110].

The urea solution that leaves the stripper bottom is further purified in HP (17.5 bar) and LP (2.5 bar) decomposers. The urea solution is first concentrated to 88.7% in a vacuum concentrator and then to the required concentration for prilling or granulating[110].

The **H**eat **R**ecycle **U**rea **P**rocess (HRUP) was developed by Urea Technologies in the 1970's and is now offered by Monsanto Enviro-Chem. This process is shown in Figure 11.5. In this process ammonia, recycled carbamate and about 60% of the make-up CO_2 feed are charged to the top of an open-ended reactor coil at 3200 psig (220 bar). Ammonium carbamate is formed within the coil, exits the coil at the bottom and flows up and around it. The exothermic heat of carbamate formation in the coil drives the endothermic dehydration of carbamate to urea on the outside of the coil. This isothermal reactor design allows the conversion rate to reach 77%. The reactor has a uniform temperature profile that eliminates corrosion in the bottom of the reactor[112].

The decomposers produce a 68 to 70 wt% urea solution that is concentrated under low vacuum in the Urea Concentrator to an 88 wt%. An atmospheric, air-swept, falling-film evaporator then evaporates the 88 wt% urea product to make a 99.7 wt% urea melt product that is ready for prilling or granulation.

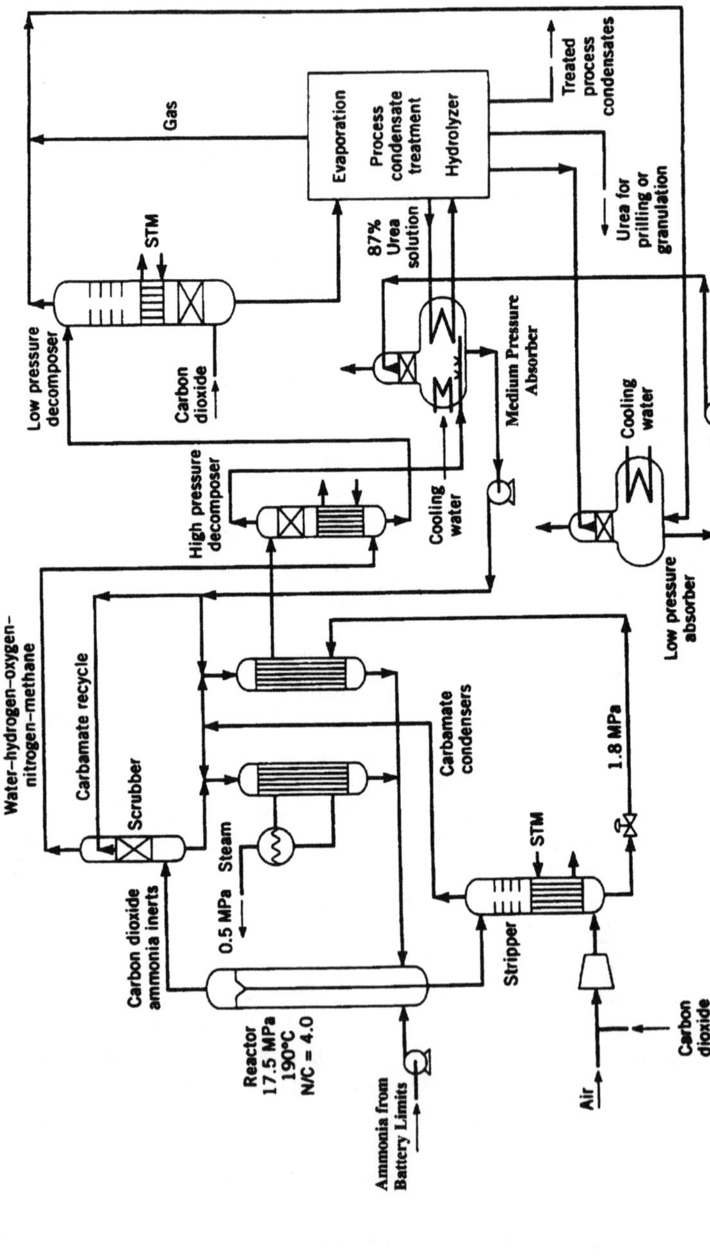

Figure 11.4. Toyo - Advanced Process for Cost and Energy Savings (ACES) process. (This mateiral is used by permission of John Wiley & Sons, Inc.)

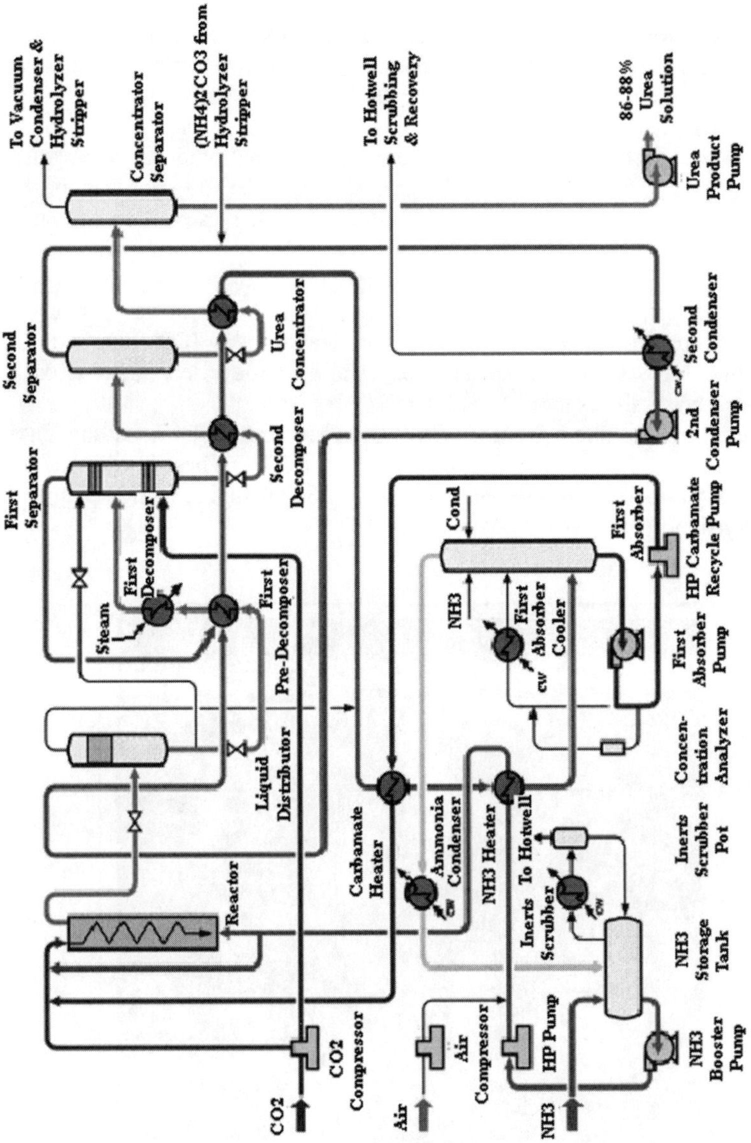

Figure 11.5. Heat Recycle Urea Process (HRUP).
(Reproduced by permission of Monsanto Enviro-Chem Systems, Inc.)

Montedison developed the Isobaric Double-Recycle (IDR) process. In this process the urea synthesis takes place at 180 to 200 bar and 185 to 190°C. The NH_3-to-CO_2 molar ratio is 3.5 to 4.0, and CO_2 conversion is about 70% per pass[110].

Most of the unconverted material in the reactor effluent is separated by heating and stripping at synthesis pressure using two strippers in series. Whereas all the carbon dioxide is fed to the plant through the second stripper, only 40% of the ammonia is fed to the first stripper. The remainder goes directly to the reactor for temperature control. The ammonia-rich vapors from the first stripper are fed directly to the urea reactor. The CO_2-rich vapors from the second stripper are recycled to the reactor via the carbamate condenser, irrigated with carbamate solution recycled from the lower-pressure section of the plant[110].

The heat of condensation is recovered as 7.0 bar steam for use in downstream process steps. Urea solution leaving the IDR loop contains unconverted NH_3, CO_2, and carbamate. These unconverted compounds are recycled to the synthesis loop[108, 110].

Several urea-producing reactors have been modernized by installing Casale High Efficiency Trays (HET) (see Figure 11.6.). These trays increase conversion, which leads to high plant capacity along with a reduction in specific energy consumption[209].

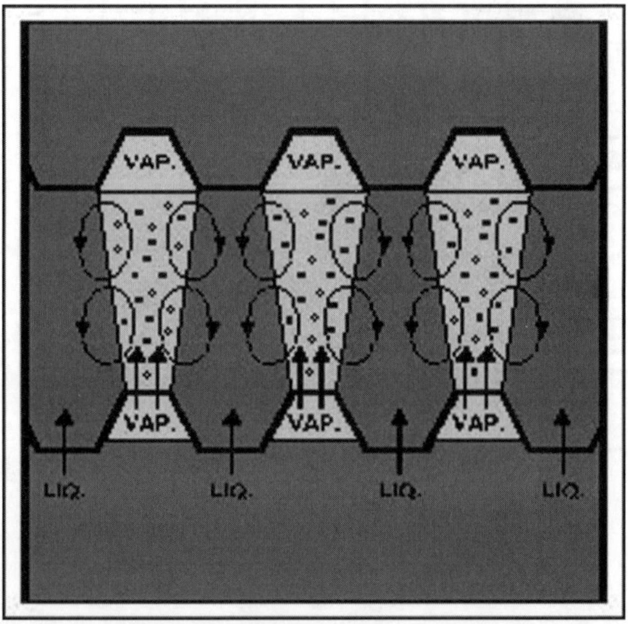

Figure 11.6. Urea Casale High Efficiency Trays (HET)
(Reproduced by permission of the Casale Group)

Typical raw material and utility consumptions for different urea processes are shown in Table 11.2. Explosive gas mixtures may form in the inerts scrubber because the off-gas contains O_2, H_2, N_2 and possible non-condensed NH_3 and CO_2. In some plants the hydrogen in the CO_2 feedstock is reduced by catalytic combustion to values below 10 ppm. This minimizes the risk of forming an explosive H_2/O_2 mixture in the scrubber[110].

Table 11.2. Typical Raw Material and Utility Consumptions in Urea Processes

Urea Process	NH_3[a] (tonne/tonne Urea)	CO_2[a] (tonne/tonne Urea)	Steam[b] (bar)		Drive Type	Cooling H_2O (m^3/tonne Urea)	Power (MJ/tonne Urea)
CO_2 Stripping	0.57	0.75	0.770	(120)	ST	70	54
			0.800	(24)	E	60	396
NH_3 Stripping	0.567	0.735	0.760	(108)	ST	80	76
IDR	0.57	0.74	0.600	(105)	ST	75	79
			0.824	(24)	E	60	425
ACES	0.57	0.74	0.700	(98)	ST	60	108
			0.570	(24.5)	E	51	436

(Reproduced by permission of European Fertilizer Manufacturers Association)

[a] The stoichiometric quantities of NH_3 and CO_2 are 0.567 tonnes per tonne of urea and 0.733 tonnes per tonne of urea.

[b] Steam pressure (in bar) is in parentheses.

[c] Assumed Cooling Water $\Delta T = 10^\circ C$.

[d] CO_2 Compressor Drive: E = electric motor, ST = steam turbine

[e] Air is needed to provide oxygen for passivation of stainless steel equipment.

[f] Crystal seeding of the melt and conditioning with formaldehyde (at 0.011 tonnes per tonne urea) may be used.

11.1.1. Prilling

The manufacture of prills is rapidly decreasing due to both environmental problems and product quality as compared to granules. In a prilling plant (see Figure 11.7) the urea solution from the recovery section is evaporated in two stages to a melt with > 99.8% strength. It is then pumped to the top of a 50 to 60 meter cylindrical concrete tower where it is fed into the prilling device (either a spinning bucket or a shower-type spray). Liquid droplets are formed that solidify and cool as they fall through the tower against a forced or natural up-draft of ambient air. The product goes from the tower base to a conveyor belt using a rotating rake, a fluidized bed or a conical hopper. The quality problem with prills is that prill size must be 1.6 to 2.0 mm to obtain proper solidification and cooling. But this must be accomplished from a fall height that is practical. Generally both the crushing and impact strength of the prill is much less than that of a granule. To improve prill strength, crystal seeding of the melt may be used. Formaldehyde can also be added to improve crushing strength by up to 25% and suppress the caking tendency in storage[108, 110].

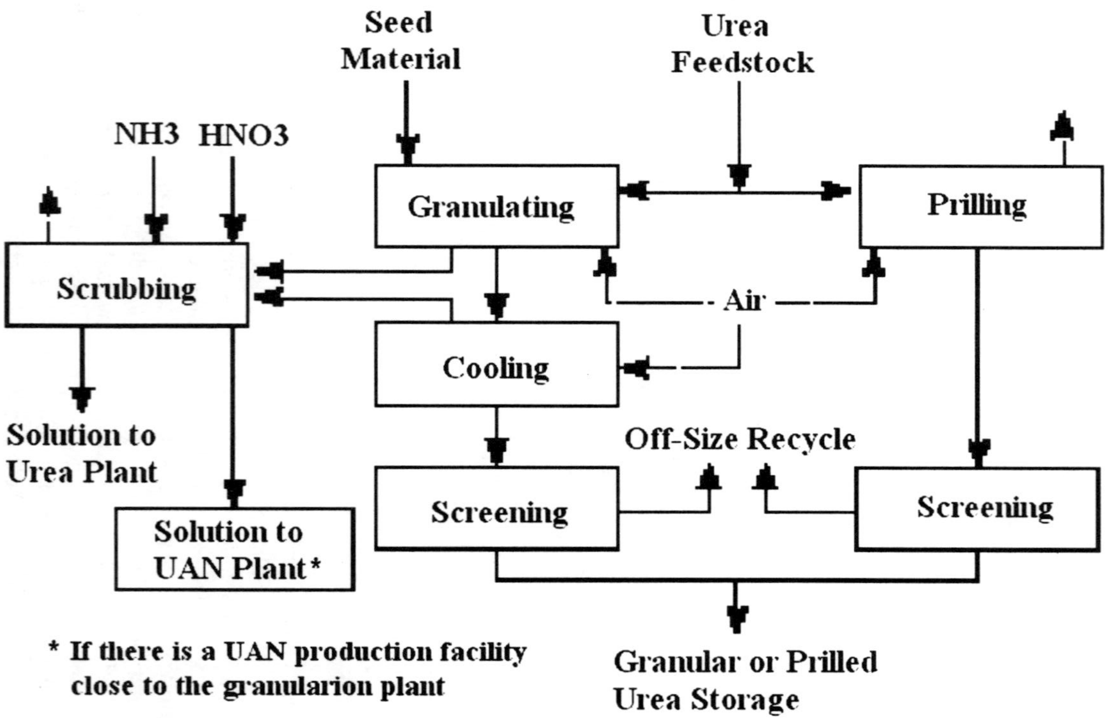

Figure 11.7. Block diagram for urea granulation and prilling processes.
(Reproduced by permission of European Fertilizer Manufacturers Association)

11.1.2. Granulation

Almost all new plants make granules and in 2003 the Hydro-Agri process, which can be built to make over 3,000 tonnes per day, is used in the majority of plants. The basic principle of the process involves the spraying of the melt onto recycled seed particles or prills circulating in the granulator. These seed particles gradually increase in size as the process continues. The heat of solidification is removed by cooling air. Since the contact time between liquid urea and air is much lower than in prilling, the dust formed in granulation processes is much coarser, and it is easier to remove[108, 109].

Another process is the C&I Girdler drum system, but it cannot compete in today's market because of size restrictions. Toyo has developed a spouting-fluid technology and by 1996 three plants were in operation. Stamicarbon will also license a fluid-bed plant that is similar to the Hydro-Agri design[108].

All granulation processes require the addition of formaldehyde or formaldehyde-containing components to the urea before it is sprayed on the crystal seed surface. Formaldehyde serves as and anti-caking agent during the granulation process and slows the release function so that the fertilizer efficiency is higher. Granulation processes also produce products that have larger diameters (2 to 4 mm vs. 1.5 to 2.0 mm) than prills. Granules may also be available with a size of 4 to 8 mm. These larger granules are used for applications to large areas such as forests and grass lands. However prilling techniques may still be used in some places because of the lower investment and lower variable costs compared to granulation[109].

Typical raw material and utility consumptions for different granulation processes are shown in Table 11.3[110].

Table 11.3. Raw Material and Utility Consumptions in Urea Granulation Processes

Granulation Process	Urea Melt Concentration (weight %)	Product Temperature (°C)	Steam (kg/ tonne Urea)	Power (MJ/ tonne Urea)
Falling Curtain	99.4 to 99.7	50	40 (LP)[a]	93
Fluid Bed Drum	95.0 to 96.0[c]	40	40 (LP)[a]	126
Spouted Bed	98.5	40	200	70
Fluid Drum	99.7	45	200 (9 bar)[b]	40

(Reproduced by permission of European Fertilizer Manufactures Association)

[a] To re-concentrate the recovered urea solution for recycle to the process.

[b] To flash-melt the over-size product for recycle to the process.

[c] The use of the higher water content in the urea solution provides estimated savings of 90 kg per tonne of LP steam in the evaporation section relative to prilling.

11.1.3. Materials of Construction

Corrosion from ammonium carbamate is a major problem. The intensity of corrosion is greatest in the reaction section and in the first recycle. These are the areas where pressures, temperatures and concentrations are highest.

The urea reactor is usually a multi-layered vessel that is made of carbon steel with a corrosion-resistant interior liner. Reactor liner materials are usually 316L stainless steel (urea grade) with a lower ferrite content (≤0.6%) for pipes and (<1%) for forgings). (see Table 11.4. for the composition of urea grade 316L stainless steel[88].

Table 11.4. Composition of 316 Stainless Steels

	Composition (%)							
Name	C (max)	Cr	Ni	Mo	Mn	P	S	Si
316	0.08	16.0-18	10-14	2-3	2	0.045	0.03	1
316L	0.03	16.0-18	10-14	2-3	2	0.045	0.03	1
316L (Urea Grade)	≤0.03	16.5-17	13-15	>2.3	2	0.045	0.03	1

In the reactor, the liquid/vapor interface is a corrosion-prone area. This section can be strengthened by attaching a sacrificial plate to the liner. Reactor outlet piping and let-down valves are subjected to erosion corrosion from the pressure reduction across the valve and from high fluid velocities in pipelines[88].

Titanium and zirconium can also be used to line urea reactors. The line must be continuously passivated with oxygen to resist corrosive action by adding air (0.5 to 0.8 vol%) in the CO_2 feed stream[88].

For low-pressure downstream equipment, 316L and 304 stainless steels are good construction materials. Aluminum alloys and piping are quite resistant to the corrosive attack by urea due to the protective oxide film. These alloys can be used in low-pressure piping, floor gratings, hand rails, etc.[88]

Dead spots and crevices - where equipment parts are not continuously wetted by oxygen-containing liquids - are prone to severe corrosion. Therefore, fabrication of this equipment should be done to avoid such vulnerable spots. An effort should also be made to reduce the potential for process contamination by corrosive agents such as sulfur (through oil in liquid NH_3), H_2S (along with CO_2) and chlorides (from cooling water) [88].

Duplex stainless steel alloys are a mixture of ferritic (400 series) and austenitic (300 series) metals. They provide 1) resistance to stress corrosion and fatigue, 2) pitting resistance, 3) are suitable for a wide temperature range (-50°C to 280°C) and 4) are cost effective. In urea plants, duplex stainless steel is used to construct strippers, decomposers, condensers and pipe lines[88].

11.2. PRODUCTION

Urea production since 1999 is listed in Table 3.1, and urea production (by region, in 2000) is shown in Figure 11.8[145]. Trends in consumption are shown in Figure 3.4 and Table 3.3. Total world urea trade is shown in Table 3.10. The distribution of urea exports (by country) is shown in Table 3.11. The regional distribution of exports and imports are shown in Figure 11.9[36] while imports to the United States are shown in Figures 3.11 and 3.12. The value of urea production is shown in Table 3.6. Urea demand in the United States is summarized in Table 11.5[230].

Table 11.5. Urea Demand in the United States

Year	Demand	Thousands of Tons Imports	Exports
2000	11,850	3,250	850
2001	12,370	4,050	850
2005	14,200	-	-

Note: Demand equals production plus imports less exports.

Urea imports to the United States have grown at 6 percent annually between 1995 and 2002. Urea use in the US grew at an average rate of 5.6 percent per year between 1997 and 2002, in part by displacing other nitrogen fertilizer materials. Annual growth between 2002 and 2005 is estimated at 3.5 percent per year[230].

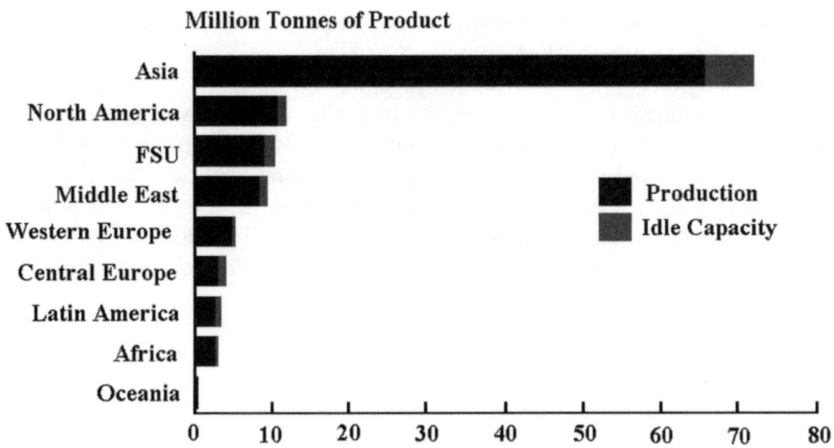

Figure 11.8. World urea production and idle capacity - 2000.
(Reproduced by permission of Fertecon)

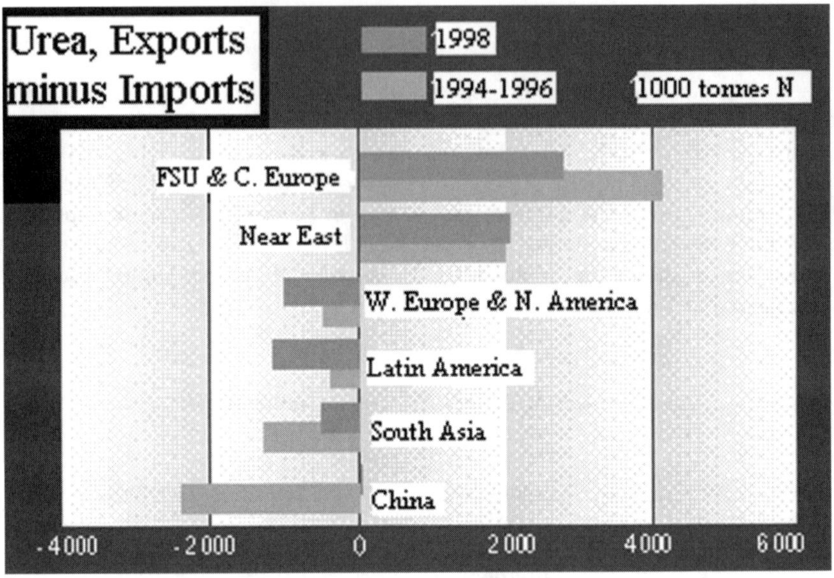

Figure 11.9. Urea imports and exports in different regions.
(Reproduced by permission of International Fertilizer Industry Association)

In the United States poor profitability resulted in the closure of five urea plants between 1999 and the end of 2002. These plants had a total capacity of 725 thousand tons per year - about 7 percent of the total US capacity in 1999. During this same period, about one-half of the operating plants were either idled for extended periods or turned down in their output because of natural gas price spikes that drove up the cost of ammonia[230].

Beginning in 1997, urea prices fell precipitously as China (who used to be a major buyer) began producing urea domestically. Between 1994 and 1997, China opened nine new urea plants and raised domestic production by 50 percent. This shifted world export product toward the US and lowered prices. At the end of 2001, China officially joined the World Trade Organization and re-entered the urea import market. The effect was to pull product away from the United States, which resulted in higher prices[230]. Historical urea prices are shown in Figure 11.10[154].

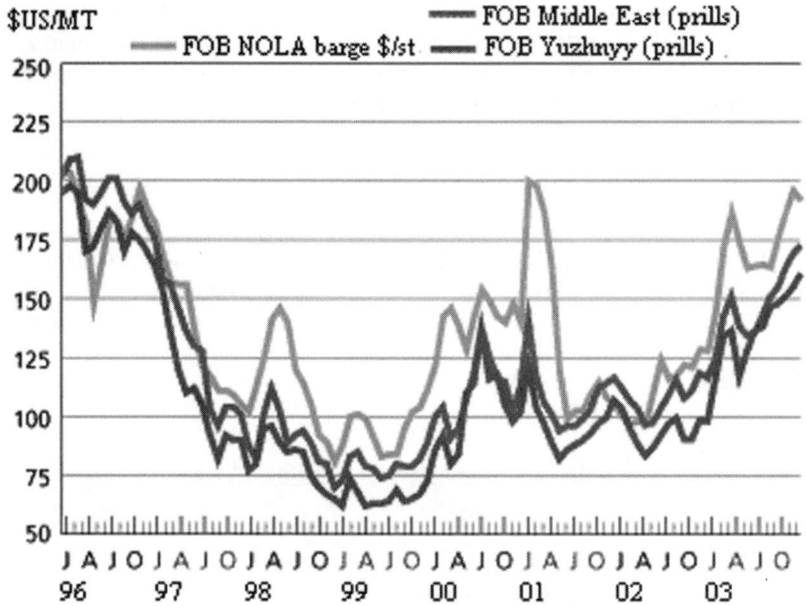

Figure 11.10. Urea Prices: January, 1996 to October, 2003.
(Reproduced by permission of Fertecon)

11.3. STORAGE & DISTRIBUTION

Urea may be supplied either in solid form or as a liquid. For liquid compound fertilizers, urea is generally used in combination with ammonium nitrate as an aqueous solution to obtain liquids containing 32% to 35% nitrogen.

Solid urea is classified as granular or prilled products because of the differences in handling properties. Prilled product is considered less suitable for bulk transportation because prills have lower crushing strength, a lower shock resistance and a higher caking tendency than granules. Because of this, prilled products are usually marginally cheaper than granulated product[109].

The majority of urea is designated as fertilizer grade, however some special forms have found other applications. Technical grade does not contain any additives. Low-Biuret grade contains less than 0.3% biuret and is used on citrus crops. Feed grade is fed directly to cattle. It is free of additives and is supplied as microprills with a diameter of about 0.5 mm. Slow-Release grades use coatings and additions in an attempt to increase the amount of nitrogen that is absorbed by crops. Urea Supergranules have a very large diameter (up to 15 mm) and are used in wetland rice and forest fertilization[109].

The shift from bagged to bulk transport and storage of prilled and granulated urea has called for warehouse designs in which large quantities of urea can be stored in bulk. Caking and subsequent product disintegration at unloading can occur due to water absorption. To avoid these problems the warehouse should be airtight and thoroughly insulated. The caking can also be reduced by adding small amounts of formaldehyde (up to 0.6 wt%) to the urea melt or by adding surfactants to the solid product[109].

11.4. USES

Urea is used in: Liquid and solid fertilizers, urea-formaldehyde resins that make adhesives and binders mostly for wood products, livestock feeds, melamine for resins, NOx control from boilers and furnaces, and numerous chemical applications[113]. Urea uses in the United States are listed in Table 11.6[230].

Table 11.6. Urea Uses in the United States

Use	
Solid Fertilizer	46%
Nitrogen Solutions	41%
Livestock Feed	5%
Urea-Formaldehyde Resins	6%
Melamine	1%
Miscellaneous	1%

Note: Miscellaneous uses include cyanuric acid for chlorinated isocyanurates, crystalline adducts, deicing agents, pharmaceutical intermediates and sulfamic acid and its ammonium salt.

12

AMMONIUM PHOSPHATES AND AMMONIUM SULFATE

12.1. AMMONIUM PHOSPHATES

Two salts are manufactured by combining ammonia with phosphoric acid: diammonium phosphate (DAP) and monoammonium phosphate (MAP).

$$2\,NH_3 + H_3PO_4 \rightarrow (NH_4)_2HPO_4 \tag{12.1}$$

$$NH_3 + H_3PO_4 \rightarrow NH_4H_2PO_4 \tag{12.2}$$

DAP is a white, crystalline compound that is completely soluble in water and, hence is 100% available to plants. MAP is a white, crystalline material that is completely soluble in water. In contrast to DAP, the solutions of which are slightly alkaline, MAP gives acid solutions with a pH of about 4.5. As in DAP all of the phosphate is 'available' since it is completely water-soluble.

Ammonium phosphates are solids between 0 and 75°C. Because ammonium hydroxide (NH_4OH) is a much weaker base than the common metal hydroxides, ammonium phosphates are comparatively unstable. For example, both triammonium phosphate ($(NH_4)_3PO_4$), and the double salt ($(NH_4)_3PO_4 \cdot 2(NH_4)_2HPO_4$) are unstable at room temperature and evolve ammonia to form diammonium phosphate. Even the commercial mono- and diammonium phosphates exhibit an ammonium vapor pressure in the solid form and in solution. Although MAP is more stable than DAP, it decomposes at high temperature to give ammonia and polyphosphoric acid. Diammonium phosphate decomposes to give ammonia and monoammonium phosphate at around 70°C[238].

The vapor pressure of a saturated solution of MAP is expressed by the following equation[238]:

$$\log_{10} P_{kPa} = -2240/T + 9.737 \quad (12.3)$$

where T is between 292 and 363°K. The vapor pressure of a saturated solution of DAP is expressed by:

$$\log_{10} P_{kPa} = -2240/T + 9.682 \quad (12.4)$$

where T is between 292 and 328°K.

12.1.1. MAP Powder Processes

MAP is produced by reaction of anhydrous ammonia and phosphoric acid in batch or continuous reactors. It is then crystallized in conventional crystallizers since the partial pressure of ammonia over this acid solution is low. Crystals are centrifuged and dried below 100°C in a rotary dryer, and mother liquor is returned to the reactor[238].

12.1.1.1. PhoSAI Process for MAP Powder

In the PhoSAI process of Scottish Agricultural Industries (Figure 12.1), ammonia reacts with phosphoric acid (minimum concentration 42 % P_2O_5) in a stirred-tank reactor to produce a slurry with an N/P ratio of 1.4 (the point of maximum solubility) under atmospheric pressure. The slurry flows to a pin mixer in which the N/P ratio is brought back to ~1.0 by the addition of more concentrated phosphoric acid. During this step the solubility decreases and more water vaporizes due to the heats of reaction and crystallization. A solid product is formed that typically contains 6% to 8 % moisture. The product is screened, the oversize particles are ground and the product is sent to storage[296].

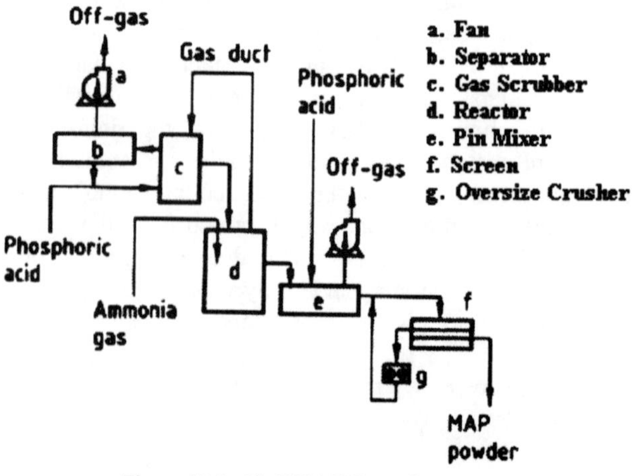

Figure 12.1. PhoSAI MAP powder process.
(Reproduced by permission of Wiley-VCH)

12.1.1.2. Hydro Minifos Process for MAP Powder

In the Minifos process of Fisons (now Hydro Fertilizers), ammonia and phosphoric acid (45 – 54 % P_2O_5) react at 0.21 MPa to give an N/P ratio of 1 (see Figure 12.2). The liquid product is flash-sprayed in a natural-draught tower, where the droplets solidify to form a powder with 6% to 8% residual moisture[296].

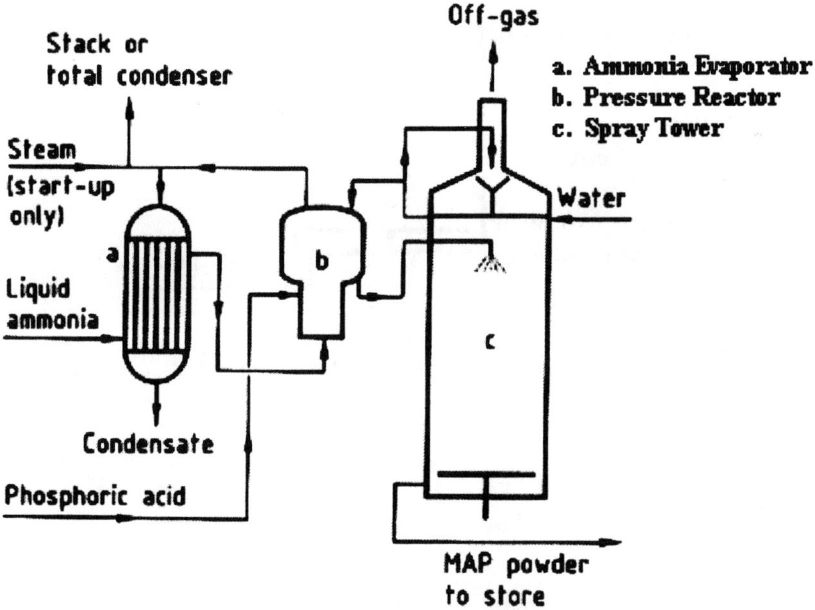

Figure 12.2. Hydro Minifos MAP powder process. (Reproduced by permission of Wiley-VCH)

12.1.1.3. Swift Process for MAP Powder

Alternative processes based on the use of a pipe reactor instead of a tank reactor were developed by Gardinier S.A., Swift Agricultural & Chemical Corporation (see Figure 12.3), and ERT – Espindesa. Product humidity is reported to be as low as 2 %, which results in less caking[296].

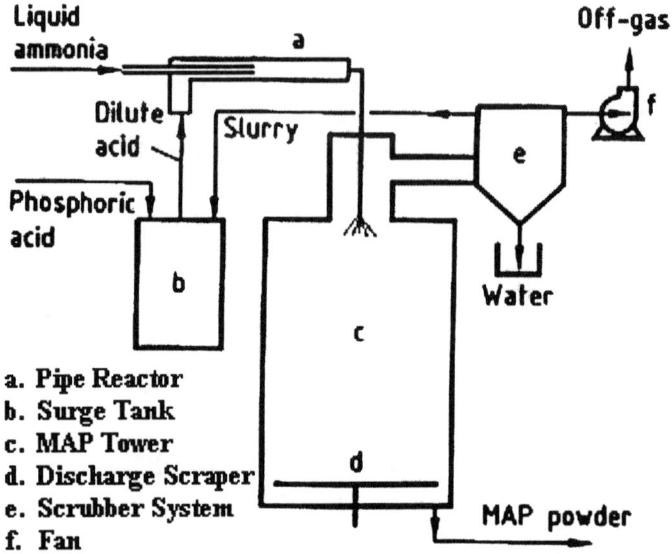

Figure 12.3. Swift MAP powder process.
(Reproduced by permission of Wiley-VCH)

12.1.2. DAP Powder Processes

DAP solutions have a high partial pressure of ammonia, and the reaction is normally carried out in a two-stage reactor system with feed acid passing countercurrent to the flow of ammonia gas. Incoming acid reacts in the scrubber with ammonia from the main reactor and may also serve as a scrubber for dryer off-gases[238].

12.1.2.1. Swift Process for DAP Powder

Based on the favorable results obtained with a pipe reactor for MAP powder manufacture, ERT – Espindesa experimented with the production of DAP powder (see Figure 12.4). The DAP product has a remarkably high water- and citrate-soluble P_2O_5 content. This is probably due to the short residence time in the reactor, which does not yield the insoluble salts that are normally formed in long-retention-time DAP plants. The same effect has been observed with MAP formed in a pipe reactor[296].

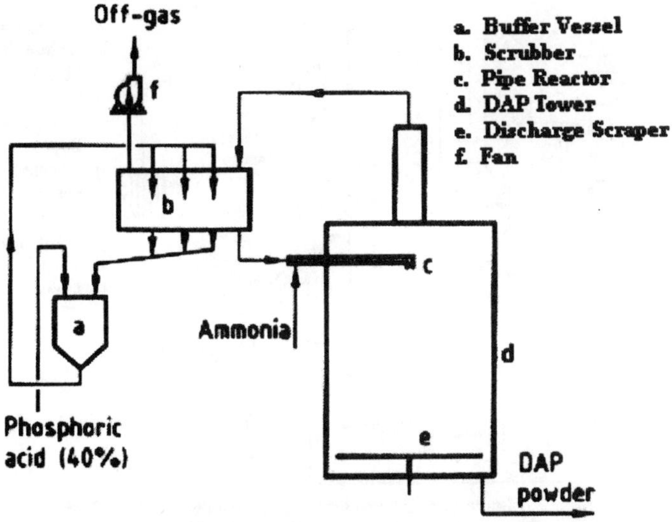

Figure 12.4. ERT- Espindesa DAP powder process.
(Reproduced by permission of Wiley-VCH)

12.1.3. Granular DAP and MAP Processes

MAP and DAP fertilizers are made in a granulation process from ammonia and wet-process phosphoric acid. The acid is partially neutralized in a tank reactor. Ammoniation and granulation are completed in a rotary drum. Drying, cooling and product screening complete the process[238].

12.1.3.1. Hydro Fertilizer Process for Granular DAP

The Hydro Fertilizer process (see Figure 12.5) that produces granular DAP has a recycle ratio of 4:1, which is lower than the 5:1 ratio used in the TVA reactor that was developed in the 1960's. This lower ratio is attained by carrying out preneutralization to an N/P ratio of 1.4 in a pressure reactor operating at 0.1 MPa overpressure. As a result of this overpressure the boiling point of the reaction mixture is elevated by about 20°C compared to operation at atmospheric pressure. Thus, the steep temperature dependence of the solubility of ammonium phosphate in water allows operation at a lower water content while maintaining the ammonium phosphate in solution. The amount of water fed to the granulator is therefore minimized[296].

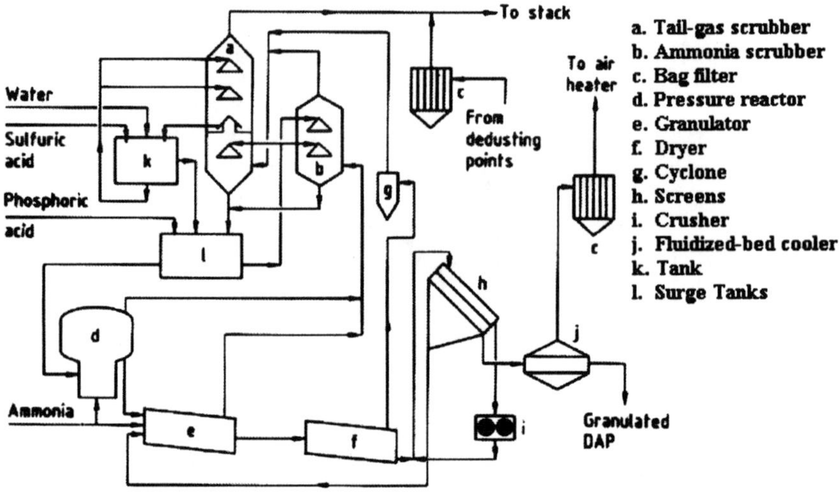

Figure 12.5. Hydro Fertilizer's granular DAP process. (Reproduced by permission of Wiley-VCH)

12.1.3.2. Hydro Fertilizer Process for Granular DAP

In an attempt to further simplify DAP plants and reduce the recycle ratio, TVA, CROS S.A., and ERT – Espindesa have tried to use pipe reactors (also called T reactors or pipe-cross reactors) directly releasing slurry in the rotary granulator, instead of applying a preneutralization tank. CROS and ERT – Espindesa report successful operation, but experiences in several U.S. plants were disappointing because of high heat input in the granulator[296].

CdF–Chimie AZF had the same experience at first, but succeeded by installing a second pipe reactor to spray solution into the dryer (see Figure 12.6).

Approximately half of the phosphoric acid is fed to a pipe reactor that operates at N/P = 1.4 and releases its product into the granulator. The remaining acid is fed into a second pipe reactor at N/P = 1.1 that sprays into the dryer. Extra ammonia is introduced underneath the rolling bed in the rotary drum granulator[296].

The pipe reactor in the dryer produces MAP powder. Part of it is carried away into the cyclones of the dedusting loop. The remainder crystallizes on the DAP product. The recycle ratio of this process is lower than with a conventional preneutralizer, and fuel consumption in the dryer is also reduced[296].

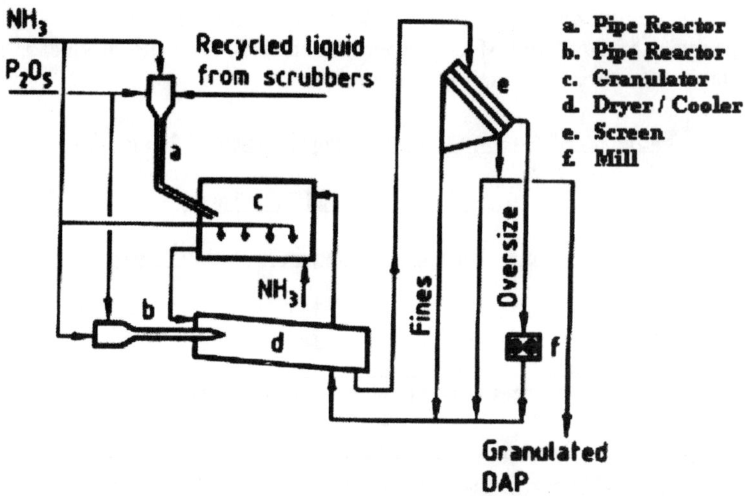

Figure 12.6. AZF granular DAP process.
(Reproduced by permission of Wiley-VCH)

12.1.3.3. Processes for Granular MAP

Granular MAP is produced in the same types of plants as DAP. The preneutralizer is operated at an N/P ratio of 0.6, (a point of high solubility) and more ammonia is added in the granulator to increase the molar ratio to 1.0[296].

In contrast to DAP production, ammonia recovery by acid scrubbing is not necessary, but all off-gases are scrubbed to recover dust and fumes. Granular plants can thus be simpler, but most units are designed so that both products can be made in the same plant[296].

The pipe reactor processes have been very successful in MAP production. They operate with a very slight ammonia loss, and little or no additional heat is required for drying. In some plants, preneutralizers have been replaced by pipe reactors[296].

12.1.4. Production of Ammonium Phosphates

DAP is a white, crystalline compound that is completely soluble in water and, hence is 100% available to plants. MAP is a white, crystalline material that is completely soluble in water. In contrast to DAP (that makes slight alkaline solutions), MAP gives acid solutions with a pH of about 4.5. As in DAP all of the phosphate is 'available' since it is completely water-soluble.

A caking inhibitor for both MAP and DAP is described in US Patent 5,578,107. The inhibitor is sodium sulfate, and it is effective at concentrations between 0.5% and 1.0%.

In 4Q02 ammonium phosphate capacity in the United States was 9,705,000 tons per year of DAP and MAP. Approximately 70% of the product is DAP. Demand is summarized in the following table[239]:

Table 12.1. Ammonium Phosphate Demand in United States

	Thousand Short Tons		
Year	Total Demand	Imports	Exports
2000	4,200	137	4,640
2001	4,080	196	4,590
2005 - Forecast	4,590	-	-

Note: Demand equals production plus imports minus exports

Total DAP/MAP demand has remained between 4.0 and 4.5 million tons since the early 1990's. Demand declined by 3% in 2001, but is expected to recover in 2002 due to an increase in corn plantings. Demand should remain near this level or increase slightly in the medium term due to the 2002 Farm Bill in the United States. This bill will support crop prices near 2002 levels through favorable loan rates. Ammonium phosphate use declined at a rate of 1.5% from 1996 through 2001. It is expected to grow at a rate of 3% from 2002 through 2005[239].

The export market was strengthening in early 2003. China resumed strong import growth and several other countries entered the DAP import market, including Pakistan and Mexico[239].

DAP utilization rates slipped from 95% in the 1990's to 75% in early 2001. Then about 10% of the industry's production capability was permanently shutdown in 2001 and 2002. The export market (typically 55% of US production) also suffered a supply shock in 2001 when two, new DAP plants started up - Oswal Chemicals in India (675,000 tons / year) and Western Mining (450,000 tons / year) in Australia[239].

The DAP price ranged from \$119/ton to \$195/ton (f.o.b. Florida works) between 1996 and 2001. The MAP price ranged from \$125/ton to \$220/ton between 1996 and 2001[239].

12.1.5. Uses of Ammonium Phosphates

Because of the thermally stable nature of ammonium phosphates, they are used in applications related to flame retarding and fire extinguishing. MAP consumption in fire retardants is second only to fertilizers. MAP is a common fire-extinguishing ingredient in ABC-type dry-powder extinguishers. The effectiveness of such extinguishers is closely related to their P_2O_5 content. Ammonium phosphates are used to prevent afterglow in matches, to control forest fires and as a flame retardant for plywood, papers and fabrics[238].

Ammonium phosphate uses are shown in the Table 12.2.

Table 12.2. Ammonium Phosphate Uses

Application	%
Fertilizers	98.0%
- DAP	70.0%
- MAP	23.0%
- Other Solid Ammonium Phosphate	5.0%
Industrial Uses	1.5%
Animal Feed	0.5%

12.2. AMMONIUM SULFATE

Ammonium sulfate (AS) is produced from many sources. The first commercial synthesis started around the time of World War I. Since then direct synthesis from ammonia and sulfuric acid has become a major production route. Ammonium sulfate is generated from several chemical processes. The nitrogen or sulfur values are recovered either to avoid pollution problems or to obtain a commercial return for these elements[241]. The basic chemical reaction is:

$$2\,NH_3 + H_2SO_4 \rightarrow (NH_4)_2SO_4 \quad \Delta H = -274\ kJ/mol \qquad (12.5)$$

Some convention is employed in the sources of AS, mainly for the benefit of statisticians, who try to follow developments in the AS business. AS that is obtained from caprolactam or acrylonitrile manufacture is normally referred to as co-product. On the other hand, AS that is produced from coke ovens, flue gas desulphurization, smelter tail gas, sewage treatment, gypsum, waste acid from methyl methacrylate and from other chemical processes is usually termed by-product[241].

Co-product and by-product AS production depends on the process that is used to make the primary product - as shown in Table 12.3. Small amounts of AS may also be recovered in other processes such as acrylonitrile, acrylamide, formic acid, hydrogen cyanide, hydroquinone and p-aminophenol, and from the neutralization of sulfuric acid used to make clays and catalysts for catalytic cracking[242].

Table 12.3. Ammonium Sulfate By-Product Ratios

Primary Product	By-Production Ratio (tons AS per ton Primary Product)
Caprolactam	0 to 4.4
Coking of Coal	0.0068
Methyl Methacrylate	3.0
Coal-Fired Electric Plant[a]	500 tons per megawatt

[a] Assumes the coal contains 3.5 weight % sulfur.

As mentioned above, AS is produced by the reaction of ammonia and sulfuric acid. The reaction is usually carried out in saturators, devices that are derived from the evaporation crystallizers[295].

In the saturator process (see Figure 12.7), neutralization and crystallization are carried out in the same vessel. The sulfuric acid is delivered to the suction side and the ammonia to the pressure side of the forced circulation pump. Crystallization of the metastable solution gives particle sizes generally between 0.5 and 3 mm. The salt is continuously discharged at the lower end of the saturator. The salt is separated in centrifuges, dried, and cooled. The mother liquor is returned to the saturator. Impurities in the sulfuric acid can adversely affect crystallization. Small quantities of phosphoric acid, urea, or inorganic salts are added to promote crystal growth[295].

Co-product AS is recovered in a crystallization process that is controlled to yield large, uniformly sized crystals. Usually by-product AS contains small amounts of organic contaminants. As a result, the crystallization process generates smaller and finer particles that are harder to handle[243].

Usually crystallizer product has a particle size that is about 90% finer than 16 mesh (about 1 mm in diameter), which is too small for satisfactory dry blending with granular fertilizer materials. Crystals of this size are suitable for use as a feed material to mixed fertilizer granulation plants, which is the main outlet for by-product ammonium sulfate[244].

Some success has been achieved in bringing the size of by-product AS up to 1 to 3.4 mm (-6 +16 mesh), which makes it suitable for dry blending with granular fertilizer materials. On approach involves increasing the retention time in the saturator-crystallizer. Larger crystals are made, but crystallizer capacity is reduced. Another approach involves agglomeration of the fine crystals in a rotary drum granulator to which some ammonia and acid are also fed to provide binding. The drawback of this method is the expense of the extra equipment and the additional raw materials. A third method for agglomeration of the fine crystals is known as compaction. In this method, which is used to some extent commercially, the crystals are fed to a roll compactor that exerts extremely high pressure to form a dense sheet of AS. This sheet is then fragmented and screened to give the desired size of AS[244].

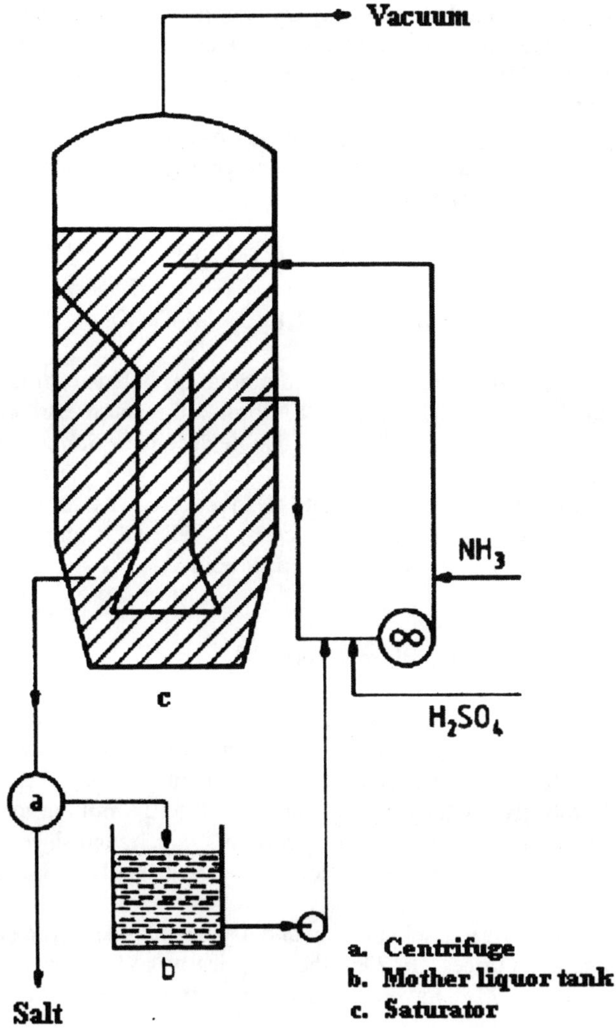

Figure 12.7. Typical Ammonium sulfate process.
(Reproduced by permission of Wiley-VCH)

In the Dakota Gasification synfuels plant in North Dakota, waste sulfur dioxide (SO_2) reacts with ammonia and air to make AS that is 99.5% pure. It has a capacity of 500 tonnes per day of granular AS. This process is referred to as the Ammonium Sulphate Forced Oxidation (ASFO) system. It was developed, patented and commercialized by General Electric Environmental Services, Inc. (GEESI). GEESI was acquired by Marsulex Inc. (North York, Ontario, Canada) in 1997[243].

A few companies have used the Merseburg process to recover sulfur values from gypsum as ammonium sulfate. This process is based on the reaction of gypsum with ammonium carbonate.

$$2\,NH_3 + CO_2 + H_2O \rightarrow (NH_4)_2CO_3 \qquad (12.6)$$

$$(NH_4)_2CO_3 + CaSO_4 \bullet 2\,H_2O \rightarrow (NH_4)_2SO_4 + CaCO_3 + 2\,H_2O \qquad (12.7)$$

In this process carbon dioxide and ammonia are bubbled through a water suspension of finely ground gypsum to form ammonium carbonate. The ammonium carbonate then goes through a double decomposition with gypsum to form insoluble calcium carbonate and AS solution. The calcium carbonate is removed by filtration and the sulfate is recovered by evaporation and crystallization. This process was developed by BASF and is only economically viable in regions with a short supply of sulfuric acid[242].

12.2.1. Production of Ammonium Sulfate

Of the total US ammonium sulfate capacity, 75% is a by-product of caprolactam production. The balance comes primarily from the sources shown in Table 12.3. Direct synthesis accounts for less than 15% of US capacity. AS demand will probably fluctuate in the range of 2.2 to 2.6 million tons per year for the foreseeable future. And it has been in this same range since 1970. Any real demand growth may be more a factor of increased supply as a by-product of the caprolactam market[243].

AS prices vary by geographic region and depend on grade. The price is also affected by ammonia prices. In 2000 the list price was $105 per ton FOB, and between 1995 and 1998 the average price varied from $US 63 to $US 84 per ton[243].

Table 12.4. Ammonium Sulfate Supply and Demand

	Thousands of Tons						
	1995	1996	1997	1998	1999	2000[b]	2002[b]
Capacity	2,936	2,984	3,200	3,195	3,250	3,250	
Production							
Non-Coke	2,648	2,662	2,711	2,787	2,875	2,935E	
Coking	230	230	210	190	175	175	
Total Prod.	2,878	2,892	2,921	2,977	3,050	3,110	
Imports	478	412	527	351	377	450E	
Exports	1,011	909	957	1,158	1,178	1,050E	
Inv. Change	+68	-67	+48	-55	+20	na	
Demand	2,277	2,462	2,443	2,225	2,229	2,450	2,400

[a] Demand equals production plus imports minus exports plus inventory change.
[b] Forecast
E Estimate

12.2.2. Uses of Ammonium Sulfate

At least 95% of the ammonium sulfate in the US is used in fertilizer applications. Over 40% of AS is applied as a direct application fertilizer while the balance is used in fertilizer mixtures. However ammonium sulfate's share of the nitrogen fertilizer market has declined because of the trend toward products that contain more nitrogen, such as ammonium nitrate and urea. AS does have advantages in that 1) it contains both nitrogen and sulfur and 2) it has a high residual acidity. These characteristics make it useful for direct application in alkaline soils, in soils low in sulfur, and for blending in fertilizer mixtures that need sulfur. However in 1999 AS represented only 2% of the world-wide nitrogen content in fertilizers[243].

The low (21.2% N) nitrogen content of ammonium sulfate translates into high shipping costs. And in most locations farmers are not willing to attribute any monetary value to the sulfur content (24.3% S) of AS. However in an extensive sulfur-deficient area of Australia, AS is synthesized in granular form and is highly regarded for its sulfur content[244].

For non-fertilizer applications AS is available from some makers in three grades: Technical grade, Anti-Caking grade and Food grade for special food applications. Some other applications include production of persulfates, food processing, flame proofing agents, fire-extinguishing powders, tanning, photographic industry, textile industry, glass industry, as a nutrient for yeast cultures, as a nutrient for bacterial cultures, particle board, and cattle feed[243, 244].

AS is a white to brownish-gray crystalline salt that is soluble in water, but is only slightly soluble in ethanol. AS has excellent storage properties in bags and in bulk, and it does not require any anti-caking conditioners. This is because AS is resistant to moisture absorption and has a critical relative humidity of 79% at 30°C[244]. AS can be irritating when in contact with the skin or mucous membranes. Normal precautions for handling dry, dusty materials should be observed. A threshold limit value for, AS has not been established[243].

At atmospheric pressure, AS will decompose before it will melt. Upon decomposition it releases ammonia and leaves bisulfate behind. Above 300°C, decomposition gives N_2, SO_2, SO_3 and H_2O in addition to ammonia. Its vapor pressure is effectively zero up to 80°C.

13

MELAMINE

Melamine is characterized by its three amino (-NH_2) groups and by its 66.6% nitrogen content. It is known by several different names: 1,3,5-triazine-2,4,6-triamine; melamine crystal; cyanurotriamide; 2,4,6-triamino-s-triazine; cyanuramide and cyanuric acid amide. Its chemical formula can be written as $C_3H_6N_6$ or $C_3N_3(NH_2)_3$. Liebig first prepared it in 1834, but it was not until 1937 that it was produced commercially.

Some of the properties of melamine are listed in Table 13.1[114]:

Table 13.1. Properties of Melamine

Appearance	Fine White Crystalline Powder
Molecular Weight	126.13
Specific Density (g/cc)	1.573 at 20°C
Melting Point	354°C
Boiling Point	Decomposes Above 300°C and Releases Ammonia Vapors
Free Bulk Density	Standard Grade: 650-750 kg/m^3 Fine Grade: 200-300 kg/m^3
Tapped Bulk Density	Standard Grade: 900-1,000 kg/m^3 Fine Grade: 500-650 kg/m^3
pH	8.2-9.4 (10% slurry in water)
Solubility	0.3 g/100 ml in water
Melamine Structure	

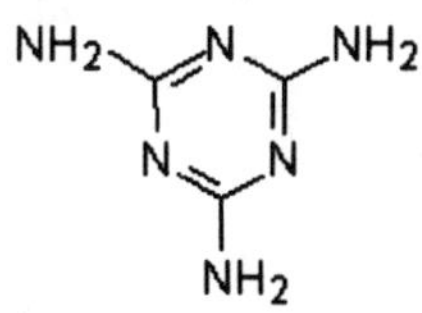

13.1. PROCESSES

Melamine can be produced from urea, dicyandiamide (a derivative of calcium cyanide) or hydrogen cyanide. In 2001 all world production was based on urea. The dicyandiamide process was the original melamine process from the 1930's, but it was phased out in the 1980's. Production of melamine from hydrogen cyanide was never commercialized[114].

Melamine is produced from urea by either a high- or a low-pressure process, and either process can consist of one or two stages. The net reaction in each process is[114]:

$$6\ H_2N\text{-}\overset{\overset{O}{\|}}{C}\text{-}NH_2 \xrightarrow[\text{Pressure}]{\text{Heat}} C_3N_3(NH_2)_3 + 6\ NH_3 + 3\ CO_2 \quad \Delta H = +\ 153\ \text{kcal/mole} \qquad (13.1)$$

Urea — Melamine — Ammonia — Carbon Dioxide

In a two-stage process, urea is heated and decomposes to make isocyanic acid and ammonia as shown in Eq. (13.2). Next this mixture is passed over a suitable catalyst to make melamine and carbon dioxide [see Eq. (13.3)] and the products are quenched with water or an aqueous mother liquor. This stream is then processed by filtration, in centrifuges or by crystallization[114].

$$H_2N\text{-}\overset{\overset{O}{\|}}{C}\text{-}NH_2 \xrightarrow[\text{Pressure}]{\text{Heat}} HN{=}C{=}O + NH_3 \quad \Delta H = \sim 800\ \text{kcal/kg Urea (endothermic)} \qquad (13.2)$$

Urea — Isocyanic Acid — Ammonia

$$6\ HN{=}C{=}O \xrightarrow{\text{Catalyst}} C_3N_3(NH_2)_3 + 3\ CO_2 \quad \Delta H = \sim 1{,}100\ \text{kcal/kg Urea (exothermic)} \qquad (13.3)$$

Isocyanic Acid — Melamine — Carbon Dioxide

High-pressure processes were first commercialized in 1963. In a typical high-pressure process, the reaction is carried out in the liquid phase, without a catalyst, and under pressures in the range of 90 to 150 bar and 380°C to 450°C. Under these conditions, urea forms cyanuric acid, which then reacts with ammonia to form melamine. The first step is the same as Eq. (13.2) while the second step is shown in Eqs. (13.4) and (13.5)[46, 114].

$$3\ HN{=}C{=}O \rightarrow C_3N_3(OH)_3 \qquad (13.4)$$

Isocyanic Acid — Cyanuric Acid

$$\underset{\text{Cyanuric Acid}}{C_3N_3(OH)_3} + \underset{\text{Ammonia}}{3\ NH_3} \rightarrow \underset{\text{Melamine}}{C_3N_3(NH_2)_3} + 3\ H_2O \qquad (13.5)$$

Figure 13.1 is a block diagram of the DSM high-pressure process, and Figure 13.2 is a block diagram of the high-pressure process that is licensed by Eurotecnica[231, 233].

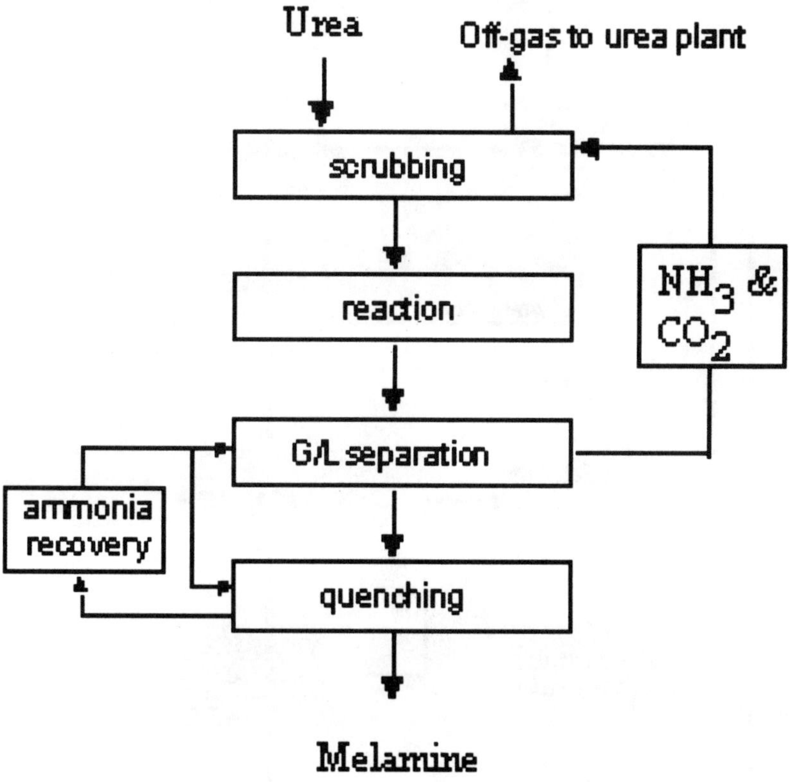

Figure 13.1. DSM High-Pressure Melamine Process
(Reproduced by permission of DSM)

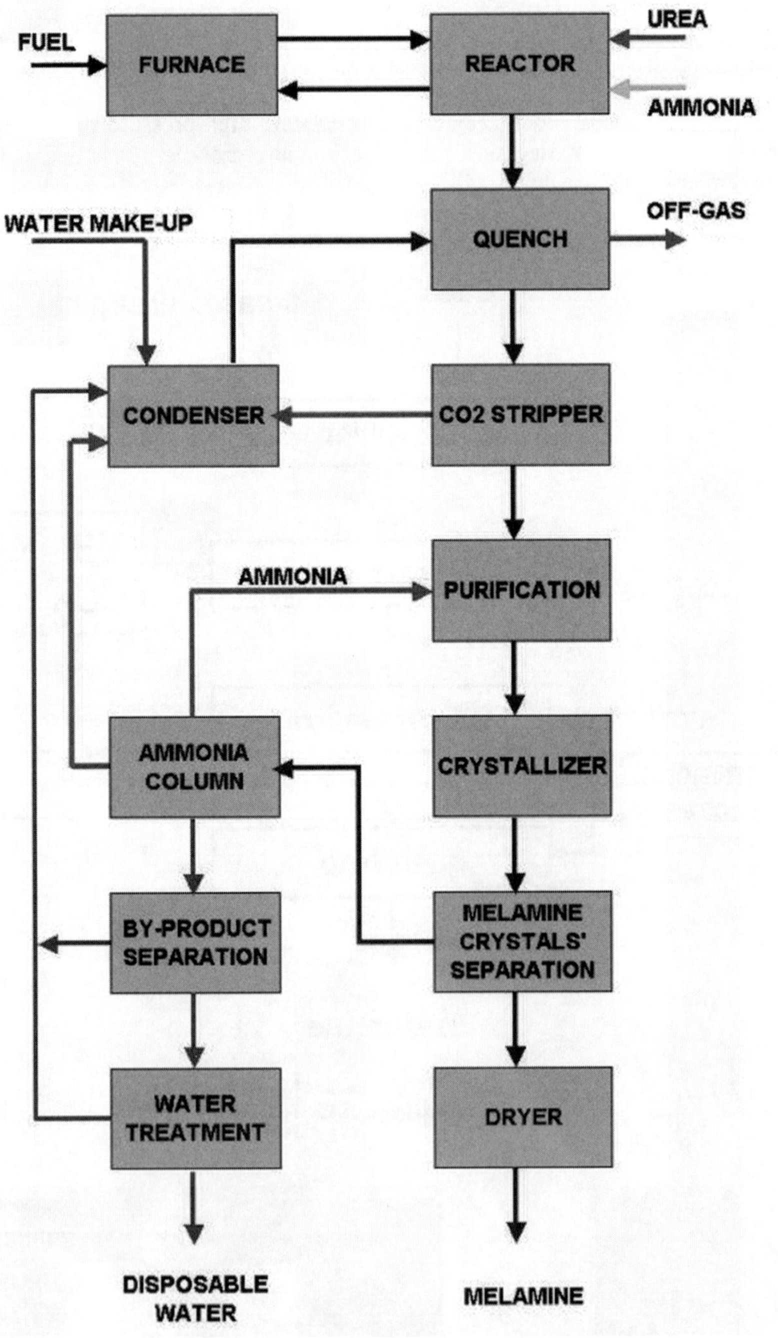

Figure 13.2. Eurotecnica High-Pressure Melamine Process (Reproduced by permission of Eurotecnica)

Low-pressure processes were first commercialized in 1966. In a typical low-pressure process, the reaction is carried out in the vapor phase at one to ten bar and 350°C to 400°C. Figure 13.3 shows a block diagram of the low-pressure process that is licensed by DSM[231].

This process uses a catalyst such as modified aluminum oxide or aluminosilicate. Under these conditions urea forms isocyanic acid, as shown in Eq. (13.2)[46, 114].

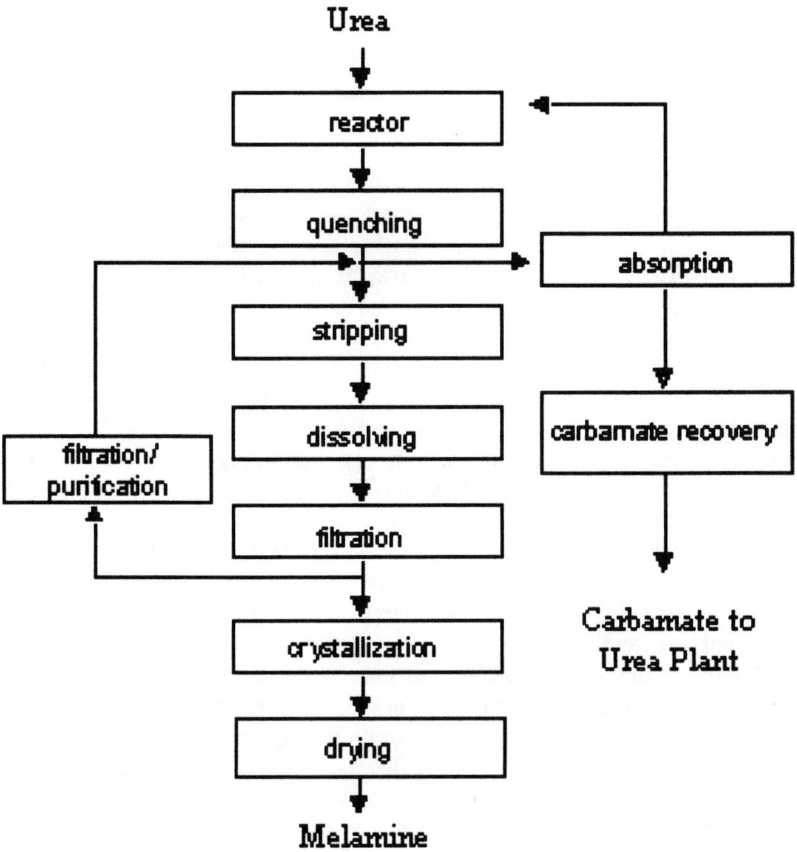

Figure 13.3. DSM Low-Pressure Melamine Process
(Reproduced by permission of DSM)

On the catalyst the isocyanic acid is converted to cyanamide or carbodiimide which is subsequently converted to melamine as shown in Eqs. (13.6) and (13.7)[46, 114].

$$\underset{\text{Isocyanic Acid}}{2\ HN{=}C{=}O} \rightarrow (\underset{\text{Carbodiimide}}{HN{=}C{=}NH} \text{ or } \underset{\text{Cyanamide}}{H_2N{-}C{\equiv}N}) + \underset{\text{Carbon Dioxide}}{CO_2} \quad (13.6)$$

$$\underset{\text{Carbodiimide}}{(HN{=}C{=}NH}\ \text{ or }\ \underset{\text{Cyanamide}}{H_2N{-}C{\equiv}N)} \rightarrow \underset{\text{Melamine}}{C_3N_3(NH_2)_3} + \underset{\text{Carbon Dioxide}}{CO_2} \quad (13.7)$$

Carbon dioxide and ammonia are obtained as by-products in both the low- and high-pressure processes. These components are usually recycled to the urea process to improve process economics. With recycling, 1.5 to 2.0 pounds of urea are used to produce one pound of melamine. This corresponds to a yield of 95% (compared to a yield of only 49% without recycle)[114].

DSM Melamine plans to start up a new 30,000 tonne per year melamine plant towards the end of 2002. The investment estimate is EUR 90 million. However part of the investment will be used to expand urea capacity at the site[115].

This plant will use the high-pressure Shortened Liquid Phase (SLP) process developed by DSM. Melamine produced with the SLP process has the same quality as melamine produced in the gas phase (or low pressure) process. The new technology is a result of DSM's further development of a process acquired from MCI (Melamine Chemical Industries) in 1997. The SLP process is expected to increase efficiency by 25%. It is also expected to enable the plant to reach a level of cost-effectiveness similar to a 100,000 tonnes/year production facility. The new process requires only 3 or 4 processing steps, in contrast to the 10 steps in conventional processes[115, 231]. The process employs the same raw materials as the low-pressure urea process, but the final melamine recrystallization step is eliminated. The company may be able to make 99% purity melamine without recrystallization[114].

Melamine Chemicals' M-II process simplifies the production of melamine over the anhydrous process in which urea is pyrolyzed to melamine. In this process melamine is recovered directly from the melamine melt as a dry powder. This process does not require the energy-intensive recrystallization steps so investment is lower[232].

The melamine that is made by this process has a purity of 96 wt% whereas other processes produce a product having greater than 99.8 wt% purity. Melamine-formaldehyde resin producers prefer high purity melamine. This is because the purity of melamine influences the physical properties of the resulting resins and affects processing times. However Melamine Chemicals' patents claim that lower-purity melamine can be substituted in the production of melamine-formaldehyde resins if the reactor conditions (e.g., pH) are carefully controlled[232].

Melamine process technologies can be licensed from DSM, BASF, Agrolinz, Allied-Eurotecnica, Nissan, Montedison, and Melamine Chemicals[114].

13.2. PRODUCTION

In 2001 the worldwide nameplate capacity for melamine was estimated to be about 1.1 million tons per year. Between 1997 and 2001 worldwide melamine capacity increased at annual rate of 13% per year. However demand growth was only 5% to 7% per year. Therefore worldwide melamine capacity will probably be in surplus after 2005[234]. In 1998, Western Europe accounted for 35% of world capacity, followed by Asia (excluding Japan) with 24% and both the United States and Japan with 16% each[114].

Between 1997 and 2001 the number of melamine producers in China increased from less than twenty (20) to at least sixty (60). During this time capacity grew from 45,000 tons per year to 197,400 tons per year (Note: Another source states that melamine capacity in China in 2001 was 222,500 tonnes per year[235]). China's melamine consumption is about 60,000 tons per year so most of the melamine enters the export market[234].

Demand for melamine is heavily influenced by construction and remodeling activities and by automobile manufacture. End-use applications for melamine vary by region. Laminates account for nearly one-half of Western European demand, while surface coatings are the largest end use in the United States. In Japan, wood adhesives and molding compounds are the main applications. One reason for this is the relative ease of establishing laminate, molding or wood adhesive production due to low technological requirements. Historically, both production and consumption of melamine in Western Europe have been higher than the United States. One reason for this is that Western Europe possesses less virgin wood resources than the United States and therefore relies more on composite wood boards and laminates. Additionally, melamine technology was discovered and developed in Western Europe[114].

During the 1980's the melamine list price in the United States was very steady at $0.50 to $0.52/lb. In 1989 and 1990 the price rose to $0.58 to $0.62/lb, but then it declined to the low $0.40's by 1994. From 1995 to 1999 the price began to rise again and reached $0.70/lb. Import prices showed similar trends during the 1980's and reached a value of $0.58/lb in 1990. Since that time import prices trended downward to less than $0.40/lb by 1995 and only recovered to about $0.50/lb by 1998[114].

13.3. USES

Melamine is supplied in 25 kg paper bags, big bags (up to 1,500 kg), and bulk (silo truck, bulk container or rail car). Some of the melamine uses are listed below[114]:

- Surface Coatings. Melamine-formaldehyde (MF) resins are used for surface coatings in automotive applications, for metal containers, for metal furniture, for coil coatings and for electric appliances.

- Laminates. Laminates are made by pressing paper or cloth saturated with MF resin under pressure and temperature onto a core material. Laminates are used in cabinets, furniture, kitchen countertops, bathroom countertops, flooring and various types of paneling.
- Paper Treating. MF resins are used in some paper coatings and as paper wet-strength resins. Both of these applications appear to be declining because alternate compounds have been identified
- Molding Compounds. Melamine molding compounds can be produced by compression or injection molding and are used to make dinnerware for the home, in food service and for institutional settings. They also have some applications in circuit breakers, mixing bowls, small appliance housings and cosmetic cases.
- Textile Treating. MF resins improve the crease-resistance of cotton and cotton-synthetic blends, stiffen synthetic fabrics such as nylon, protect fabrics from mildew and aid in water repellency.
- Wood Adhesives. Melamine-modified wood adhesives are used in various wood products such as plywood, particle board, medium-density fiberboard (MDF), truck and railcar flooring, furniture doors and wooden salad bowls and trays.
- Concrete Additives. Melamine-formaldehyde resins can be sulphonated and then used in fresh concrete to create a "plasticizing" effect. The resin makes 1) concrete more flowable, thus improving concrete's workability and 2) allowing the water content to be reduced for faster drying times. The end result is an improved, stronger material that is structurally superior.
- Other uses include ceiling tiles, tire cord, flame retardants and many other small-volume applications.

In the first half of 2002, an Austrian chemical company (OMV) increased melamine sales volume by 57% by targeting markets for new applications. Most of the increased sales occurred in the automotive and construction industries. The melamine is being used in reinforced products where it is mixed with polymers. Melamine provides components and materials with an extra hard surface. Unfortunately OMV's melamine earnings actually decreased because of intense competition from Asian producers[236].

14

ALIPHATIC AMINES

14.1. ALIPHATIC AMINES

Aliphatic amines are derivatives of ammonia. In these compounds monovalent hydrocarbon radicals have displaced the hydrogen atoms in the ammonia. Depending on the number of hydrogen atoms displaced, amines are classified as Primary (RNH_2), Secondary (R_2NH) or Tertiary (R_3H). The amines can be categorized further according to the type of hydrocarbon radicals that substitute for the hydrogen atoms: Aliphatic (saturated or unsaturated), Aromatic, Heterocyclic, Alicyclic, or any combination of these.

Imines are one group of compounds that are similar to amines. Imines contain an ammonia molecule in which two hydrogen atoms are displaced by bivalent hydrocarbon radicals (R=NH). Nitriles are another group of compounds that are similar to amines. In nitriles all the hydrogen atoms in ammonia are displaced by a trivalent hydrocarbon radical ($RC\equiv H$).

The three methylamines and ethylamine are gaseous at room temperature. Diethylamine, triethylamine and higher amines (up to about twelve carbon atoms per alkyl chain) are liquid, and long-chain amines containing still higher alkyl groups are solid. All alkylamines have a characteristic ammonia odor. - especially the lower amines, which have high vapor pressures[245].

14.1.1. Methylamines

Methylamines are colorless liquids that are volatile at normal atmospheric conditions. They have threshold odor limits of less than 10 ppm, and at low concentrations they have a fishy smell. At high concentrations they smell like ammonia. The physical properties are given in Table 14.1 and Table 14.2.

Table 14.1. Physical Properties of Anhydrous Methylamines

	MMA	DMA	TMA
Chemical Formula	CH_3NH_2	$(CH_3)_2NH$	$(CH_3)_3N$
Molecular Weight	31.06	45.08	59.11
Freezing point, °C	-93.5	-92.02	- 117.3
Boiling point, °C	-6.3	6.9	2.9
Vapor Density (1 atm, 25°C) g/liter	1.3	1.9	2.5
Specific gravity, air = 1	1.1	1.6	2.0
Liquid density (25°C), g/ml	0.6562	0.6496	0.6270
lb/gal	5.48	5.42	5.23
Liquid Viscosity (25°C), cP	--	0.190	0.175
Critical temperature, °C	156.9	164.5	160.1
Critical pressure, atm	73.6	52.4	40.2
Enthalpy of Formation, ΔH^o_f (25°C), kcal/mol			
Gas	- 5.49	-4.41	- 5.81
Liquid	- 11.3	- 10.5	- 11.0
Solution (1000 moles $H_2$0)	- 16.78	- 17.3	- 18.6
Vapor pressure, psia			
25 °C	50	30	32
100 °C	400	230	205
Heat Capacity, $C_{p.}$ (25°C), cal,/°C mol			
Gas	12.7	16.9	21.9
Liquid	--	32.9	32.31
Heat of Fusion (m.p.), cal/g	47.20	31.50	26.46
Heat of Vaporization (b.p.), cal/g	198.6	140.4	92.7
Autoignition temperature, °C	430	400	190
Flammable Limits, vol. %			
Lower	4.9	2.8	2.0
Upper	20.7	14.4	11.6

Source: Du Pont Methylamines Bulletin

Notes: MMA = Monomethylamine
DMA = Dimethylamine
TMA = Trimethylamine

Table 14.2. Physical Properties of Aqueous Methylamines

Properties	40% MMA	40% DMA	60% DMA	25% TMA
Boiling Point (°C)	48	54	36	43
Freezing Point (°C)	-38	-37	-74.5	6
Liquid Density (25°C)				
g/ml	0.897	0.892	0.829	0.930
lb/gal	7.49	7.44	6.92	7.76
Vapor Pressure (25°C)				
psia	5.8	4.2	9.7	6.6
mm Hg	300	215	500	340
kPa	40	29	67	45
Flash Point, closed cup (°C)	-12	-18	-52	6

14.1.1.1. Conventional Processes

Various modifications of the reaction of an alcohol with ammonia provide the most common commercial routes to alkylamines. Some others routes that are used to make certain individual amines include aldehyde/amine additions, nitrile reduction, the Ritter reaction, amination of isobutylene, and hydrogenation of anilines. Capacities of many plants depend on the product mix of mono/di/tri products as well as the variety of amines (ethyl, propyl and butyl). One must know the product mix that the capacity is based upon and the actual scheduled output of product(s) to determine what amounts can actually be manufactured. Capacities are often in excess of anticipated demand to satisfy seasonal demands for pesticide uses[116].

In Alcohol Amination, methanol and excess ammonia react at 350°C to 500°C and 15 to 30 bar in the presence of aluminum oxide, silicate or phosphate catalysts according to the following reactions[46]:

$$NH_3 + CH_3OH \xrightarrow{\text{Catalyst}} CH_3NH_2 + H_2O \quad \Delta H = -5.358 \text{ kcal/mol} \tag{14.1}$$

$$CH_3NH_2 + CH_3OH \xrightarrow{\text{Catalyst}} (CH_3)_2NH + H_2O \quad \Delta H = -9.598 \text{ kcal/mol} \tag{14.2}$$

$$(CH_3)_2NH + CH_3OH \xrightarrow{\text{Catalyst}} (CH_3)_3N + H_2O \quad \Delta H = -14.098 \text{ kcal/mol} \tag{14.3}$$

$$NH_3 + (CH_3)_3N \xleftrightarrow{\text{Catalyst}} CH_3NH_2 + (CH_3)_2NH \quad \Delta H = +8.740 \text{ kcal/mol} \tag{14.4}$$

$$NH_3 + (CH_3)_2NH \xleftrightarrow{\text{Catalyst}} 2\,CH_3NH_2 \quad \Delta H = +4.240 \text{ kcal/mol} \tag{14.5}$$

$$CH_3NH_2 + (CH_3)_3N \xleftrightarrow{\text{Catalyst}} 2\ (CH_3)_2NH \quad \Delta H = +4.500 \text{ kcal/mol} \qquad (14.6)$$

The reactions shown in Eqs. (14.1) through (14.3) are known as the Alkylation reactions. They are exothermic and highly irreversible, except for Eq. (14.3). The reactions in Eqs. (14.4) through (14.6) are known as Disproportionation reactions. They are reversible and are endothermic. The alkylation reactions dictate the rate of consumption of methanol and are somewhat faster than the disproportionation rates that govern the selectivity of the three amines.

The mono- and dialkylation [Eqs. (14.1) and (14.2)] are favored 1) by excess ammonia, 2) by the addition of water and 3) by the recycling of the trimethylamine. Changing the alcohol/ammonia ratio and recycling amine mixtures can be used to control the product mixture exiting the reactor. At 500°C with an ammonia/methanol ratio of 2.4 to 1, the reactor discharge stream contains 54% mono-, 26% di- and 20% trimethylamine. The typical ratio of mono/di/tri methylamines without recycle is 15%/45%/50%. Side reactions form small amounts of CO, CO_2, CH_4, H_2 and N_2. The total selectivity to methylamines reaches about 94%, and yields based on alcohol typically exceed 95%. Due to azeotrope formation, the reaction products are separated in a combination of pressure and extractive distillation[46, 116]. Figure 14.1 shows a typical methylamines process diagram[117]. The raw material and utility use are summarized in Table 14.3[17].

Table 14.3. Methyl Amine Production: Raw Material and Utility Use

	(Tonne per Tonne of Product Leaving the Reactor)		
	MMA	DMA	TMA
Methanol	1.053	1.451	1.660
Ammonia	0.559	0.385	0.294
Steam	3.0		
Cooling Water (m^3)	150		
Electricity (kWh)	150		

Although companies can produce a range of C_2 to C_6 amines on a campaign basis in a higher alkylamine plant, methylamine plants are usually designed and dedicated solely for production of a given chain-length product. The methylamine plants also use a different catalyst from that in a higher amines plant[116].

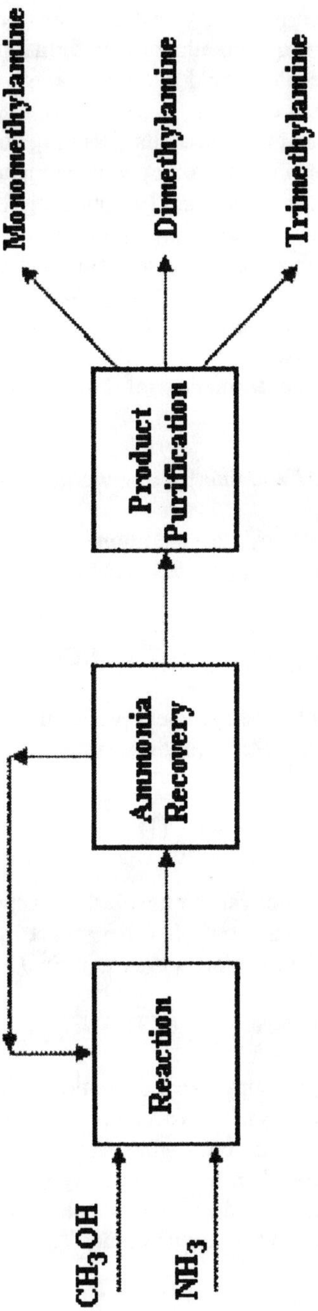

Figure 14.1. Typical methylamines process.

Mitsubishi Rayon (formerly Nitto Chemical) has operated a process since 1984 in which the equilibrium of methylamine formation is shifted to make more dimethylamine by use of an acid zeolite catalyst with a particular pore structure. The product stream contains 7 mol% MMA, 86 mol% DMA and 7 mol% TMA, and its investment as well as its operating costs are lower than the conventional Leonard process that is used by most companies. Since DMA has the highest sales volume, the process might be appealing[246].

In US Patent 5,569,785 an attrition-resistant zeolite catalyst is described that can be used for the production of methylamines in fluidized bed reactors. The technology claims to provide improved temperature control because of better heat transfer and more efficient solids handling in the fluidized bed. The process also offers more precise temperature control to maintain the activity of the catalyst and eliminate the formation of hot spots that lead to catalyst deactivation.

14.1.1.2. Aldehyde-Amine & Ketamine Hydrogenation Process

Imines can be produced by the addition of ammonia to a carbonyl compound, and the imines are hydrogenated make amines per Eq. (14.7):

$$RCHO \quad + \; NH_3 \;\rightarrow\; RCH{=}NH \;\xrightarrow{H_2}\; RCH_2NH_2 \qquad (14.7)$$

Commercial use of this route has become less prevalent with the development of alcohol-based syntheses. Some ethylamines are produced in Japan by this route due to the alcohol tax situation[116].

14.1.1.3. Nitrile Reduction

The choice of catalyst and variations of the temperature, pressure and concentration of ammonia can control the product mix of amines from hydrogenation of a nitrile. The reaction is shown in Eq. (14.8):

$$RC{\equiv}N \; + \; H_2 \;\rightarrow\; RCH_2NH_2 \; + \; (RCH_2)_2NH_2 \; + \; (RCH_2)_3N \qquad (14.8)$$

A large excess of ammonia suppresses formation of di- and trialkylamines. With a rhodium-on-carbon catalyst in process conditions of 75°C and 50 psig plus continuous removal of ammonia, the reduction of propionitrile with hydrogen proceeds with greater than 85% selectivity for di-n-propylamine, with 2 to 5% mono-n-propylamine coproduct. Several companies produce amines in the C_{12} to C_{18} range from fatty nitriles by this route[116].

14.1.2. Other Alkyl Amines

14.1.2.1. Ritter Reaction

Mitsubishi Rayon and Sumitomo Chemical in Japan make t-butylamine via the addition of hydrogen cyanide to isobutylene in acidic media, followed by hydrolysis of the resulting t-butylformamide [see Eq. (14.9)]. Average yields are about 85% based on isobutylene. Rohm and Haas in the United States makes t-octylamine from hydrogen cyanide and diisobutylene by a similar route[116].

$$\begin{array}{ccccc} \overset{\displaystyle CH_3}{\overset{|}{H_2C{=}C{-}CH_3}} + HCN & \xrightarrow[H_2O]{H^+} & \overset{\displaystyle O\quad CH_3}{\overset{\|\quad\;\; |\quad}{\underset{\displaystyle \quad\;\; CH_3}{\underset{\quad\;\; |}{HC{-}N{-}C{-}CH_3}}}} & \xrightarrow{OH^-} & \overset{\displaystyle CH_3}{\overset{|}{\underset{\displaystyle CH_3}{\underset{|}{CH_3{-}C{-}NH_2}}}} \end{array} \qquad (14.9)$$

14.1.2.2. Direct Amination

Direct amination of olefins is a special case and has been successfully developed for only a few amines. BASF has developed proprietary technology for the direct amination of isobutylene to yield t-butylamine [see Eq. (14.10)], and has been operating a commercial plant in Western Europe since 1993[116].

$$\overset{\displaystyle CH_3}{\overset{|}{H_2C{=}C{-}CH_3}} + NH_3 \rightarrow \overset{\displaystyle CH_3}{\overset{|}{\underset{\displaystyle CH_3}{\underset{|}{CH_3{-}C{-}NH_2}}}} \qquad (14.10)$$

14.1.2.3. Hydrogenation of Anilines

A manufacturing process that is specific for cyclohexylamines is the catalytic hydrogenation of anilines or phenols in the presence of ammonia. The catalytic hydrogenation of aniline is the classical method for the manufacture of cyclohexylamine. Hydrogenation of phenol in the presence of ammonia produces predominantly cyclohexyl or dicyclohexylamine depending upon catalysts, reaction conditions and reactant ratios[116].

14.1.3. Production

In the United States in 2001 the overall outlook for methylamines is very good. Production should remain strong as applications in the largest market segments are at or above the growth rates for GDP[247]. Raw material supplies of ammonia and methanol are abundant, but the price for natural gas has been significantly higher compared to prices prior to 2001. As a result some producers have instituted a surcharge based on the price of natural gas.

Worldwide growth in alkylamines demand should average a modest 2% to 3% annually from 2000 through 2004[116]. In the United States growth is expected to increase slightly to 3.3% per year between 2003 and 2006. Demand in the United States is shown in Table 14.4[247].

Table 14.4. Methyl Amine Demand in the United States

	(million pounds)		
	2001	2002	2006 Forecast
Production	318	320	-
Imports	94	100	-
Less Exports	10	8	-
Total Demand	402	412	470

In 2002 the demand was split among the three methylamines as follows: monomethylamine - 21%, dimethylamine - 63% and trimethylamine - 16%.

The United States has usually maintained a net export balance for alkylamines, but the trade surplus for methylamines has eroded since the mid-1980's as a result of increasing imports of methylamines from Canada. Exports of methylamines reached a peak of 34 million pounds in 1993 and then fell to 8 million pounds in 2002 as overseas capacity has grown. The 1999 trade balance for methylamines is one of net imports[116].

Most applications will grow at or above the Gross Domestic Product growth rate. Some segments that are expected to grow even faster are shown in Table 14.5[247].

Table 14.5. Fastest Growing Applications of Methyl Amines

Monomethylamine Uses	Growth Rate (%/year)
Pesticides	2.5
Alkylalkanolamines	2.7
Dimethylamine Uses	**Growth Rate (%/year)**
Water Treatment Polymers	5.9
Dimethylaminoethanol	7.5
Dimethylaminopropylamine	13.5

The major world producers of methylamines include BASF, Air Products, DuPont and UCB. The major producers of C_2 to C_6 alkylamines include Air Products, BASF and Celanese[116].

Manufacturing costs for alkylamines depend on the specific amine and the process. Methylamines are commonly sold as anhydrous gases or as 25%, 40%, 50% and 60% aqueous solutions. The other alkylamines are usually sold on a 100% basis although some are also marketed as aqueous solutions. The published prices for most alkylamines were reasonably steady between 1990 and 1999 except for methylamines. Their prices increased steeply from mid-1993 to early 1995 because methanol prices soared. Methylamine list prices went up to $0.50 per pound in mid-1994 and increased another $0.08 per pound at year-end 1994. In 1997 MMA and TMA prices increased to $0.73 per pound and DMA went up to $0.77 per pound[116]. In 2003 market prices have declined and are substantially below the list prices[247]. From 1995 to 2000 the price of cyclohexylamine has varied from a low of $1.40 per pound to a high of $1.52 per pound. In 2001 its price was at the high end of the price range[118].

14.1.4. Uses

The three methylamines are important intermediates for the manufacture of solvents, insecticides, herbicides, pharmaceuticals, and detergents. The demand for the individual methylamines has developed independent of one another. DMA is the most important due to its use in the manufacture of N,N-dimethylformamide and N,N-dimethylacetamide, which find wide application as solvents. It is also used in water treatment chemicals, to make dimethylaminoethanol, fatty tertiary amines, dimethylaminopropylamine, and pesticides[46, 116].

MMA ranks second in terms of demand. It is mainly used in the to make dimethyl urea and N-methylpyrrolidone, as well as for methyltaurine which is employed in CO_2 washes or as a raw material for detergents. Other uses include pesticides, alkylalkanolamines, surfactants, pharmaceuticals, and explosives[46, 116].

TMA plays only a minor role. It is used in the manufacture of choline chloride, quaternary salts and epichlorohydrin[46, 116].

The different uses for methylamines are listed in Table 14.6[247].

Table 14.6. Uses for Methyl Amines

Monomethylamine Uses	(%)
n-methylpyrrolidone (NMP)	34
Pesticides	37
Metam-sodium - based	21
Methyl isocyanate-based	16
Alkylalkanolamines	23
Surfactants	3
Miscellaneous (explosives & pharmaceuticals)	3

Table 14.6. Uses for Methyl Amines (continued)

Dimethylamine Uses	(%)
Solvents - including dimethylformamide (DMF) and dimethylacetamide	29
Water Treatment Polymers	28
Dimethylaminoethanol (DMAE)	18
Fatty Tertiary Amines	8
Dimethylaminopropylamine	8
Pesticides	3
Miscellaneous (including rubber processing chemicals)	6

Trimethylamine Uses	(%)
Choline Chloride	48
Miscellaneous (including cationic starch reactants)	52

Some of the other alkyl amines and their uses are:

- Ethylamines: pesticides, dithiocarbamates, thioureas, semisynthetic penicillin
- n-Propylamines: herbicides (trifluralin), Epram, pharmaceuticals
- iso-Propylamines: glyphosate salt for Round-up herbicide, triazine, Basagram
- n-Butylamines: herbicides, rubber treatment chemicals
- iso-Butylamines: thiocarbamate herbicide "Sutan"
- sec-Butylamines: herbicides, rubber treatment chemicals
- Cyclohexylamines: corrosion inhibitors in steam systems, catalysts, antibiotics, herbicides, cyclamate sweeteners
- 2-Ethylhexylamine: Intermediate for pharmaceuticals, dyes, insecticides, rubber accelerators, flotation agent , corrosion inhibitor and antioxidants.
- Alkyl alkanolamines Acid gas scrubbing agents, pharmaceutical intermediates, imidazolines, and surfactants.
- Aminoethylethanolamine lube oil additives, wet strength resins, chelating agents, epoxy curing agents and surfactants.

15

ETHANOLAMINES AND SECONDARY PRODUCTS

Ethanolamines are industrially valuable products whose main use is for the manufacture of detergents by reaction with fatty acids. The principal secondary products of ethanolamines are morpholine, ethylenimine and ethylenediamine.

15.1. ETHANOLAMINE PROCESS

Commercial production of ethanolamines (EOA) is by reaction of ethylene oxide with aqueous ammonia. The ethylene oxide reacts exothermically with 20% to 30% aqueous ammonia at 60 to 150°C and 30 to 150 bar in a tubular reactor to form the three possible ethanolamines (mono-ethanolamine - MEA, di-ethanolamine - DEA and tri-ethanolamine - TEA) with high selectivity. The product stream is then cooled before entering the first distillation column where any excess ammonia is removed overhead and recycled. In the second column, ammonia and water are removed and the EOA's are separated in a series of vacuum distillation columns.

$$\underset{\diagdown\;\diagup\atop O}{H_2C\text{-}CH_2} + NH_3 \rightarrow H_2NC_2H_4OH + HN(C_2H_4OH)_2 + N(C_2H_4OH)_3 \qquad (15.1)$$

Ethylene Oxide   Ammonia   MEA   DEA   TEA

The composition of the reaction product stream can be influenced by temperature and pressure, but it is especially sensitive to the ratio of ammonia to ethylene oxide. Higher ratios of excess ammonia lead to a higher monoethanolamine content as shown in Table 15.1. The high triethanolamine content that results from an equimolar mixture of reactants indicates that the primary reaction with ammonia is slower than the secondary reactions[46].

Table 15.1. Effect of Excess Ammonia on Ethanolamine Selectivity

Molar Ratio (NH_3 to EO)	Selectivity Ratios Mono-	Di-	Tri-ethanolamine
10:1	75	21	4
1:1	12	23	65

A typical ethanolamine flowsheet is shown in Figure 15.1. Nippon Shokubai has developed technology that uses a new zeolite catalyst. This new catalyst suppresses the formation of TEA and produces more MEA and DEA[120]. A process that was proposed by some students at Rice University is shown in Figure 15.2. They claim that a single column with multiple product streams (instead of the typical series of three columns) can achieve high purity separation while reducing material and energy costs. Their separation unit makes 99.7% MEA, 97.5% DEA and 99.6% TEA. They also claim that the design has the ability to handle varying production levels as well as different feed compositions[121].

Himtek Engineering in St. Petersburg, Russia offers ethanolamine (EOA) technology that operates at 40 to 70°C and 15 to 35 atmospheres. The ammonia and part of the MEA are recycled. Three vacuum stills are used to separate the three products. This technology is protected by Russian patents No. 2,141,475 and No. 2,063,955[122].

EOA's are corrosive and will attack some plastics and rubbers. They are a moderate fire hazard. MEA and TEA are clear, viscous liquids with a mild ammonia-like odor. EOA's absorb water and carbon dioxide from air. DEA is crystalline or a viscous liquid. All are soluble in water and ethanol. Vapor is irritating to the eyes, skin and respiratory tract and depression of the central nervous system can occur. It can also be absorbed by the skin in toxic amounts[120].

Physically, EOA's are clear, viscous liquids at room temperature and white crystalline solids when frozen. Diethanolamine has the highest freezing point of the group, 28.0°C, and monoethanolamine the lowest, 10.5°C. They all have comparatively high boiling points, although at elevated temperatures they are subject to decomposition, particularly diethanolamine.

Their odors are ammoniacal to varying degrees, with monoethanolamine being the most marked. The ethanolamines are hygroscopic and miscible with water, most alcohols, and polyols. They and their aqueous solutions are alkaline. Being bifunctional, ethanolamines react with acids to form esters or salts, and most of their industrial applications are dependent to some degree on these features.

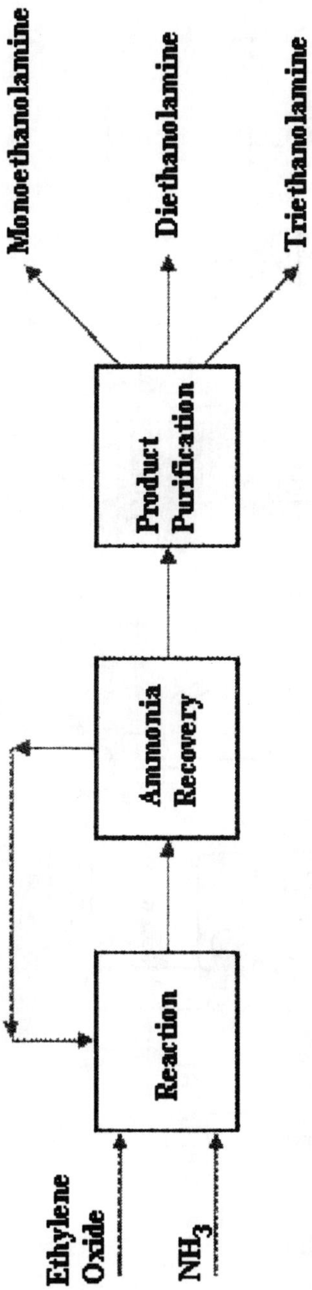

Figure 15.1. Typical Ethanolamine Process

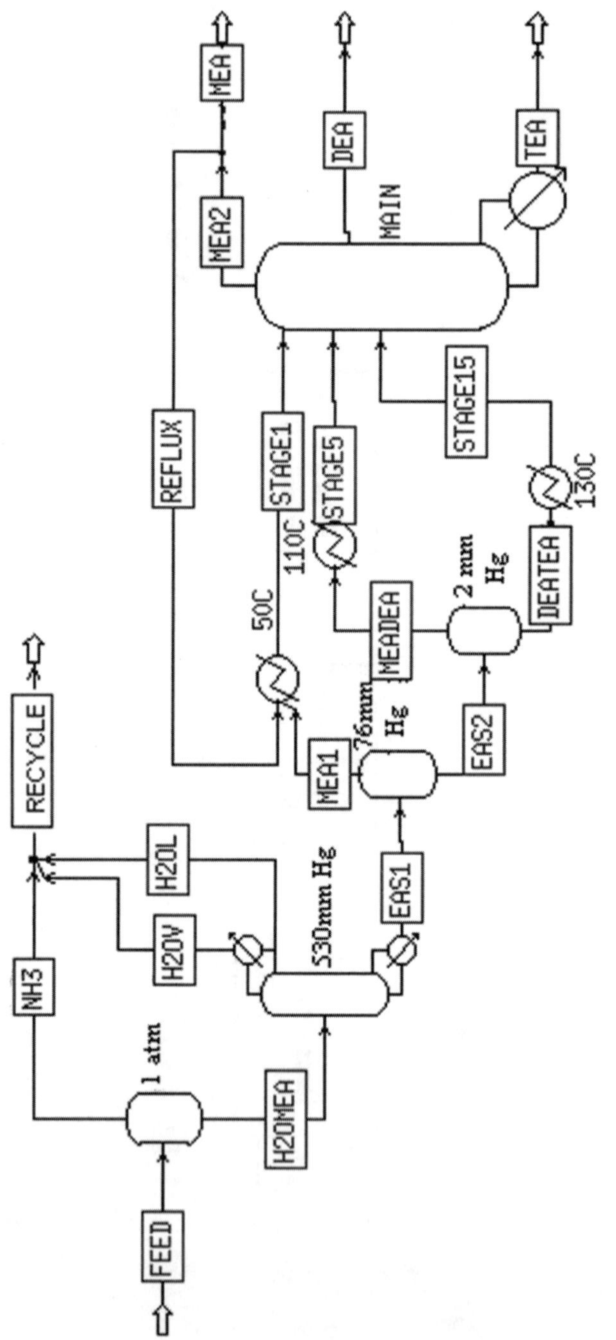

Figure 15.2. Separation Process for Dischagre Stream from Ethanolamines Reactor

15.2. PRODUCTION

All EOA plants make a mixture of products. The most important is MEA, which accounts for about half of total production. It is followed by DEA with 30% to 35% of production. As of April, 2001 INEOS plc is the largest global producer followed by BASF and Huntsman. EOA demand is growing between 4% and 6% annually with MEA and DEA growth rates substantially exceeding that for TEA. In 2001 EOA supply exceeds demand, and the oversupply situation will continue through 2002 with startups of more new plants. In 2000 European operating rates were at 86% of capacity. The European operating rates were expected to fall to 75% in 2003, but then recover to 85% by 2005. About 50% of EOA capacity is in the Americas, more than 30% is in Europe and the balance is in the Pacific Rim and the Middle East. Table 15.2 summarizes world consumption and growth rates for EOA[38, 120].

Table 15.2. World Consumption of Ethanolamines and Growth Rates[120]

	Consumption (thousands of tonnes)			
Region	1995	2000	2002	2005
Americas	279	358	388	424
Western Europe	260	329	377	417
Asia	99	157	196	297
Middle East / Africa	27	60	30	36
Totals	665	904	988	1,105
		Growth Rates (%)		
Americas	-	5.1	-	3.1
Western Europe	-	4.8	-	3.2
Asia	-	9.7	-	4.6
Other	-	17.3	-	8.4
Totals	-	6.3	-	3.8

The historical prices (1995 to 2000) for different ethanolamines are listed below[249]:

- MEA: High = \$0.59/lb, Low = \$0.56/lb
- DEA: High = \$0.61/lb, Low = \$0.57/lb
- TEA: High = \$0.67/lb, Low = \$0.67/lb

15.3. USES

EOA production is consumed in the following areas: Detergents (32%), Ethyleneamines (20%), Gas Purification (17%), Metal Cleaning (10%), Herbicide Intermediates (10%), Textiles (5%) and Miscellaneous (6%)[249].

At 140 to 160°C, the fatty acid can react 1) with the amino group on the EOA to form a fatty acid ethanolamide, 2) with the hydroxyl group on the EOA to give a fatty acid aminoethyl ester or 3) with both groups to form a di-fatty acid ethanolamide ester[46].

MEA is used as an intermediate for cosmetics, surface-active agents, emulsifiers and plasticizing agents; as a gas-scrubbing agent for absorption and removal of H_2S and CO_2 from refinery and natural gas streams; and in CO_2 and ammonia manufacturing[123].

DEA is used as an intermediate for cosmetics; as surface-active agents in household detergents and textile specialties; in insecticides and herbicides; in petroleum demulsifiers; as a gas scrubber in refinery and natural gas operations; in waxes, polishes and coatings emulsifiers; in soluble oils; and as corrosion inhibitors[123].

TEA is used as an intermediate for surface-active agents used in textile specialties, waxes and polishes, herbicides, petroleum demulsifiers, toilet goods, cement additives, cutting oils, and photograph-film developers; as corrosion inhibitors; as a dispersant for dyes, casein, shellac, and rubber latex; as sequestering agents; and as a rubber chemicals intermediate[123].

15.4. SECONDARY PRODUCTS OF ETHANOLAMINE

15.4.1. Morpholine

Morpholine is a solvent and intermediate for optical brighteners and rubber chemicals. It is also a corrosion inhibitor, especially in steam boiler systems. Some other uses include:

- An emulsifier for cosmetics, rubless waxes, and polishes,
- A separating agent for volatile amines,
- An intermediate for textile lubricants, softening agents, adjuvants in medical applications, whitening agents, sizing emulsifiers, rubber vulcanization accelerators, antioxidants, surface-active agents, plasticizers, viscosity improvers, insecticides, fumigants, herbicides, dyes, and catalysts.

In one process morpholine is obtained from DEA by dehydration with 70% H_2SO_4 to close the ring:

$HN(C_2H_4OH)_2 \xrightarrow[-\ H_2O]{[H_2SO_4]}$  or C_4H_9O (15.2)

In the USA a newer production route has eclipsed the cyclization of diethanolamine. In this method, diethylene glycol, ammonia and a little hydrogen react at 150°C to 400°C and 30 to 400 bar over a catalyst containing Ni, Cu, Cr or Co to give morpholine. The process is operated by Huntsman Chemical Company in Port Neches, Texas, but more details are not known[46].

Morpholine is a colorless, oily, hygroscopic, volatile liquid with a characteristic amine odor. It is heavier than air and, as a result, the vapor can travel a significant distance to a source of ignition and "flash back".

Morpholine can undergo a diversity of chemical reactions. It is an amino ether. The ether function of the molecule is typically inert and most of the reactions involve the secondary amine group.

Morpholine is distributed as an anhydrous liquid and as 40% and 88% solutions with water. It is marketed as a product with approximately 99% purity.

The physical properties of morpholine are shown in Table 15.3[248].

Table 15.3. Physical Properties of Morpholine

Chemical Formula:	C_4H_9NO
Molecular Weight:	87.1
Boiling point at 101.3 kPa	129°C
Melting point	-5°C
Relative density (water=1)	1.0
Solubility in water	miscible
Vapor pressure, kPa at 20°C	1.06
Relative vapor density (air=1)	3.00
Relative density of the vapor / air mixture at 20°C (air=1)	1.01
Flash point (open cup)	38-43°C
Autoignition temperature	310°C
Explosive limits, vol% in air	1.8-15.2
Octanol/water partition coefficient as log P_{ow}	-0.86 (at pH 5)

Morpholine is a colorless hygroscopic liquid, with a characteristic odor; It decomposes on heating and produces nitrous oxides. It is a base that reacts violently with acids, and it is corrosive to aluminum, zinc and copper. Morpholine reacts with oxidants and nitrites. In addition, the carcinogenic N-nitrosomorpholine can be formed from nitrosation of morpholine[248].

15.4.2. Ethylenimine

Ethylenimine can be made from MEA. BASF and Hoechst developed an industrial process that has two steps.

$$H_2NC_2H_4OH + H_2SO_4 \rightarrow H_2NC_2H_4OSO_3H + H_2O \qquad (15.3)$$

$$H_2NC_2H_4O\,SO_3H + 2\,NaOH \rightarrow \begin{array}{c} H_2C\text{—}CH_2 \\ \backslash\ \ / \\ N \\ | \\ H \end{array} + Na_2SO_4 + 2\,H_2O \qquad (15.4)$$

If the imine formation is conducted in a flow tube, residence times of 4 to 10 seconds can be used to suppress secondary reactions such as the polymerization of ethylenimine. The imine selectivity then increases to 80% to 85% (based on MEA)[46].

Dow makes ethylenimine by reacting 1,2-dichloroethane with ammonia in the presence of CaO at about 100°C:

$$ClC_2H_4Cl + CaO + NH_3 \rightarrow \begin{array}{c} H_2C\text{—}CH_2 \\ \backslash\ \ / \\ N \\ | \\ H \end{array} + CaCl_2 + H_2O \qquad (15.5)$$

Most ethylenimine is converted into polyethylenimine, a processing aid in the paper industry. It also serves as a reaction intermediate when ureas are formed with isocyanates[46].

16

ETHYLENEDIAMINE AND CHELATING AGENTS

16.1. ETHYLENEDIAMINE

Commercial ethylenediamine (EDA) is produced by two different routes:

- From 1,2-dichloroethane (EDC) and ammonia by Cl/NH_2 substitution.
- From monomethylamine (MEA) and ammonia by OH/NH_2 substitution.

In the first process EDC reacts with an excess of aqueous ammonia at about 180°C and 10 to 20 atmospheres. Diethylenetriamine and triethylenetetramine are also formed in secondary reactions of the ethylenediamine with EDC. Caustic is added to neutralize the crude product and free the ethylenediamine, the higher amines and the ammonia from the amine hydrochloride and ammonium chloride. The chloride salts are a waste that must be properly handled. The crude product is refined by fractionation. The unreacted ammonia is recycled, and the amines are purified. The yield of EDA is about 60%[46].

Low reaction temperature and low pressure are more selective for EDA, but conversion is low. High temperature and high pressure increase conversions, but with less selectivity for EDA. The reaction is:

$$ClC_2H_4Cl + 2\,NH_3 \rightarrow NH_2C_2H_4\,NH_2 + \text{Higher Amines} + NH_4Cl + \text{Amine Chlorides} \quad (16.1)$$

In 1969 BASF developed a process to make EDA by the reaction of monoethanolamine (MEA) and ammonia as shown in Eq. (16.2):

$$NH_2C_2H_4OH + NH_3 \xrightarrow[H_2]{Ni} NH_2C_2H_4NH_2 + H_2O \qquad (16.2)$$

This gas-phase reaction is run at less than 300°C and under 250 bar in the presence of hydrogen and nickel catalysts with various promoters such as Co, Fe, and Re. Excess ammonia determines the selectivity, and secondary reactions can give higher amines such as piperazine. One of the key advantages of this process is the absence of the chloride by-products[46].

The properties of EDA and some of the higher amines are given in Table 16.1.

Table 16.1. Properties of Ethylenediamine and Higher Amines

	EDA[a]	DETA[b]	TETA[c]	TEPA[d]	PEHA[e]
Molecular Weight	60.11	103.17	146.24	189.31	232.36
Melting point, °C	11	- 39	-35	-30	-26
Boiling point, °C	117	207	277	decomp @ 340°C	
Density, lb/gal	7.67	7.96	8.19	8.33	8.39
Flash pt. (O.C.), °C	38	102	143	185	185
Specific Gravity	0.898				
pH	11.9 (25% solution in water)				
Vapor Density	2.07				
Nitrogen Content, wt %.	46.6				
Viscosity, mPa-s	2.0 (at 20°C)				
Autoignition Temp, °C	385				
Physical State	Colorless liquid with amine-like odor				

[a] EDA: Ethylenediamine
[b] DETA: Diethylenetriamine
[c] TETA: Triethylenetetramine
[d] TEPA: Tetraethylene Pentamine
[e] PEHA: Penta-Ethylenehexamine

16.1.1. Ethylenediamine Production

In 1999 the EDA capacity in the USA was 390 million pounds per year of ethyleneamines. Demand was expected to grow as follows:

1998: 310 million pounds per year
1999: 320 million pounds per year
2003: 370 million pounds per year.

These estimates include exports that average 70 million to 80 million pounds per year and imports that total about 10 million to 12 million pounds per year. From 1989 to 1998 growth averaged 4% per year, but it is expected to slow to 3% per year through 2003[124]. Prices are kept confidential because the USA has very few major producers. One industry source placed the 2001 US EDA price at $0.80 to $0.85 per pound[118].

16.1.2. Ethylenediamine Uses

Ethylenediamines are used in the following products: Lube Oil & Fuel Additives (20%), Chelating Agents (16%), Wet-Strength Resins (15%), Epoxy Curing Agents and Polyamides (13%), Surfactants (10%), Oil Field Chemicals (7%), and Miscellaneous (19%). EDA is used in the production of TAED (tetra-acetylethylenediamine) and EDTA (ethylenediamine tetra-acetic acid) as well as in fungicides. DETA is mainly used as an epoxy curing agent and in the production of wet strength resins for paper.

16.2 CHELATING AGENTS

A chelating agent is a molecule that can form several coordinate bonds with a single metal ion. That is, a chelating agent is a polydentate ligand. It can be either an organic or inorganic chemical. The most common and most widely used chelating agents are those that coordinate to metal ions through oxygen or nitrogen donor atoms, or through both. Chelating agents that coordinate through sulfur in the form of -SH (thiol or mercapto) groups are not as common in commercial applications, but they perform a significant role in complexing metal ions in biological systems. Three widely used chelating agents are ethylenediamine, ethylenediaminetetraacetic acid and dimercaprol[251].

Chelating agents can coordinate with metal ions at a minimum of two sites. Typically this bidentate coordination solubilizes or otherwise inactivates these metal ions. The effect is to counter any deleterious actions these metals might have on the system in which the chelates are used. The chelating agents form ring structures that incorporate the metal ion within the molecule[250].

The choice of a particular chelant is dependent on several factors, including pH range of the using system, which metal ions are to be controlled, and overall economics of the system. The active portion of the chelant is the anionic organic moiety. Large variations in the structure of chelating agents make possible the control of a wide range of metal ion concentrations and redox tendencies.

16.2.1. Manufacturing Processes for Chelating Agents

All of the processes used to make aminopolycarboxylic acid chelating agents are similar. They can be illustrated by the route to the tetrasodium salt of

EDTA. The manufacturing processes are all variations of the conversion of ethylenediamine to a cyanomethyl ($-CH_2CN$) derivative and hydrolysis of this intermediate. Two processes go directly from ethylenediamine to the final product in a single step. They differ in that one used sodium cyanide (see Eq. 16.3) while the other used hydrogen cyanide and caustic soda (see Eq. 16.4)[250].

$$4\,NaCN + 4\,HCHO + H_2NCH_2CH_2NH_2 + 4\,H_2O \rightarrow (NaOOCCH_2)_2NCH_2CH_2N(CH_2COONa)_2 + 4\,NH_3 \quad (16.3)$$

$$4\,NaOH + 4\,HCHO + H_2NCH_2CH_2NH_2 + 4\,HCN \rightarrow (NaOOCCH_2)_2NCH_2CH_2N(CH_2COONa)_2 + 4\,NH_3 \quad (16.4)$$

Several by-products are formed in these processes. They include ammonia, sodium salts of NTA (nitrilotriacetic acid), iminodiacetic acid (IDA), glycolic acid and some partially substituted EDTA compounds.

A third, two-step synthesis route produces the cyanomethyl derivative in a distinct step. This process uses formaldehyde, hydrogen cyanide and ethylenediamine[250].

$$4\,HCHO + H_2NCH_2CH_2NH_2 + 4\,HCN \rightarrow (NCCH_2)_2NCH_2CH_2N(CH_2CN)_2 + 4\,H_2O \quad (16.5)$$

The resulting cyanomethyl compound is insoluble in water and is therefore easily separated from excess reactants and by-products that remain in the water. The ammonia by-product is removed by distillation. The purified cyanomethyl compound is then hydrolyzed with caustic soda in a separate step to make pure tetrasodium EDTA[250].

$$4\,NaOH + (NCCH_2)_2NCH_2CH_2N(CH_2CN)_2 + 4\,H_2O \rightarrow (NaOOCCH_2)_2NCH_2CH_2N(CH_2COONa)_2 + 4\,NH_3 \quad (16.6)$$

Other aminopolycarboxylic acids are produced by substituting organic polyamines (e.g., diethylenetriamine) or ammonia for the ethylenediamine. NTA is made via the two-step process using ammonia (instead of ethylenediamine), formaldehyde and HCN as starting materials. The resulting cyanomethyl intermediate is subsequently hydrolyzed in a separate step to the free acid or sodium salt[250].

A type of hydroxycarboxylic acid that is used as a chelating agent is gluconic acid. It is produced on a commercial scale by bacterial fermentative oxidation of glucose. The reaction can be represented by Eq. 16.7:

$$CH_2OH(CHOH)_4CHO + {}^1/_2\,O_2 \xrightarrow[\text{oxidase}]{\text{glucose}} CH_2OH(CHOH)_4CHOOH \quad (16.7)$$

Sodium gluconate is produced by the neutralization of gluconic acid with NaOH, or it can be fermented directly from glucose. Sodium glucoheptonate is made directly from the reaction of NaCN with glucose[250].

16.2.2. Production and Uses of Chelating Agents

The consumption of aminopolycarboxylic chelating agents (excluding NTA) in the US, Western Europe and Japan during 1998 was approximately 343 million pounds (156 thousand tonnes). These chelating agents are used as components or process chemicals in five main areas - cleaning compounds, pulp and paper manufacture, water treatment, photography and agriculture. These five areas account for about 80% of the worldwide consumption. Chelates (combinations of chelating agents and metal ions) are used in pharmaceuticals, agriculture and feed additives[250].

Overall market growth is mature and heavily dependent on the economy and environmental regulations. In the US, growth from 1998 to 2003 was expected to be about 2% per year[250].

16.2.2.1. Ethylenediaminetetraacetic Acid (EDTA)

EDTA is a versatile chelating agent. It can form four or six bonds with a metal ion. It forms chelates with both transition-metal ions and main-group ions. It is a colorless crystalline material that decomposes when heated above 240°C. It is only slightly soluble in water. However its sodium salts are quite soluble in water. EDTA is frequently used in soaps and detergents because it forms complexes with calcium and magnesium ions. These ions are in hard water and interfere with the cleaning action of soaps and detergents. The EDTA binds to them, sequestering them and preventing their interference[251].

EDTA is also used as a stabilizing agent in the food industry. EDTA deactivates enzymes that cause food to spoil by removing metal ions from them and forming stable chelates with them. It promotes color retention in dried bananas, beans, chick peas, canned clams, pecan pie filling, frozen potatoes and canned shrimp. It improves flavor retention in canned carbonated beverages, salad dressings, mayonnaise, margarine and sauces. EDTA salts are used in foods at levels ranging from 33 to 800 ppm[251].

EDTA is used in the separation of the rare earth elements. EDTA complexes have a stability that varies slightly from the rare earth elements. This slight variation allows EDTA to effectively separate rare-earth ions. EDTA is used as an anti-coagulant in blood. It prevents coagulation by sequestering the calcium ions required for clotting[251].

The sodium salt of EDTA is the primary product of the aminopolycarboxylic chelants. In 1998 it accounted for over 75% of the total chelant demand in the US, Western Europe and Japan[250].

16.2.2.2. Diethylenetriaminepentaacetic Acid and Salts (DTPA)

DTPA tends to be a weaker chelator than EDTA. However it is very stable and has been found to be especially useful in peroxide bleach systems. DTPA works in a pH range similar to that of EDTA for most ions, but tends to have a slightly higher active range in controlling ferrous ions[250].

16.2.2.3. Hydroxyethylethylenediaminetriacetic Acid and Salts (HEDTA)

HEDTA was developed specifically for chelating ferric ions in alkaline systems. It chelates with maximum effectiveness in a pH range of 6 to 12. This prevents precipitation of iron compounds (e.g., ferric hydroxide and oxide) in this alkaline range. It is not as effective as EDTA for calcium and magnesium ions. It also tends to be less stable[250].

16.2.2.4. Nitrilotriacetic Acid (NTA)

The trisodium salt of nitrilotriacetic acid (NTA) adds another 74 million pounds (34 thousand tonnes) to the demand for aminopolycarboxylic chelants. Canada had been a major user of NTA. When Proctor & Gamble stopped using NTA in Canadian detergents in 1996, the North American market became much smaller[250].

NTA is an efficient, general-purpose chelating agent. It has the greatest chelating action on a weight-for-weight basis of all aminopolycarboxylates. However, it is the least stable of the aminopolycarboxylics and tends to work in a narrower pH range[250].

Generally, sequestering agents produce chelates with a particular metal with stabilities decreasing in the following sequence[250]:

$$DTPA > EDTA > HEDTA > NTA$$

The sequestering power (i.e., the quantity of metal chelated per unit weight of chelating agent) varies in the inverse order. This chemical behavior is an important reason for the extensive use of NTA for cleaning applications. These applications need a high sequestering power but a low stability toward oxidation is acceptable. In oxidative bleaching (e.g., in the pulp and paper industry), where maximum stability against oxidation is imperative, DTPA is the chelating agent of choice[250].

16.2.2.5. Hydroxycarboxylic Acids

Gluconic acid is a good general chelating agent for di- and trivalent metal ions (e.g., iron, copper, calcium, zinc and aluminum). It is particularly effective as a chelating agent in alkaline and free caustic soda solutions. Glucoheptonic acid, particularly its sodium salt, is a good general chelating agent for polyvalent metals. Sodium gluconate and sodium glucoheptonate are generally interchangeable in many applications. It is the ability to chelate ions effectively in strongly alkaline solutions (which citrates and EDTA do not do as well) that makes gluconates and glucoheptonates important in industrial uses. These chelants are also easily biodegradable (98% within two days)[250].

17

HEXAMETHYLENETETRAMINE (HEXAMINE)

Hexamine is also known as hexamethylenetetramine, aminoform, crystamine, methenamine or formin. It was first prepared in 1859 by Butlerov of Russia. It is a white, crystalline powder with a slight amine odor. It is soluble in water, alcohol, and chloroform, but it is insoluble in ether. However the aqueous solutions exhibit inverse solubility, i.e., less hexamine dissolves as the temperature increases. The hydrate, $(CH_2)_6N_4 \cdot 6H_2O$ can be crystallized from the aqueous solution at temperatures below 14°C. Some additional properties are listed in Table 17.1.

Table 17.1. Properties of Hexamine/Hexamethylenetetramine

Molecular Weight	140.19
Sublimation Temperature	285 – 295°C
Flash Point	250°C
Density of solid @ 20°C	1.33 g/cm^3
Bulk Density	700 to 800 g/L
Particle Size	700 micron - Maximum
Specific Heat	36.5 cal/°C
Heat of Formation @ 25°C	28.8 kcal/mol
Heat of Combustion @ 25°C	1,003 kcal/mol
Solubility in Water	
20°C	874 g/L
25°C	867 g/L
60°C	844 g/L
pH of 10% Aqueous Solution	8 to 9
Vapor Pressure @ 20°C	0.0035 mbar

17.1. HEXAMINE PROCESSES

Hexamine is made by the liquid phase reaction of ammonia and formaldehyde:

$$4\ NH_3 + 6\ HCHO \rightarrow (CH_2)_6N_4 + H_2O \quad \Delta H = -55\ \text{kcal/mol Hexamine} \quad (17.1)$$

The ammonia and formaldehyde are added simultaneously to the reaction vessel to produce hexamine in a water solution. Anhydrous ammonia is used to reduce the amount of water that enters the reactor. The reaction is controlled at a pH between 7 and 8 because formic acid and carbon dioxide are formed under acidic conditions[125].

Adding aqueous formaldehyde solution to a reaction kettle can also make Hexamine. Then ammonia gas (in a 3:2 formaldehyde/ammonia mole ratio) is added. During the addition of the reactants, the temperature is maintained at about 20 to 30°C. The reaction mixture is then fed to a vacuum evaporator where it is maintained at a temperature between 30 and 50°C and at a pH of 7 to 8[252].

The ammonia-formaldehyde reaction is highly exothermic. Therefore the reactor is operated under a vacuum so that the heat of reaction can be used to evaporate some of the water of reaction. This helps concentrate and crystallize the hexamine in the water solution. If dry solids are desired, the hexamine-water solution is pumped through a crystals receiver and then separated in a centrifuge. The crystals pass through a rotary dryer and are packaged for shipment. In some cases a liquid solution may be shipped to customers[125].

The temperature in the final drying step is kept below 50°C to avoid yield losses due to decomposition. The overall yield is in the range of 95% to 97% based on formaldehyde[125, 252].

17.2. PRODUCTION

Hexamine production is listed in Table 17.2[252].

Table 17.2. Hexamine Supply and Demand

	Million of Pounds						
	1995	1996	1997	1998	1999	2000	2001[b]
Capacity	69	69	69	61	61	61	61
Total Prod.	48	46	43	38	38	35	32
Imports	2	2	2	2	3	4	5
Demand[a]	50	48	45	40	41	39	37

[a] Demand equals production plus imports.

[b] Forecast

In 2000 Mexico accounted for 63% of the US imports. In 2001, 58% of imported material came from Germany while 31% came from Mexico. In 2001 two producers, Borden Chemical and Wright Chemical, served the US market. As the above table shows, capacity is much greater than demand[252].

No long-term demand growth is expected unless a significant new end use is found. There has been growth in some smaller volume applications, (such as commercial explosives and steel pickling solutions). However the growth in these markets will not be great enough to offset the slow decline in the use of phenolic resins that contain hexamine[252].

The price of hexamine can generally be expected to track the price of methanol, which is the raw material for formaldehyde. As a rule of thumb, the realized cost of hexamine moves 1 cent for every 4 cent move in the methanol price. The price also moves one-quarter of a cent for every $5/ton move in ammonia costs. Between 1995 and 2000, the price of imported hexamine dropped from $0.54/lb to $0.34/lb[252].

17.3. USES

Hexamine is produced as a granular and free-flowing powder as well as a 42.5% solution. The solution is shipped in tank trucks, railcars and drums. Solid forms are packed in bags, fiber drums and super sacks. Hexamine is sensitive to moisture. Therefore it should be stored in an atmosphere with a relative humidity below 60%[126].

Hexamine is used in the following areas[127]:

- Rubber Industry: Vulcanization accelerator and rubber blowing agent
- Explosives Industry: Cyclonite (RDX), octogen (HMX), hexamethylene triperoxide amine (HMTA)
- Synthetic Resin Industry: Liquid resin stabilizer, molding powder, carbohydrate resins, vulcanization of vinyl resins and copolymers, aniline shellac resins
- Pharmaceutical Industry: Disinfectant (formin, urotropin, crystazol, helmitol), urinary antiseptic
- Photographic Industry: Stabilizer for developers
- Organic Synthesis Industry: Additives in deodorizing powder, absorption of phosgene gas, preservation of fresh products
- Metal Industry: Inhibitor against acids and hydrogen sulphide
- Leather Industry: Conservation of furs and skins
- Paper Industry: Surface treatment during manufacture of water repellent papers and cardboards
- Lubricant Industry: Stabilizer for greases and oils
- Fertilizer Industry: Anticaking agent for prilled urea

- Other: Dyeing and artificial aging of wood, preservative for cosmetics, treatment of cholera in chickens

An estimate five (5) million pounds per year of hexamine are consumed to make commercial explosives. Hexamine demand from the phenolic resin segment has declined due to increased competition from formaldehyde-free resins and other resins that offer performance advantages.

The production of nitrilotriacetic acid (NTA) may be the largest application for hexamine (40 to 60 million pounds per year). However, since hexamine for this use is manufactured as a captive intermediate (in solution) this segment is usually not included in the production statistics.

The hexamine uses are shown in Table 17.3[252].

Table 17.3. Hexamine Uses

Application	%
Phenolic Resin Curing Agent	60%
Explosives	13%
Pharmaceuticals & Bactericides	10%
Pickling Solutions	4%
Miscellaneous	13%

(Reproduced by permission of Mannsville Chemical Products Corp.)

18

HYDRAZINE

Hydrazine, NH_2-NH_2, is the simplest diamine. Lobry de Bruyn first prepared anhydrous hydrazine in 1894. In 1953 Arch Hydrazine (formerly Olin Chemical) pioneered U.S. production of hydrazine to fuel the first Titan rocket.

Hydrazine is a clear, hygroscopic fuming liquid with the odor of ammonia. It is a mild base that is miscible in polar solvents such as water, ammonia, amines and alcohols. Additional properties are shown in Table 18.1[135].

TABLE 18.1. Physical Properties of Hydrazine

Molecular weight	
Anhydrous	32
Hydrate	50
Boiling point (°C)	113.5
Melting point (°C)	1.4
Specific gravity (g/ml)	
0°C	1.025
15°C	1.014
25°C	1.004
50°C	0.982
Critical temperature (°C)	380
Critical pressure (atm)	145
Vapor pressure (mm Hg)	
25°C	14
31°C	20
36°C	100

TABLE 18.1. Physical Properties of Hydrazine - Continued

Viscosity (cp)	
5°C	1.2
25°C	0.9
Heat of vaporization (kcal/mol)	9.6
Heat of solution (kcal/mol) @ 25°C	-3.9
Heat capacity (J/mol °K) @ 25°C	98.87
Heat of combustion (kcal/mol)	- 146.6
Heat of formation (kcal/mole)	
Liquid	12
Gas	23
Flash point (°C)	52
Explosive limits in air by vol. (%)	4.7 to 100
Azeotrope Boiling Point (68% Hydrazine) (°C)	120.5

	Hydrazine Concentration (wt %)[132]					
	100	64	51.2	35.2	22.4	15.4
Melting Point, °C	2.0	-51.7	-59.8	-64.6	-26	-14
Boiling Point, °C	113.5	120.5	117.2	108	107	103
Density (25°C), g/ml	1.0045	1.0320	1.0281	1.0209	1.0132	1.0083
Viscosity (20°C), μPa – s	0.974	1.5	1.44	1.10	1.08	1.04
pH		12.75	12.10			10.5

18.1. PROCESSES

Commercial production of hydrazine from its elements has not been successful. However three processes are available for the commercial production of hydrazine: 1) The Raschig Process, 2) The Raschig/Olin Process, 3) The Hoffmann (urea) Process, 4) Bayer Ketazine Process, and 5) the Peroxide process from Produits Chimiques Ugine Kuhlmann (of France).

18.1.1. Raschig Process

The Raschig process was discovered in 1907 and then modified into the Olin process. The chemical reactions take place in the liquid phase and involve three steps:

$$NaOH + Cl_2 \rightarrow NaOCl + HCl \qquad (18.1)$$

$$NH_3 + NaOCl \rightarrow NH_2Cl + NaOH \qquad (18.2)$$

$$NH_2Cl + NH_3 + NaOH \rightarrow NH_2\text{-}NH_2 \bullet H_2O + NaCl \qquad (18.3)$$

Hydrazine is produced in the hydrated form with one mole of water added. Although a significant fraction of hydrazine is used as the hydrate, numerous applications (such as rocket propulsion) require anhydrous hydrazine. Because of the azeotrope at 68% hydrazine, reactive distillation or extractive distillation must be used to produce pure hydrazine.

In Eq. (18.1), sodium hypochlorite is produced by feeding chlorine into a 30% aqueous caustic solution in a circulating reactor/cooler system. To avoid sodium chlorate formation, the reaction temperature is kept below 30°C and NaOH concentration is kept below 1 g/liter. Typical reaction temperature is 5 °C[132].

In Eq. (18.2), the reaction rate for chloramine (NH_2Cl) formation is rapid relative to the formation of hydrazine in Eq. (18.3). In this step dilute ammonia solution (5% to 15%) is added to the NaOCl at a ratio of 3:1. Use of NaOCl is estimated to be 3.5 pounds per pound of hydrazine.

Because the decomposition rates are relatively insensitive to temperature, Eq. (18.3) is operated at 130°C to 150°C and 3.0 MPa[132] to speed up the rate-determining step. Excess ammonia, at a ratio of 40:1, is used to minimize the hydrazine-chlorine decomposition. Synthesis efficiency favors a dilute system although the increase in operating cost due to the low concentration may ultimately become inhibiting. The Raschig process is shown in Figure 18.1[132].

At the reactor outlet, the reaction liquor contains 1% hydrazine hydrate and 4% NaCl. The pressure is reduced to atmospheric in a battery of evaporators. Ammonia is condensed, concentrated and recycled. The liquor from the bottom of the stripping columns is freed from salt in a conventional, forced-circulation salting evaporator. The distillate is then concentrated to 100% hydrazine hydrate[132].

The Raschig process can also be used to react amines with chloramine to make monosubstituted or unsymmetrical disubstituted hydrazines.

$$NH_2Cl + RNH_2 \rightarrow RNHNH_2 \tag{18.4}$$

$$NH_2Cl + R_2NH \rightarrow R_2NNH_2 \tag{18.5}$$

18.1.2. Raschig/Olin Process

The Raschig/Olin process (see Figure 18.2) is used to make anhydrous hydrazine. In this process the NaOCl production occurs at a low temperature to prevent decomposition and chlorate formation. The excess NaOH is kept to a low level[132].

The NaOCl solution is mixed with a threefold excess of ammonia at 5°C to form chloramine, which is then rapidly added to a 30-fold molar excess of anhydrous ammonia under pressure (20-30 MPa) and heated to 130°C[132].

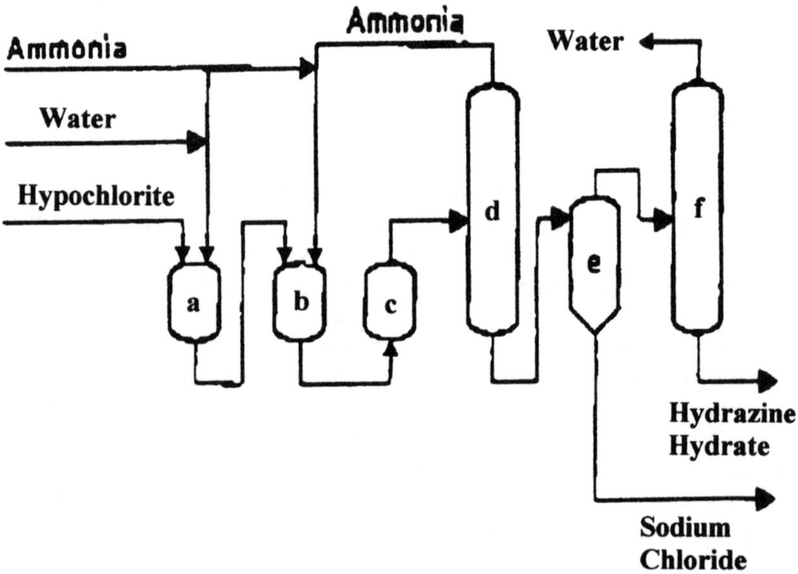

a. Chloramine Reactor
b. & c. Hydrazine Reactors
d. Ammonia Evaporator
e. Hydrazine-Sodium Chloride Separator
f. Hydrazine Hydrate Concentrator

Figure 18.1. Raschig Process for Hydrazine Production. (Reproduced by permission of Wiley-VCH)

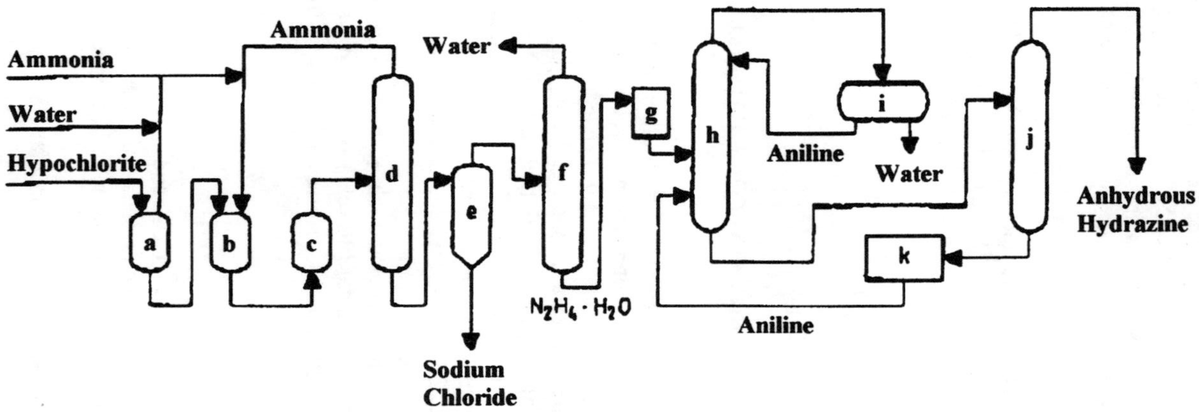

a. Chloramine Reactor
b. & c. Hydrazine Reactors
d. Ammonia Evaporator
e. Hydrazine-Sodium Chloride Separator
f. Hydrazine Hydrate Concentration
g. Hydrazine Hydrate Storage
h. Hydrazine Hydrate Dehydration
i. Aniline-Water Decantation
j. Anhydrous Hydrazine Distillation
k. Aniline Storage

Figure 18.2. Raschig / Olin Process for Hydrazine Production.
(Reproduced by permission of Wiley-VCH)

The refining area has five steps: 1) ammonia removal, 2) NaCl concentration and removal, 3) hydrazine-water distillation to reach azeotropic concentration, 4) extractive distillation with aniline to break the azeotrope and 5) distillation to separate hydrazine from aniline. Sometimes 50% caustic replaces aniline in the extractive distillation. If ultra pure hydrazine is needed, freeze crystallization is used to remove the supernatant fluid. Assays between 99.5% and 99.99% have been achieved.

The overall yield based on chlorine is 65%. The combined yield for Eqs. (18.1) and (18.2) is about 95%, whereas the yield for Eq. (18.3) is 70%. Japanese patent SHO-62-83308 claims to increase hydrazine yield by almost 10% by high frequency heating immediately after the reactants are mixed in Eq. (18.2).

18.1.3. Hoffmann (Urea) Process

The overall reaction for the Hoffmann process is:

$$\underset{\text{Urea}}{CO(NH_2)_2} + \underset{\text{Hypochlorite}}{NaOCl} + 2\,NaOH \rightarrow \underset{\text{Hydrazine}}{NH_2\text{-}NH_2} + NaCl + Na_2CO_3 + H_2O \quad (18.6)$$

As in the Raschig process, aqueous caustic reacts with chlorine to make sodium hypochlorite solution. The urea solution is prepared by dissolving urea in water with the addition of steam to provide the heat needed for the endothermic dissolution. The temperature is kept at about 5°C for 43 percent urea solution. Glue is added at a ratio of 0.5g/liter of solution to inhibit side reactions. The urea and hypochlorite solutions are added to the hydrazine reactor at a ratio of 1:4, and the reaction temperature is allowed to rise to 100°C. The crude product contains approximately 35 g N_2H_4/liter and can be refined in the same steps as used for the Raschig process.

This process is not being operated in 2001 although it has operated commercially in the past. Compared with the standard Raschig process, it was the most economical method for low production levels. However rapid growth in plant size made it obsolete[132].

18.1.4. Peroxide Process

This process (see Figure 18.3) was invented by PCUK and is operated by AtoFina in France. The reaction is carried out in the presence of methyl ethyl ketone (MEK) at atmospheric pressure and 50°C. The molar ratio in the feed of hydrogen peroxide (H_2O_2) : MEK : NH_3 is 1:2:4.

The reaction pathway involves the formation of an intermediate that is able to oxidize ammonia to a hydrazine derivative. The methyl ethyl ketazine is insoluble in the reaction mixture and is separated by decantation and is then purified by distillation. The purified ketazine is hydrolyzed under pressure (0.8 to 10 MPa) to give concentrated aqueous hydrazine and MEK that is recycled[132].

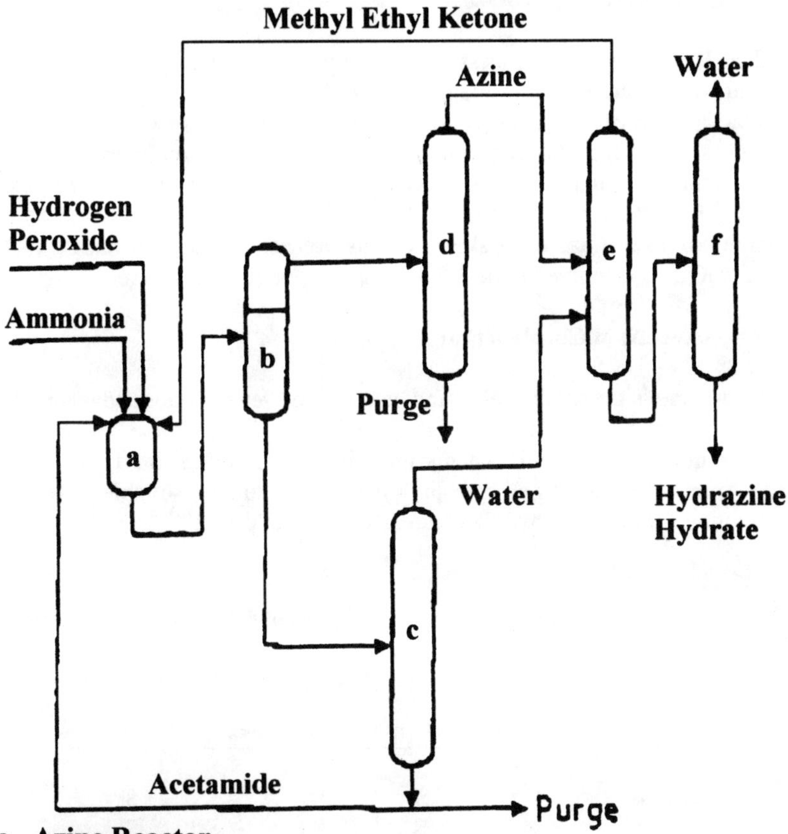

a. Azine Reactor
b. Phase Separator
c. Aqueous Phase Concentration
d. Azine Purification
e. Azine Hydrolysis
f. Hydrazine Hydrate Concentration

Figure 18.3. Peroxide Process for Hydrazine Production.[132]
(Reproduced by permission of Wiley-VCH)

The Peroxide process has many advantages compared to other processes: no salt by-product, high yields, low energy consumption, low molar excess and no aqueous effluent treatment[132].

18.1.5. Bayer Ketazine Process

The Bayer Ketazine process is based on the reaction of chloramine with ammonia in the presence of acetone at pH 12 to 14. NaOCl, acetone and a 20% aqueous solution of ammonia (at a mole ratio of 1:2:20, respectively) are fed to a reactor at 35°C and 200 kPa to make the aqueous dimethyl ketazine solution. Excess ammonia and acetone are removed in a series of columns and recycled to the reactor. The ketazine solution is distilled to make a hydrazine hydrate containing 64% hydrazine. A sketch of this process is shown in Figure 18.4[132]. Use of NaOCl is estimated to be 3.5 pounds per pound of hydrazine.

18.1.6. Materials of Construction

Hydrazine is thermally stable and storable for years without adverse effects either to the product or the storage container. However the recommended materials must be used; all systems must be clean; and an inert gas (such as nitrogen) must be maintained over the system at all times. Suitable materials of construction for handling hydrazine are given in Table 18.2[253].

TABLE 18.2. Materials of Construction for Aqueous Hydrazine Solutions[a]

	Hydrazine Concentration (wt %)			
Material	< 10	35	54.4	64
Stainless Steel	S	S	S	S
304L	S	S	S	S
347	S	S	S	S
316[b]	S	S	S	S
Cold-Rolled Steel	S	NR	NR	NR
Copper	NS	NS	NS	NS
Brass	NS	NS	NS	NS
Aluminum	NS	NS	NS	NS
PTFE[c]	S	S	S	S
Polyethylene	S	S	S	S
Polypropylene	S	S	S	S

(Reproduced by permission of John Wiley and Sons)

[a] S = generally satisfactory; NR = not recommended; NS = not suitable owing either to decomposition or to adverse effects of the solution on the material of construction.

[b] Only up to 65°C

[c] PTFE = poly(tetrafluoroethylene)

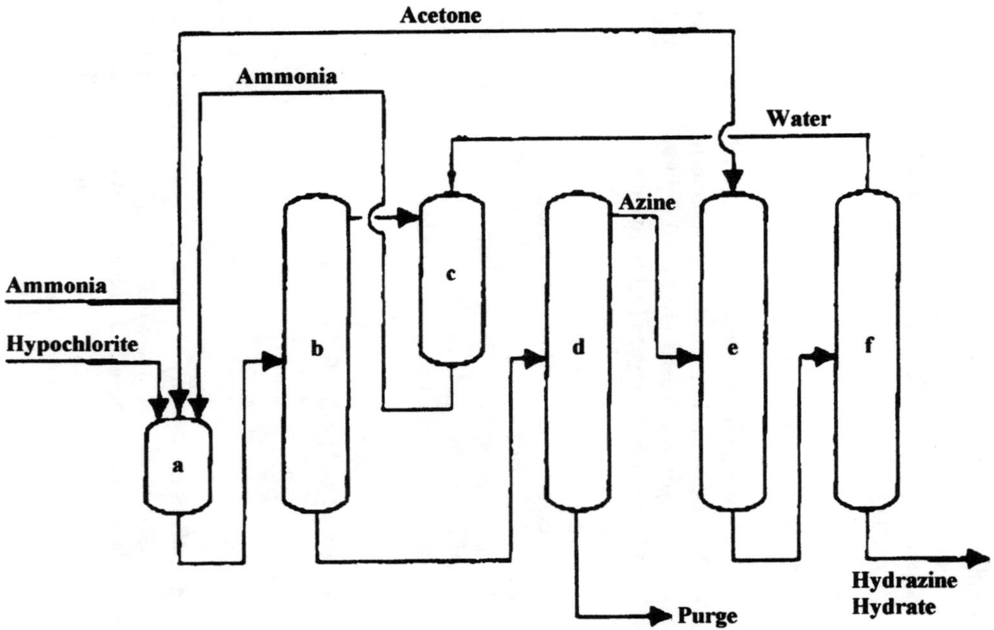

a. Azine Reactor
b. Ammonia Stripping
c. Ammonia Quenching
d. Azine Distillation
e. Azine Hydrate Concentration

Figure 18.4. Bayer Hydrazine Process.
(Reproduced by permission of Wiley-VCH)

For solutions that contain less than 10% hydrazine, cold-rolled steel is satisfactory. PVC is not recommended. Ethylene-propylene-diene monomer (EPDM) rubber, polyketones and polyphenylene sulfides are suitable for use with anhydrous hydrazine[253].

In some cases testing may be needed to identify a suitable material. For example, low concentrations of CO_2 (250 ppm) in anhydrous hydrazine accelerate the decomposition of hydrazine in stainless steel. Long-term storage tests of hydrazine propellants in 17-7 PH stainless steel and AM350 precipitation-hardened stainless steel at 50°C for about three years showed no pressure rise or hydrazine decomposition[253].

18.2. PRODUCTION

Hydrazine is marketed as anhydrous hydrazine, as the monohydrate (64% hydrazine), and as a 54.5% aqueous solution (85% hydrazine monohydrate). The largest U.S. producers of hydrazine based on a 2000 report[134] are: Arch Chemical (23 Million Pounds per Year), Bayer (17 Million Pounds per Year), and Fairmont Chemical (1 Million Pounds per Year).

Total U.S. Production averages around 36 million pounds per year. About 29 million pounds are sold commercially while the manufacturers retain the remainder for internal use. The space industry uses only 5% of all hydrazine produced in the U.S[134]. Total capacity in the western world in 1988 was estimated to be about 130 million pounds per year[132].

The Permissible Exposure Limit (PEL) for hydrazine hydrate on an 8 hour, time-weighted average (TWA) is 0.1 ppm. The ammonia-like odor of hydrazine hydrate normally can be detected at concentrations above 3 to 5 ppm, which does not provide adequate warning of potentially hazardous concentrations[135].

18.3. USES

Hydrazine is used directly as an oxygen scavenger and as an energy source. As an intermediate, it is used in the production of a number of important chemicals and polymers. Hydrazine is consumed in the following applications: Blowing Agents (33%), Pesticides (32%), Water Treatment (18%) and Miscellaneous (17%). The miscellaneous applications include pharmaceuticals, production of urethanes, lubricant additives, in anti-oxidant compounds, as a reducing agent for noble metals, in the production of mirrors, in the recovery of precious metals, as a raw material for flame-retardants, in the purification of muriatic acid and in the production of iodine compounds[128].

19

HYDROGEN CYANIDE

Hydrogen cyanide, HCN, is also known as hydrocyanic acid, prussic acid and formonitrile. Although HCN is a colorless liquid, iron cyanide is blue (prussian blue) and its presence as a slight contaminant usually gives HCN a bluish tint that led to the German name Blausäure – blue acid. Selected physical properties of hydrogen cyanide are shown in Table 19.1.

Table 19.1. Physical Properties of Hydrogen Cyanide

Molecular Formula	HC≡N
Molecular Weight	27.03
Melting Point	-13.24°C
Triple Point	-13.32 °C
Boiling Point (1 atm)	25.70 °C
Density, g/ml	
0 °C	0.7150
10 °C	0.7017
20 °C	0.6884
Specific Gravity of Aqueous Solutions	
(@18 °C compared to water @ 18 °C)	
10.04%	0.9838
20.29%	0.9578
60.23%	0.8290
Vapor Specific Gravity (@31 °C, air = 1)	0.947
Vapor Pressure, kPa	
-29.5 °C	6.697
0.0 °C	35.24
27.2 °C	107.6
Liquid Viscosity @ 20.2 °C	0.2014

Table 19.1. Physical Properties of Hydrogen Cyanide - continued

Specific Heat, J/mol	
-33.1 °C, Liquid	58.36
16.0 °C, Liquid	70.88
27.0 °C, Gas	36.03
Heat of Formation (@ 25 °C, 1 atm), ΔH_f, kJ/mol	
Gas @ 25 °C	-130.5
Liquid @ 25°C	-105.4
Heat of Fusion @ -14 °C, kJ/mol	7,100
Latent Heat, kJ/mol	23.8
Heat of Combustion, net, kJ/mol	642
Conductivity, S/cm	3.3 x 10-6
Heat of Vaporization, kJ/mol	25.2
Heat of Polymerization, kJ/mol	42.7
Flash cup, closed cup, °C	-17.8
Explosive Limits in Air at 100 kPa & 20 °C, vol%	6 to 41
Autoignition Temperature, °C	538

19.1. SAFETY

Cyanides are non-cumulative, protoplasmic poisons. That means they can be detoxified readily. Cyanide combines with those enzymes at the blood tissue interfaces that regulate oxygen transfer to the cellular tissues. Unless the cyanide is removed, death results through insufficient oxygen in the cells[129].

The warning signs of cyanide poisoning include dizziness, numbness, headache, rapid pulse, nausea, reddened skin and bloodshot eyes. More prolonged exposure can cause vomiting and labored breathing followed by unconsciousness; cessation of breathing; rapid, weak heat beat; and then death. Sever exposure by inhalation can cause immediate unconsciousness. This rapid knockdown power without an irritating odor makes hydrogen cyanide very dangerous[129].

Many people can detect hydrogen cyanide by odor or taste sensation at a concentration of 1 ppm in air while most people can detect 5 ppm. But HCN does not have an offensive odor, and a few people cannot smell it even at toxic levels. OSHA has set 4.7 ppm as the maximum, average safe exposure limit for a 15-minute period. Exposure to 20 ppm of HCN in air causes slight warning symptoms after several hours; 50 ppm causes disturbances within an hour; 100 ppm is dangerous for exposures of 30 to 60 minutes; and 300 ppm can be rapidly fatal unless prompt, effective first aid is administered. A small concentration of cyanide (0.02 to 0.04 mg/L) always exists in a person's body, and the body has a mechanism for continuous removal of small amounts of cyanide[129].

Besides toxicity, HCN presents other hazards. HCN undergoes an exothermic polymerization at a pH of 5 to 11 to form solid black compounds. This polymerization can become explosively violent, especially if it is confined. The polymerization reaction is between HCN and cyanide ions, and the presence of water and heat contribute to the onset of polymerization. Therefore stored HCN should contain less than 1.0 wt.% water; it should be kept cool; and it should be inhibited with sulfuric, phosphoric or acetic acid[129].

The presence of any contaminant or surface that depletes the acid stabilizer can cause polymerization. This is first indicated by a yellow-brown color in the HCN and is followed by the development of heat. To avoid further polymerization, the solution should be promptly re-acidified and cooled[129].

Explosively violent hydrolysis can occur if an excess of a strong acid (H_2SO_4, HNO_3 or HCl) is added to HCN. The reaction is fastest at or near stoichiometric ratios, e.g., 1 to 2 moles H_2SO_4 per mole HCN and can cause severe equipment damage if confined[129].

Although HCN is a weak acid and is normally not corrosive, it has a corrosive effect under two special conditions: 1) water solutions of HCN cause transcrystalline stress cracking of carbon steels under stress even at room temperature and in dilute solution and 2) water solutions of HCN containing H_2SO_4 as a stabilizer severely corrode steel above 40°C and stainless steels above 80°C[129].

An important environmental issue is the fate of cyanide. HCN, if spilled, evaporates quite readily. That which does not evaporate is soon decomposed or rendered nonhazardous by complexing with iron in the soil, by biological oxidation or by degradation from photolysis[129]. Several methods for treatment and disposal of cyanide compounds are described in references 129 and 133.

19.2. MANUFACTURE

Hydrogen cyanide can be manufactured from almost any form of hydrogen, carbon, and nitrogen if sufficient energy is provided. However, only six processes are presently of economic significance:

1) The Andrussow process is currently the principal HCN manufacturing process in the world,
2) The BMA process is practiced by two companies and exhibits high yield and selectivity by using a complex reaction system,
3) The Fluohmic process is of interest in locations where electricity is inexpensive,
4) The Formamide process is useful for sites with inexpensive carbon monoxide,
5) The Sohio acrylonitrile process produces HCN as a by-product, and
6) The Methanol process.

19.2.1. Andrussow Process

The Andrussow process was patented by Dr. L. Andrussow of I.G. Farben AG in Germany in 1933 (U.S. Patent 1,934,838). Its main advantages are low converter investment, low maintenance costs and high natural gas yields when the waste gas is used as a boiler fuel. The Andrussow process produces HCN by the reaction of ammonia, air and natural gas at 1,000°C to 1,200°C in the presence of a platinum/rhodium catalyst. The reaction is:

$$NH_3 + CH_4 + 1.5\,O_2 \xrightarrow[\text{Pt/Rh}]{1200^{\circ}\text{C}} HCN + 3\,H_2O \quad \Delta H = -115.2 \text{ kcal/mol HCN} \quad (19.1)$$

The reaction generally takes place at a pressure of less than two bar with a very short residence time. The converter discharge gas is rapidly quenched to less than 400°C to avoid decomposition of HCN. The reaction gases are usually quenched in a steam-generating, waste-heat boiler that is directly below the catalyst[129]. The heat of reaction is recovered in the waste-heat boiler to generate about five (5) pounds of steam per pound of HCN produced.

The precious metal catalyst is usually 90% platinum and 10% rhodium in gauze form. Platinum alone does not have adequate mechanical strength to endure extended exposure to the high reaction temperature. When platinum is alloyed with rhodium, the catalyst life can range from 4,000 hours to as much as 10,000 hours. A catalyst pack is usually replaced because it has been contaminated by minor impurities in the feed gases. Very little catalyst metal is consumed or lost during the life of a catalyst.

The reaction takes place under fuel-rich conditions to maintain a non-flammable feed mixture. Typical feed composition is 13% to 15% ammonia, 11% to 13% methane and 72% to 76% air on a volumetric basis. Control of feed composition is essential to guard against deflagrations as well as to maximize the yield. The yield from methane is approximately 60% of theoretical. Conversion, yields, and productivity of the HCN synthesis are influenced by the extent of feed gas preheat, purity of the feeds, reactor geometry, feed gas composition, contact time, catalyst composition and purity, converter gas pressure, quench time and materials of construction.

The reactions in the Andrussow process are more complex than that shown in equation 72[130]. Most of the heat required for HCN formation is supplied by combustion of methane. This results in an overall reaction that is exothermic even though the endotherm of the methane-ammonia reaction is 60 kcal per mole of HCN[129]. The converter off-gas typically has the following composition:

Product	Mole %
HCN	6 – 12
NH_3	1.5 – 3
H_2	7 – 11
N_2	50 – 55
H_2O	20 – 25
CO	3 – 4
CO_2	0.3
CH_4	0.1

Two processes can be used to recover HCN from the converter off-gases. Figure 19.1 shows a process that recovers unreacted ammonia for recycle to the converter while the process in Figure 19.2 uses sulfuric acid to scrub the ammonia out of the gas and makes ammonium sulfate as a by-product. Ammonium sulfate can create a disposal problem, but the recycle system is capital and energy intensive. Ammonia must be removed from the off-gas before HCN can be recovered because the ammonia promotes polymerization of the HCN[129].

After ammonia is removed from the converter off-gases, the remaining gas stream enters an absorber where HCN is absorbed in cold water. The dilute HCN solution is stripped and fractionated by conventional means to produce HCN of 99.5% or higher purity. To avoid polymerization, the process streams in the HCN absorber and downstream equipment contain a small concentration of acid. All HCN recovery equipment is made of austenitic stainless steels to minimize corrosion problems and maintain product purity[129].

The waste gas that leaves the HCN absorber contains enough hydrogen and carbon monoxide to give it a heating value that makes it a valuable fuel. It is usually used to displace other fuel in boilers[129].

19.2.2. BMA Process

Degussa developed the BMA (Blausäure-Methan-Ammoniak, or HCN-methane-ammonia) process around 1949 in cooperation with Heinrich Koppers GmbH. In the process, ammonia reacts with methane in the absence of oxygen in a gas-fired tubular reactor. US patent 5,785, 942 describes the BMA process and improvements that were made during the mid 1990's. The reaction is[46]:

$$NH_3 + CH_4 \xrightarrow[\text{Pt/Ru}]{1300^\circ C} HCN + 3\,H_2 \qquad \Delta H = +60 \text{ kcal/mol HCN} \qquad (19.2)$$

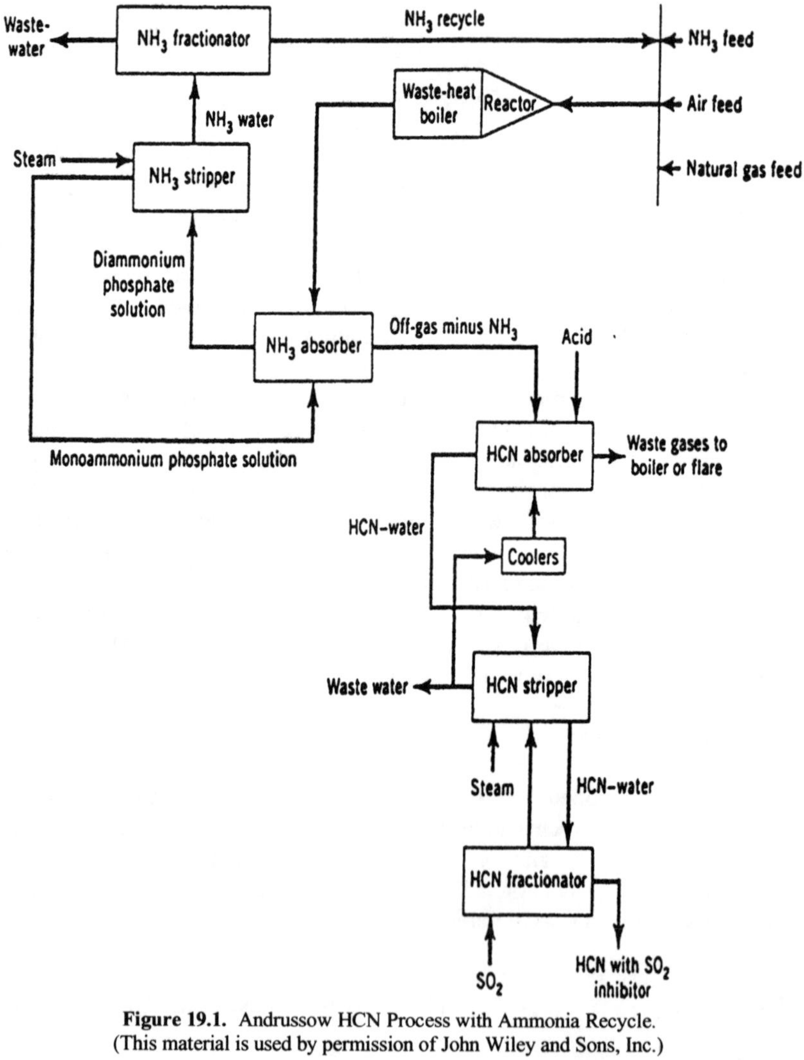

Figure 19.1. Andrussow HCN Process with Ammonia Recycle. (This material is used by permission of John Wiley and Sons, Inc.)

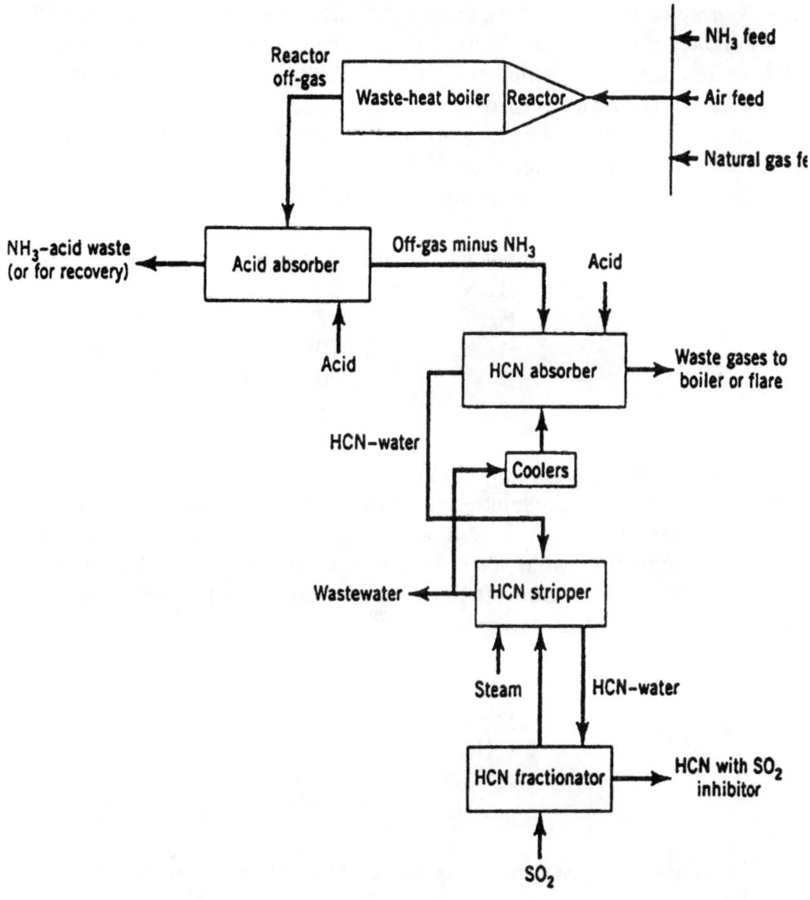

Figure 19.2. Andrussow HCN Process with Ammonia Removal. (This material is used by permission of John Wiley and Sons, Inc.))

In a reactor that is similar to a reformer, the reaction occurs in tubes that are heated externally to supply the endothermic heat of reaction[129]. Sintered corundum (α-Al_2O_3) tubes with an internal layer (~15 microns thick) of platinum/ruthenium catalyst are used. In some cases a platinum/aluminum catalyst may be used. To achieve adequate heat transfer, the tubes may be only ¾" in diameter and 6½ feet long. Selectivities of 90-91% for methane and 83-84% for ammonia are reached at 1200°C to 1300°C reaction temperatures.

A typical converter is made up of multiple furnaces, each of which contains 8 to 10 reactors. Each reactor contains 10 to 30 tubes with a catalyst lining. Catalyst life is approximately 10,000 hours[129].

The converter off-gas typically has the following composition:

Product	Mole %
HCN	20 – 23
NH_3	2 – 3
H_2	70 – 72
N_2	0.5 -- 1
CH_4	1 – 2.4

The unreacted ammonia is removed from the reactor off-gas by scrubbing with sulfuric acid to make ammonium sulfate. Just like the Andrussow process, ammonia must be removed from the off-gas before HCN can be recovered because the ammonia promotes polymerization of the HCN. After ammonia is removed from the converter off-gases, the remaining gas stream is processed in a way similar to the Andrussow process[129].

In the BMA process the vent gas from the HCN absorber has the following composition:

Product	Mole %
H_2	96
N_2 & O_2	1.5
CH_4	2.5

The vent gas is used either as a chemical feedstock or as fuel gas in the HCN furnaces.

The advantages of the BMA process are high ammonia yields, high natural gas yields and the useful hydrogen vent gas. But the high investment and high maintenance costs for the converter are a disadvantage. The complexity of the reaction system also makes the process less attractive for large-scale production[129]. In 2003 the BMA process was being used at four plants[131].

19.2.3. Fluohmic Process

Shawinigan Chemicals of Canada developed the Fluohmic process around 1960. This process involves the reaction of ammonia with hydrocarbons (usually propane or butane) in an electrically-heated, fluidized bed of coke. The reaction is:

$$3\ NH_3 + C_3H_8 \xrightarrow{1370^\circ C} 3\ HCN + 7\ H_2 \quad \Delta H = +151\ \text{kcal/mol HCN} \qquad (19.3)$$

The fluidized bed is heated to 1350°C to 1650°C by passing an electric current between carbon electrodes immersed in the bed, and no catalyst is used to initiate the reaction. The performance of the reaction system is reported to be:

	Conversion (Mole %)	Yield (Mole %)	Yield (lb/lb HCN)
C_3H_8	88%	88% to 90%	0.60 to 0.62
NH_3	86%	86% to 90%	0.70 to 0.73

The reactor off-gas typically has the following composition:

Product	Mole %
HCN	25
NH_3	0.25
H_2	72
N_2	3

The low ammonia concentration in the off-gas eliminates the need for recovery. The HCN is separated from the off-gas for refining, and the hydrogen leaves the system as a by-product together with a small amount of nitrogen and unreacted ammonia. The high electrical consumption (~3 kWh/lb HCN) makes the process attractive only in locations where the cost of electricity is extremely low.

In 2003 the Fluohmic process was being used in Spain, Australia and South Africa.

19.2.4. Formamide Process

Several companies developed the Formamide process, but in 2001 it was only practiced by BASF. The process involves the following steps[46]:

1. Methyl Formate Synthesis:

$$CO + CH_3OH \xrightarrow[\text{20 to 200 bar}]{\text{70°C, Sodium Methylate}} \underset{\text{Methyl Formate}}{HC(=O)\text{-}OCH_3} \qquad (19.4)$$

With excess methanol the conversion to methyl formate is 95% with nearly 100% selectivity.

2. Formamide Synthesis

$$\underset{}{HC(=O)\text{-}OCH_3} + NH_3 \xrightarrow[\text{4 to 6 bar}]{80^{\circ}C \text{ to } 100^{\circ}C} \underset{\text{Formamide}}{HC(=O)\text{-}NH_2} + CH_3OH \qquad (19.5)$$

The conversion of ammonia to formamide is 60% to 85%.

3. HCN Formation

$$HC(=O)\text{-}NH_2 \xrightarrow[\text{Acidic, Fe}]{400^{\circ}C\text{, Vacuum}} HCN + H_2O \quad \Delta H = +18 \text{ kcal/mol HCN} \qquad (19.6)$$

The dehydration of formamide is conducted in iron catalyst tubes at 380°C to 430°C under reduced pressure. The tubes are filled with Fe or Al phosphate catalyst that also contains Mg, Ca, Zn or Mn as promoters. The selectivity to HCN is 92% to 95%[46].

The overall performance of the reaction system is listed below:

	Yield	
	(Mole %)	(lb/lb HCN)
CH_3OH	--	0.20 to 0.30
NH_3	79% to 90%	0.70 to 0.80
CO	52% to 94%	1.10 to 2.00

The reactor off-gas contains 60% to 70% HCN and is suitable for direct liquefaction.

BASF has obtained European patent 1,110,913, which describes HCN production by the thermolytic reaction of gaseous formamide in the presence of a fine-particle, solid catalyst in a fluidized bed. They claim that the process is more economical and results in less equipment damage. They also received patent DE 10,144,891 for production of HCN from formamide in a flow-type reactor with electric resistance heating.

Elements of the existing formamide process are included in a methyl methacrylate process that Mitsubishi Gas Chemical started up in 1997. The formamide is obtained by reacting hydroxyisobutyramide with methyl formate. Therefore the external requirements for HCN are reduced to the replacement of yield losses[131]. The process is described in US patent 6,075,162 and the catalyst is described in European patent 1,086,744.

19.2.5. Methanol Process

In this process HCN is produced when methanol reacts with ammonia and oxygen in the presence of an oxide catalyst that contains iron, antimony, phosphorous and vanadium. The reaction occurs in the vapor phase in a fluidized bed reactor with an oxygen-to-methanol molar ratio in the gas phase that is less than 1.6. The process and the catalyst are described in patents that were issued to Nitto Chemical (now Mitsubishi Rayon) during the late 1990's (European Patent 864,532; Japanese patents 10-167,721, 10-251,012, 11-043,323; US Patent 5,976,482).

In 2003 Mitsubishi Rayon was believed to be operating a commercial-scale, methanol-based HCN process and had offered to license the technology to other companies. The technology might provide a low-cost way to convert an acrylonitrile plant to HCN-only production. However methanol is a higher cost source of carbon compared to natural gas so the methanol process probably has a higher operating cost than the Andrussow process.

19.2.6. Acrylonitrile Process

The Standard Oil Company of Ohio or SOHIO (now BP Amoco) developed and commercialized in 1960 a fluidized bed process in which the catalytic oxidation of a mixture of propylene and ammonia produced acrylonitrile (ACRN). By-products from this reaction are HCN and acetonitrile. The yields of HCN depend on the process conditions and on the catalyst system[131]. The reactions are:

$$\underset{\text{Propylene}}{CH_2{=}CHCH_3} + NH_3 + {}^3/_2\,O_2 \rightarrow \underset{\text{Acrylonitrile}}{CH_2{=}CHCN} + 3\,H_2O \qquad (19.7)$$

$$\underset{\text{Propylene}}{CH_2{=}CHCH_3} + {}^3/_2\,NH_3 + {}^3/_2\,O_2 \rightarrow {}^3/_2\,\underset{\text{Acetonitrile}}{CH_3CN} + 3\,H_2O \qquad (19.8)$$

$$CH_2{=}CHCH_3 + 3\,NH_3 + 3\,O_2 \rightarrow 3\,HCN + 6\,H_2O \qquad (19.9)$$

Depending on the type of ACRN process and the operating conditions, HCN production will be equivalent to between 10% and 24% of the ACRN that is produced[46]. US Patent 5,840,648 describes a catalyst that permits more HCN production.

During the 1990's several patents were issued in relation to the addition of methanol to the feed stream of an ACRN plant to increase HCN production. The benefits of methanol addition are described in European patent 878,464, in US patent 5,288,473 and US patent 6,204,407.

BP Amoco and Asahi have also been running tests related to the ammoxidation of propane to produce ACRN. This process is believed to yield a higher level of HCN than the optimized oxidation of propylene[131].

19.2.7. Other HCN Processes

Some alternate HCN processes that have been studied during the 1990's are:

- Production of HCN from Acetonitrile (see Japanese patent 10,167,721),
- Production of HCN by use of a corona (German patent 1,054,982),
- Production of HCN by use of microwave heating (US patents 5,393,393; 5,470,541; and 5,529,669), and
- Production of HCN in an electrically-heatable fixed bed reactor (EP-1,213,052).

None of these processes are believed to be used in commerical production in 2003.

19.2.8. Materials of Construction

Dilute cyanide-water solutions at ambient temperature can exhibit a unique destructive action on stressed carbon steel. Cracks can develop in the metal in a very short time. The characteristic feature of the cracks is their transcrystalline course without formation of slip planes[254].

This is rather unique and is completely different from stress-corrosion cracking of unalloyed steel in alkaline solutions. In alkaline solutions the cracks always run between the crystals and usually only after plastic deformation. Normalized, stress-relieved steel is not usually subject to cracking[254].

Teflon® is satisfactory for valve packing and gaskets in cold or ambient temperature HCN service. Above 45°C, HCN will penetrate the surface of Teflon®, polymerize and destroy the material. Hypalon® is probably the best material for hose linings for handling HCN at ambient temperature. Viton® is better at somewhat higher temperatures[254].

Globe valves are used in HCN service. Gate valves should not be used since they can trap HCN in the bonnet cavity, which can polymerize violently[254].

HCN lines should be sloped a minimum of 1/8" per foot to drain when not in use. More slope is preferred. No low spots or sags should be permitted because lines will not drain and the HCN can polymerize in the low spots[254].

19.3. PRODUCTION

In 1999 a total of 34 companies were operating 47 HCN production facilites in the USA, Western Europe and Japan. Capacity in these three regions was about 3.6 billion pounds (1.6 million tonnes). Direct production accounts for 75% of total capacity, and the balance is co-product from ACRN production. Worldwide capacity is about 1.8 million tonnes. DuPont owns or has a financial interest in about 36% of this capacity while Degussa-Hüls and Rohm and Haas have about 6% each[131]. In 2000 HCN production in North America was about 1.6 billion pounds, and it is projected to reach 1.8 billion pounds by 2004[261].

Between 1994 and 1999 HCN consumption increased about 2.2% per year. Consumption is expected to increase at around 2.8% per year from 1999 through 2004[131]. In the United States strong exports of adiponitrile and sodium cyande (NaCN) were the key growth factors for HCN in the late 1990's. Several acrylonitrile (ACRN) plants have either been built or expanded in the Asia / Pacific region during the early years of the new century. Most of the HCN that is made in these ACRN plants has been used to make NaCN. Therefore, NaCN exports from the United States have fallen as these ACRN plants increase their HCN and NaCN production.

19.4. USES

HCN is usually consumed at its production site, and most consumers without on-site HCN production capability are believed to be developing plans to eliminate HCN shipments. For example, BASF in the United Kingdom produces by-product HCN in an ACRN plant and ships HCN railcars to Lucite. This HCN is used to produce acetone cyanohydrin (ACH) in a methyl methacrylate plant. To eliminate these shipments, BASF and Lucite plan to 1) install an ACH plant on the BASF plant site and 2) transfer the ACH to Lucite via a 9 km pipeline[136].

ACRN plants are highly dependent on the actions of the acrylic fiber industry, the major outlet for ACRN. As a result, HCN consumers may have to operate at reduced rates if the ACRN plant is not producing an adequate supply of HCN. In 2001 many consumers of HCN from ACRN plants are believed to be evaluating options that will give them a more consistent supply of HCN.

Some of the uses for HCN are:

- Adiponitrile (ADN). ADN is made when HCN reacts with butadiene. It is a key ingredient in production of nylon.
- Acetone Cyanohydrin (ACH). ACH is made when HCN reacts with acetone. It is a key ingredient in many methyl methacrylate (MMA) plants. MMA is used to make acrylic sheeting and related products.
- Sodium Cyanide (NaCN). NaCN is made when HCN reacts with sodium hydroxide (NaOH). NaCN is used in precious metals (gold and silver) mining, metal plating and some chemical production.
- Methionine. Methionine is an essential amino acid that is used in animal feed. The key ingredients are HCN or NaCN, methyl mercaptan, acrolein and NH_4HCO_3. In 2003 global capacity was 450,000 to 500,000 tonnes per year.
- Cyanuric Chloride (CYC). CYC is made when HCN reacts with chlorine and is then rearranged into a ring compound. It is used to make herbicides, pharmaceuticals, and explosives.

- Chelating Agents. Chelating agents are made when HCN reacts with formaldehyde and amines to make aminocarboxylic acids, which are then saponified to produce the acid salts. Their major uses are in soaps, water treating agents and various cleaning agents. For more details see Chapter 16.
- Other uses include ferrocyanides (for blue jeans), acrylates, lactic acid, nitriloacetic acid (NTA), pharmaceuticals and specialty chemicals.

The distribution of HCN uses in 2001 is shown in Table 19.2.

Table 19.2. Hydrogen Cyanide Uses[261]

Application	%
Adiponitrile	47%
Acetone Cyanohydrin	27%
Sodium Cyanide	8%
Methionine	6%
Chelating Agents	2%
Cyanuric Chloride	2%
Miscellaneous	8%

20

ANILINE AND NITROBENZENE

20.1. ANILINE

Aniline (also know as aminobenzene and benzamine) was first produced in 1826 by Unverdorben by the dry distillation of indigo, the oldest known vat dye. Fritsche also obtained aniline from indigo by heating it with potash (K_2O), and he named it aniline. Hofmann obtained aniline by the reduction of nitrobenzene in 1843 and was able to prove the structure. Aniline is a colorless, oily, flammable liquid that is slightly soluble in cold water and infinitely soluble in alcohol and ether. Its physical properties are summarized in Table 20.1.

Table 20.1. Physical Properties of Aniline

Property	Value
Molecular Weight	93.12
Boiling point. °C	
101.3 kPa (760 mm Hg)	184.4
4.4 kPa (33 mm Hg)	92
1.2 kPa (9 mm Hg)	71
Melting point, °C	- 6.15
Density, *d*	
at 20/4°C	1.02173
at 20/20°C	1.022
Viscosity at 20°C, mPa•s (= cP)	4.423-4.435
Dissociation constant, *pK*	
at 20°C	4.60
at 60"C	8.88
Enthalpy of dissociation, kJ/mol (kcal/mol)	21.7 (5.19)

Table 20.1. Physical Properties of Aniline - continued

Property	Value
Heat of combustion,	
kJ/ mol (kcal/mol)	3389.72 (810.55)
Specific heat, 20-20°C	0.518
Latent heat of vaporization	
J/g (cal/g)	476.3 (113.9)
Flash point (closed-cup), °C	76
Autoignition Temperature, °C	615
Flammable Limits in Air (% by Volume)	
LEL	1.3
UEL	11.0
Vapor Density (Air = 1)	3.22
Solubility (g in 100 g Water @ 20°C)	3.5
Odor Threshold, ppm	1.1

20.1.1. Aniline Processes

The aniline processes that are being used in 2003 include:

1) Hydrogenation of nitrobenzene,
2) Nitrobenzene reduction with iron filings, and
3) Ammonolysis of phenol.

Almost 97% of the nitrobenzene produced in the United States is converted to aniline between the hydrogenation process and the iron filings process[137].

20.1.1.1. Hydrogenation of Nitrobenzene

The primary aniline production process in the world is the hydrogenation of nitrobenzene. BASF, DuPont, ChemFirst (purchased by DuPont in 2002) and Rubicon use this process in the United States. This technology is also used by all Western European aniline producers and all but one Japanese aniline producer.

$\Delta H = -117$ kcal/mol (20.1)

The catalytic gas-phase hydrogenation processes for nitrobenzene can be carried out using a fixed-bed or a fluidized bed reactor. Bayer and Allied work with nickel sulfide catalysts at 300°C to 475°C in a fixed bed. The selectivity to aniline is more than 99%. The catalytic activity slowly decreases due to carbon deposition. However, the catalyst can be regenerated with air at 250°C to 350°C and subsequent H_2 treatment. Similar processes are operated by ChemFirst which uses a copper on pumice catalyst from Lonza; ICI with Cu, Mn or Fe catalysts with various modifications involving other metals; and by Sumitomo with a Cu-Cr system[46]. A typical process diagram is shown in Figure 20.1[137].

The gas-phase hydrogenation of nitrobenzene with a fluidized-bed catalyst is used in a process from BASF. The BASF catalyst consists of Cu, Cr, Ba and Zn oxides on a SiO_2 support. The hydrogenation is conducted at 270°C to 290°C and 1 to 5 bar in the presence of a large excess of hydrogen (H_2:Nitrobenzene = 9 to 1). The high heat of reaction is removed by a cooling system that is built into the fluidized bed. The selectivity to aniline is 99.5%, and the nitrobenzene conversion is quantitative. The catalyst must be regenerated with air periodically[46].

20.1.1.2. Nitrobenzene Reduction with Iron Filings

The older Bechamp method for iron oxide pigment production gives aniline as a co-product and is operated by Bayer in West Virginia. Nitrobenzene is reduced by reaction with iron filings in the presence of a hydrochloric acid catalyst. The iron is oxidized to the ferrous or ferric state, and the aniline-water mixture is separated from the iron-hydroxide sludge. The heavier aniline layer is removed and vacuum distilled to yield pure aniline. The yield is 90% to 95% of theoretical. The reactions are represented as follows[138, 255]:

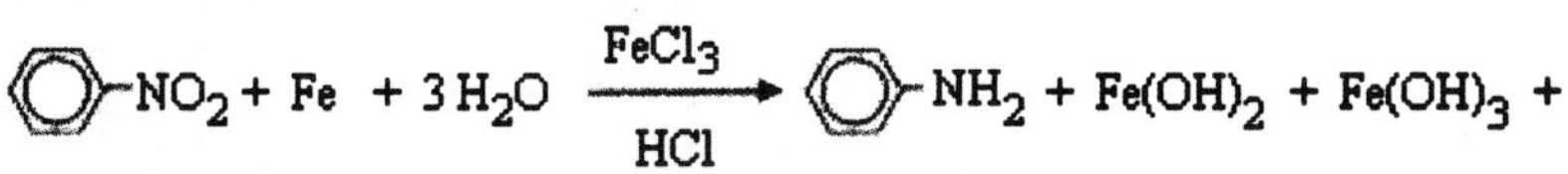

$$FeO + \tfrac{1}{2} H_2 \quad \Delta H = -117 \text{ kcal/mol} \qquad (20.2)$$

$$2\,Fe(OH)_3 + Fe(OH)_2 \rightarrow Fe_3O_4 + H_2O \qquad (20.3)$$

This process would have been replaced much earlier by more economical reduction methods if it had not been possible to obtain valuable iron oxide pigments from the iron oxide sludge. However the increasing demand for aniline has far surpassed the market for iron pigments, so this process is no longer preferred[46].

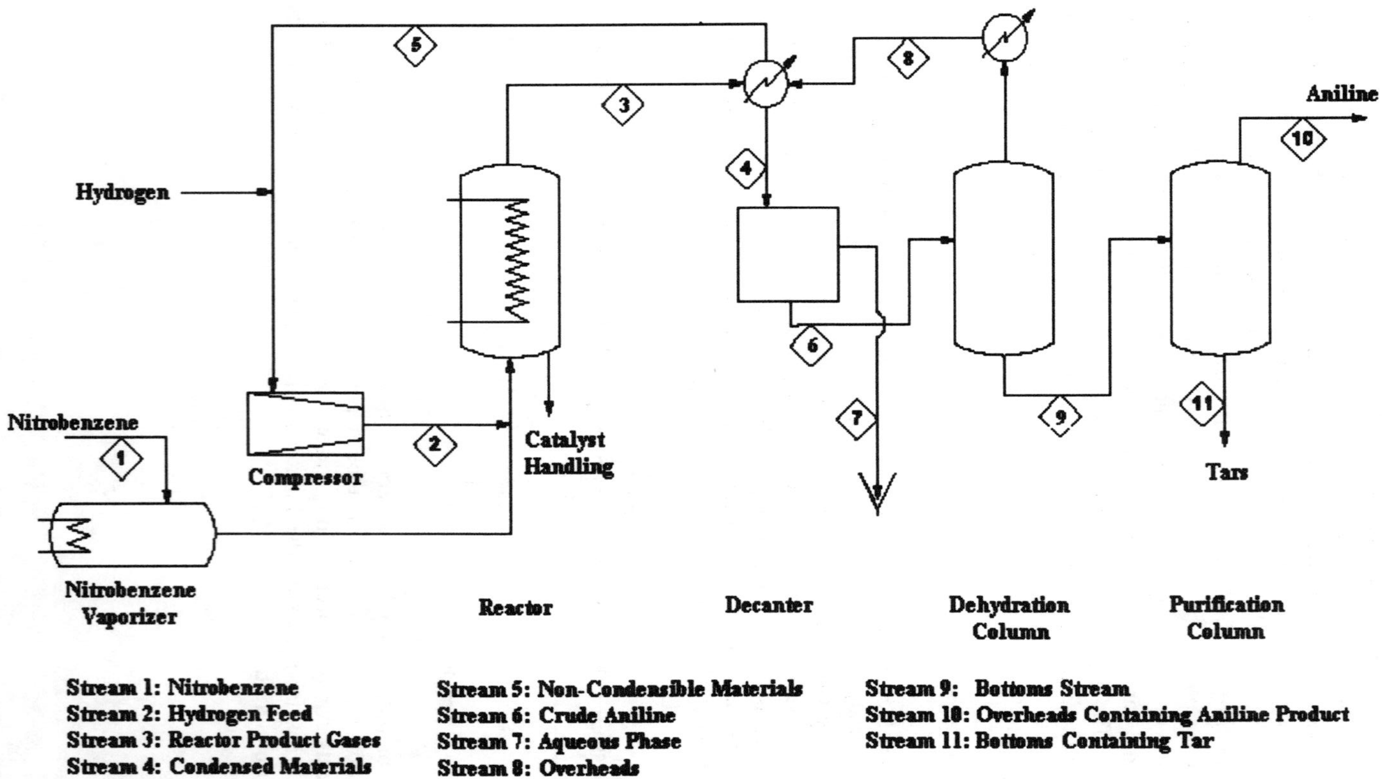

Figure 20.1. Aniline production process: hydrogenation of nitrobenzene

20.1.1.3. Ammonolysis of Phenol

Aniline can also be produced when phenol is subjected to gas-phase ammonolysis at 200 bar and 425°C in an adiabatic, fixed-bed reactor. This is the Halcon/Scientific Design process. The chemistry is:

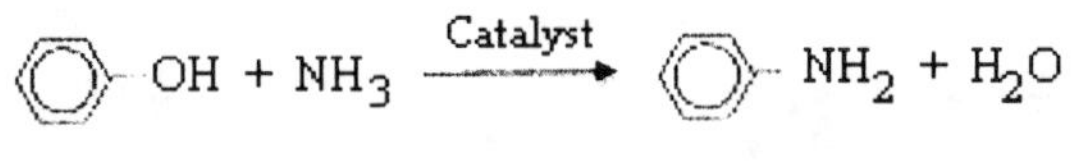

(20.4)

The catalysts are $Al_2O_3{\bullet}SiO_2$ (possibly as zeolites) and oxide mixtures of Mg, B, Al and Ti. These can be combined with additional co-catalysts such as Ce, V or W. With a large excess of ammonia, the selectivity to aniline is 87% to 90% at a phenol conversion of 98%. The by-products are diphenylamine and carbazole. This technology is used at one plant by Sunoco (previously Aristech Chemical) in Ohio and at another plant in Japan. The economics of this process are favorable if low-cost phenol is available, and high-purity aniline is desired. Capital costs are low because benzene nitration is avoided. A typical process sketch along with a material balance is shown in Figure 20.2[139].

20.1.2. Aniline Production

MDI [4,4'-methylenebis (phenylisocyanate)] has been the driving force behind the recovery of the aniline business since 1982 when the industry had a capacity utilization rate of less than 50%. By 1996, capacity utilization had approached 95% in some regions[138].

Aniline's global production capacity in 1999 was 2.9 million tonnes, and demand in 2000 was estimated to be 2.68 million tonnes per year. The capacity in 1999 was found in these regions: Western Europe - 47%, North America - 30% and Asia / Pacific - 19%[140,256]. In 2000 China had over 20 aniline producers with a total capacity that exceeded 200,000 tonnes per year. Aniline production in 1999 in China was 142,700 for a utilization rate of around 70%. Historical production in the United States is summarized in Table 20.2[255].

Table 20.2. Aniline Supply and Demand

	Million of Pounds						
	1995	1996	1997	1998	1999	2000	2001[b]
Capacity	1380	1420	1535	1745	1745	2310	2310
Total Prod.	1388	1395	1339	1545	1588	1866	1975
Imports	–	24	60	42	26	11	12
Exports	67	41	19	43	38	58	65
Demand[a]	1321	1378	1380	1544	1574	1819	1922

[a] Demand equals production plus imports minus exports.

[b] Forecast

(Reproduced by permission of Mannsville Chemical Products Corp.)

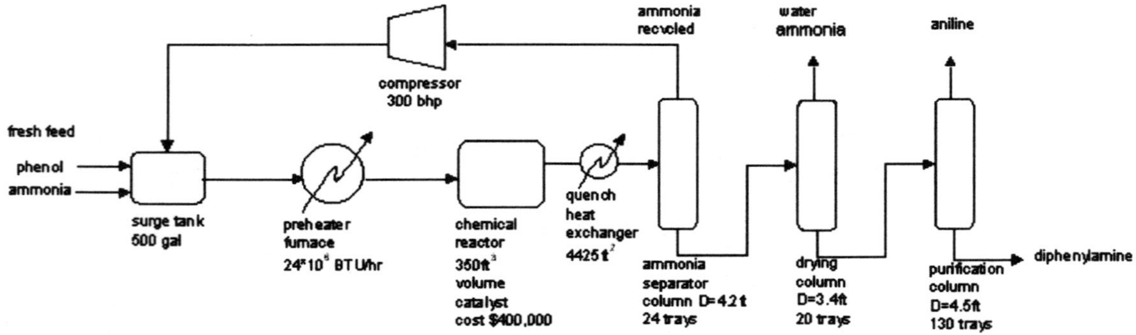

Production rate = 100 million pounds per year

Utilities:

Fuel: $24*10^{6}$ BTU/hr

Steam: 600psig - $10.2*10^{6}$ BTU/hr 150psig - $10*10^{6}$ BTU/hr

Cooling water: $9*10^{6}$ BTU/hr for $\Delta T=20^{\circ}F$

Electricity to drive the motor for the 300 bhp compressor

Preliminary Material Balance on Process

mass flow rates (lb/hr)	fresh feed	ammonia recycle	reactor feed	reactor product	ammonia separator column overhead	ammonia separator column bottoms	drying column overhead	drying column bottoms	purification column overhead	purification column bottoms
phenol	13,159	0	13,159	13	0	13	0	13	13	0
ammonia	4,199	43,070	47,269	44,900	43,070	1,830	1,830	0	0	0
aniline	0	10	10	12,890	10	12,880	315	12,565	12,565	0
water	0	11	11	2,528	11	2,517	2,505	12	12	0
diphenylamine	0	0	0	118	0	118	0	118	0	118
total	17,358	43,091	60,449	60,449	43,091	17,358	4,650	12,708	12,590	118
	60449				60449		17358		12708	

	fresh feed	ammonia recycle	preheater		quench exchanger		ammonia separator overhead	ammonia separator bottoms	drying column overhead	drying column bottoms	purification column overhead	purification column bottoms
temperature (°F)	100	142	118	670	673	110	112	248	141	289	278	491
pressure (psig)	260	260	260	250	240	230	220	220	6	6	5	5

Figure 20.2. Aniline production process: ammonolysis of phenol

Aniline consumption grew at a rate of 5.5 percent per year from 1995 to 2000. It is expected to grow at 3.9 percent per year from 2001 through 2004[262].

Aniline pricing tends to track benzene pricing rather closely. In general the aniline price will move 1.2 cents per pound for each 10 cents per gallon movement in benzene price. Between 1995 and 2001, the aniline sales price varied from $US 0.35 per pound to $US 0.40 per pound[255].

20.1.3. Aniline Uses

MDI [4,4'-methylenebis (phenylisocyanate)] accounted for almost 85% of the worldwide demand for aniline in 2000. MDI is used primarily to make rigid polyurethane foam and polyurethane elastomers. MDI growth is expected to be 6% to 8% per year during the 2000's as its use continues to increase in the construction industry (the largest user of rigid polyurethane foam) and the auto industry (the largest user of reaction-injection molding plastics). The aniline uses are shown in Table 20.3[255].

Table 20.3. Aniline Uses[255, 262]

Application	%
Isocyanate	85%
Rubber Chemicals	9%
Agricultural Chemicals - Pesticides	3%
Dyes & Pigments	2%
Specialty Fibers	1%
Miscellaneous[a]	1%

[a] Miscellaneous includes explosives, epoxy curing agents and pharmaceuticals .

The use of aniline in rubber chemicals will closely track the Gross Domestic Product growth in the United States, while its use in agricultural chemicals is expected to slowly decline. Aniline-based dyes and pigments are forecast to have little growth[255].

Aniline is considered a suspect carcinogen. It is absorbed through the skin, lungs or mucous membranes, and it can produce serious burns. In 2001 the ACGIH recommended exposure limit is 2 ppm. The odor of aniline can usually be detected in concentrations as low as 0.5 ppm. Aniline is combustible, but it is not explosive or spontaneously flammable in air[255].

20.2. NITROBENZENE

Nitrobenzene, also known as oil of mirbane, is a pale yellow liquid with an odor resembling that of bitter almonds. It was first synthesized in 1834 by E.

Mitscherlich by treating benzene with fuming nitric acid and has commercial importance as the major precursor to aniline[46, 258].

20.2.1. Nitrobenzene Processes

Nitrobenzene is made by three processes: direct nitration, dehydrating nitration and adiabatic nitration. The first process involves the direct nitration of benzene using a mixture of nitric acid and sulfuric acid[46, 258].

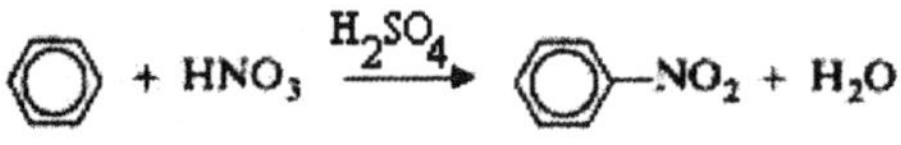

$\Delta H = -28$ kcal /mol (20.5)

The sulfuric acid promotes the formation of the nitrating agent (the nitronium ion) and prevents the dissociation of nitric acid into an oxidizing NO_3^- ion by binding water as a hydrate. It also enhances the solubility between the aqueous and organic phases[46].

This process may be carried out by a batch or a continuous process. The continuous process is more effective for large plants. Both processes wash the crude nitrobenzene with water. The crude nitrobenzene is fed to a distillation system to separate the water, benzene and dinitrobenzene. This process requires reactor cooling and a system for sulfuric acid reconcentration. Yields by the continuous process are 95% to 96% of theoretical. Bayer used this process in the United States until it discontinued nitrobenzene production in 1994[258].

In the batch process, the nitrating acid (a mixture of 32 to 39 wt % HNO_3, 53 to 60 wt % H_2SO_4 and 8 wt % H_2O) flows into the benzene, and the chemicals react in cast-iron nitrating vessels at 50 to 55°C. The process requires a lot of agitation due 1) the formation of two phases and 2) the required mass and heat exchange. After a residence time of several hours, the nitrating acid is largely exhausted and the denser nitrobenzene is separated, washed and distilled. The selectivity reaches 98 to 99%[46].

The continuous nitration plant consists of a cascade of stirred vessels (usually three nitrating vessels) with a stepwise, slowly increasing temperature (35 to 40°C in the first vessel, 50°C in the second and 55 to 60°C for the final reaction. The largest units are the final settling and washing vessels[46].

DuPont developed a dehydrating nitration process by which water produced in the mixed acid nitration reaction is removed by humidification of the inert gas stream. This eliminates the energy-intensive sulfuric acid process[258].

American Cyanamid and ICI Canada jointly developed a process to adiabatically produce mononitrobenzene. This process uses the heat of reaction to reconcentrate the spent acid stream. The process development was completed and commercialized by Kvaerner Chemetics and ICI Canada. As of 2003, a total of seven (7) adiabatic mononitrobenzene plants have been constructed, with the largest being rated at 227,000 tonnes per year. A process sketch is shown in Figure 20.3[290].

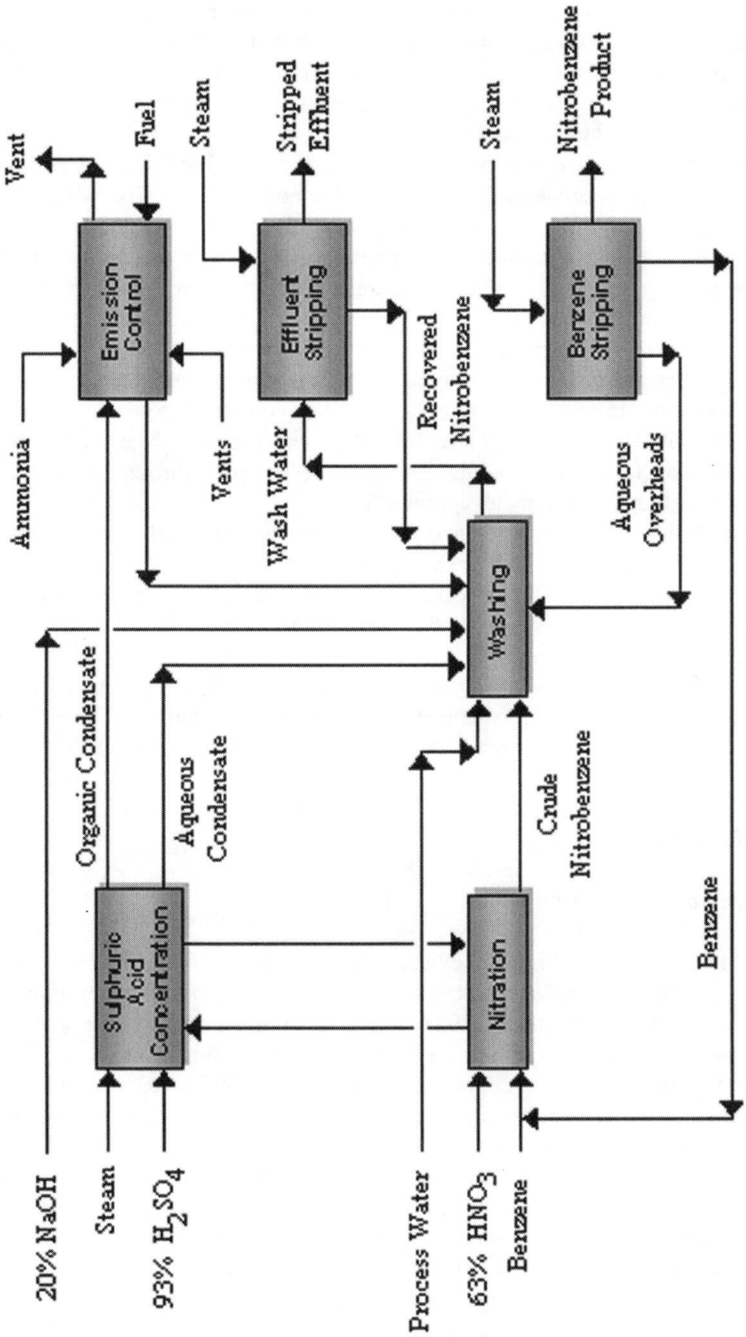

Figure 20.3. Adiabatic Nitrobenzene Production Process (Reproduced by permission of Chemetics, a division of Aker Kvaerner Canada Inc.)

In the adiabatic mononitrobenzene (MNB) process, benzene is nitrated in the presence of a large volume of mixed acid, such that the total heat of nitration and dilution is absorbed by the acid as sensible heat. The process utilizes this energy in a vacuum sulphuric acid evaporator to reconcentrate the spent acid, which is then returned to the nitration section.

Nitration is carried out in a series of stages. Mixed sulphuric and nitric acids together with benzene are introduced at the bottom of the first stage. The reaction takes place while the fluids flow upwards through the reactors.

The organic and acid phases are separated in a static separator located at the top of the process building. From the separator the acid phase discharges to the vacuum concentrator, while the organic phase passes to the washing section.

The vacuum sulphuric acid concentrator consists of a single stage evaporator. The vacuum is maintained by a two-stage steam ejector system utilizing shell and tube inter-and after-condensers. Most of the heat required to reconcentrate the sulphuric acid is recovered as sensible heat from the reaction and is contained in the spent acid. The balance is added through a tantalum bayonet heater immersed in the evaporator body.

In the washing process, acids and nitrophenols are removed from the crude MNB by means of an alkaline wash. Condensate from the acid concentrator may be used as wash water. The washing section of the plant employs specially designed equipment for space saving, minimum use of wash water and elimination of circulation pumps.

In the subsequent product stripping section, the product is stripped free of benzene and water using a live steam stripper or a reboiler column. The effluent discharged from the washing area is steam-stripped to recuperate MNB.

There is a single vent stack in the plant. The stack gases are treated in a benzene scrubber to remove benzene vapours prior to incineration or other means of treatment[290].

20.2.2. Nitrobenzene Production

MDI has been the driving force behind the recovery of the aniline business, and consequently nitrobenzene, since the early 1980's. Nitrobenzene production capacity in the United States in 2002 was 2.96 billion pounds. Demand has grown from 2.235 billion pounds in 1999 to 2.526 billion pounds in 2000. Production grew at a rate of 6.5 percent per year from 1995 to 2000. And from 2000 through 2004 production is expected to continue growing, but at a slower growth rate of 3.9 percent per year. The slowdown in growth is due to a downturn in the United States economy. However the forecast growth rate is still higher than the rate of growth in the gross domestic product. The supply side of nitrobenzene is in excellent shape. Plant operating rates are about 85 percent, and feedstock supplies of benzene, nitric acid and sulfuric acid are adequate to meet demand. Foreign trade is negligible[263].

Between 1996 and 2001, the nitrobenzene sales price varied from $US 0.30 per pound to $US 0.35 per pound[263].

20.2.3. Nitrobenzene Uses

Aniline accounted for almost 97% of the United States demand for nitrobenzene in 2001. The other three percent is used in p-aminophenol, dyes and pigments and as a chemical intermediate and solvent.

21

CAPROLACTAM

Caprolactam is a white, hygroscopic, crystalline solid with a characteristic odor. Its basic properties are summarized in Table 21.1. Most caprolactam is used to make nylon 6, which is the starting material for fibers that have many uses in textile manufacture and in the industrial sector.

Table 21.1. Caprolactam Properties[264]

Chemical Formula	$C_6H_{11}ON$ or $C_6H_5NHCH_2CN$
Molecular Weight	113.16
Melting Point (°C)	69.2
Boiling Point (°C)	
at 101.3 kPa	268.5
at 6.7 kPa	174
at 1.3 kPa	134
at 0.4 kPa	111
Density (kg/L)	
at 120°C	0.9829
at 100°C	0.9983
at 80°C	1.0135
Viscosity (mPa • s)	
at 120°C	2.93
at 100°C	4.87
at 80°C	8.82
Specific Heat (kJ / kg - °K)	
at 150°C	2.345
at 80°C	2.135
Heat of Fusion (kJ/kg)	123.5
Heat of Polycondensation, k(J/kg)	138

Table 21.1. Caprolactam Properties (continued)

Heat of Vaporization (kJ/kg)	
at 268°C	481
at 168°C	574
at 105°C	628
Vapor Pressure (kPa)	
at 268°C	101.3
at 168°C	5.3
at 105°C	0.25
Flash Point (°C)	139.5
Ignition Temperature (°C)	375
Lower Explosion Limit	1.4 vol% at 135°C
Upper Explosion Limit	8.0 vol% at 180.5°C

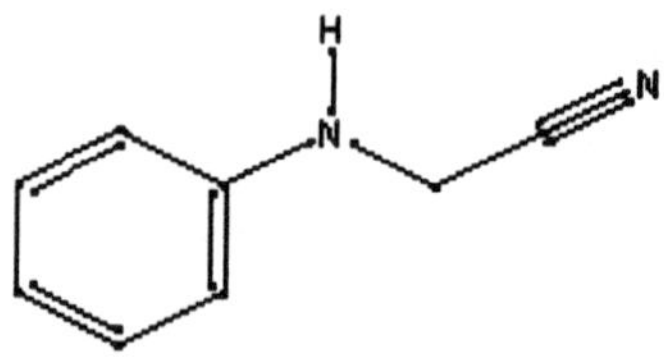

Polymerization is caprolactam's most important chemical property. The ring is hydrolyzed at 260 to 270°C. Liner polymer chains are formed by polycondensation. Caprolactam also reacts directly by polyaddition with the polymer chains. These reactions lead to an equilibrium between the polymer and caprolactam that favors a 90% conversion to polymer[264].

21.1. PROCESSES

All commercial caprolactam processes are based on benzene or toluene. These processes employ cyclohexanone, cyclohexane or toluene as starting materials. Large-scale industrial processes are multistage processes in which ammonium sulfate and sometimes organic compounds are formed as by-products. At least seven processes are used commercially[264, 266]. And each of these processes still has room for improvement.

1. Conventional,
2. DSM/HPO,
3. BASF,
4. Inventa-NO Reduction,
5. PNC or Toray,
6. Snia Viscosa, and
7. Recovery from Nylon-6 Waste.

Several other processes have been evaluated in the past, and some have reached pilot scale evaluation. A few technologies operated commercially for a while but were then shut down. Most of these processes are mentioned in References 46, 264, 265 and 266.

In 2001 DSM and Chiyoda were developing the ALTAM™ process. This process is based on butadiene and carbon monoxide and involves no by-products. Production costs are expected to be $US 1,300 per tonne ($0.59/lb)[268].

About 90% of the caprolactam is produced by the conventional cyclohexanone process. Cyclohexanone is obtained by catalytic oxidation of cyclohexane with air, or by hydrogenation of phenol and dehydrogenation of the cyclohexanol byproduct. The conversion of cyclohexanone to cyclohexanone oxime followed by Beckmann rearrangement gives caprolactam. About 10% of caprolactam is produced by photonitrosation of cyclohexane or by nitrosation of cyclohexanecarboxylic acid in the presence of sulfuric acid[264].

21.1.1. Conventional Process

The Conventional Process includes three 'organic' steps and one 'inorganic' step[46]:

1. Manufacture of cyclohexanone,
2. Manufacture of hydroxylamine,
3. Oxime formation from cyclohexanone with hydroxylamine, and
4. Beckmann rearrangement of cyclohexanone oxime to caprolactam.

The yield of caprolactam in the conventional process is approximately 70% of theoretical based on cyclohexane. The process makes about 0.94 pounds of caprolactam per pound of cyclohexane consumed. The yield is approximately 92% of theoretical based on phenol and yields about 1.11 pounds of caprolactam per pound of phenol consumed[266].

21.1.1.1. Manufacture of Cyclohexanone

Most cyclohexanone is made from cyclohexane as shown in Eq. (21.1):

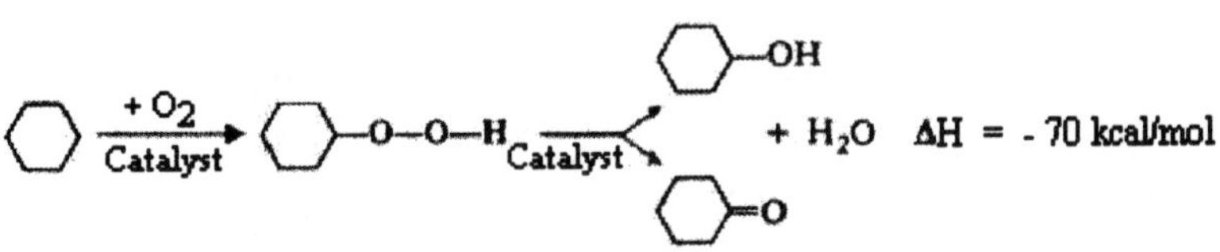

(21.1)

Cyclohexane is oxidized to a ketone/alcohol (KA) mixture at 125 to 165 °C and 8 to 15 bar. The reaction is conducted in the liquid phase with air and Mn- or Co-salts as catalysts. Cyclohexyl hydroperoxide, the primary product of this radical reaction, reacts further by means of a catalyst to the alcohol and ketone. Cyclohexane conversion is limited to 10 to 12% in order to increase the alcohol/ketone selectivity to 80 to 85%. The unreacted cyclohexane is distilled off and recycled to the oxidation step. Cyclohexanone and cyclohexanol (at a ratio of 1:1) are obtained in 99.5% purity by distillation[46].

Cyclohexanone is distilled from the cyclohexanone/cyclohexanol mixture and the cyclohexanol portion is catalytically dehydrogenated at 400 to 450°C and atmospheric pressure over Zn or Cu catalysts [(see Eq. (21.2)]. The cyclohexanol conversion is about 90% and selectivity to cyclohexanone is about 95%[46].

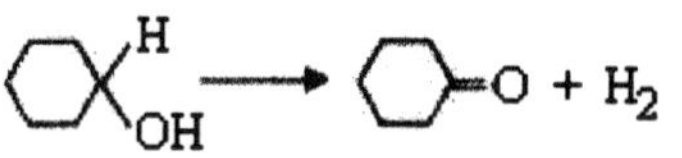

$\Delta H = -15$ kcal/mol (21.2.)

In earlier processes phenol could only be converted into cyclohexanone in a two-step process: 1) First the ring in the phenol was hydrogenated with a nickel catalyst at 140 to 160°C and 15 bar and then 2) the dehydrogenation was conducted as shown in Eq. (21.1)[46].

The production of cyclohexanone from phenol was simplified when selective hydrogenation with Pd catalysts was made possible ([see Eq. 21.3)]. In this process, phenol is completely converted in the gas phase at 140 to 170°C and 1 to 2 bar using a supported Pd catalyst containing alkaline earth oxides (e.g., Pd-CaO/Al_2O_3). The selectivity to cyclohexanone is greater than 95%[46].

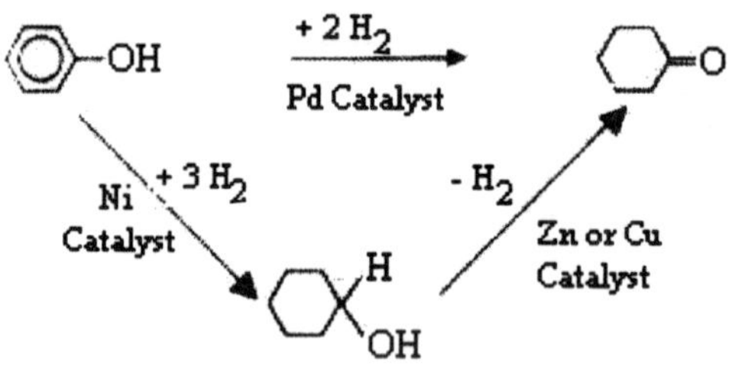

(21.3)

21.1.1.2. Hydroxylamine Formation

Hydroxylamine sulfate is usually prepared in a modified Raschig method that starts with air oxidation of ammonia. The nitrogen oxides are reacted with ammonium carbonate to produce ammonium nitrite and carbon dioxide [see Eq. 21.3)]. The ammonium nitrite solution then reacts with sulfur dioxide to give hydroxylamine diammonium sulfate, which is hydrolyzed to hydroxylamine sulfate and by- product ammonium sulfate[46, 266].

$$NH_3 + \text{air} \rightarrow N_2O_3 \xrightarrow{+ (NH_4)_2CO_3} NH_4NO_2 + CO_2 \quad (21.4)$$

The steps in Eqs. (21.5) and (21.6) consist of the reduction of ammonium nitrite with SO_2 at about 5°C to the disulfonate, which is then hydrolyzed at 100°C to hydroxylamine sulfate. One mole of $(NH_4)_2SO_4$ is formed for each mole of hydroxylamine sulfate[46].

$$NH_4NO_2 + NH_3 + 2\ SO_2 + H_2O \rightarrow HON(SO_2NH_4)_2 \qquad (21.5)$$

$$HON(SO_2NH_4)_2 + H_2O \rightarrow NH_2OH \cdot H_2SO_4 + (NH_4)_2SO_4 \qquad (21.6)$$

21.1.1.3. Oxime Formation from Cyclohexanone with Hydroxylamine

Oxime formation with cyclohexanone takes place with a hydroxylamine salt (either sulfate or phosphate) at 85°C[46]:

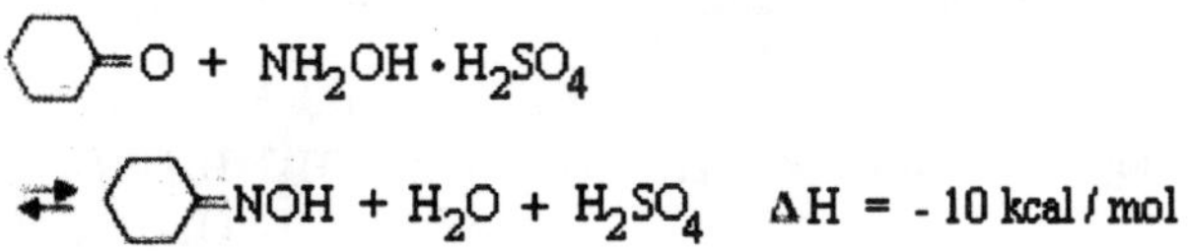

(21.7)

In order to displace the equilibrium, ammonia must be continuously introduced to maintain a pH of 7. Therefore another mole of ammonium sulfate is produced[46].

21.1.1.4. Beckmann Rearrangement of Cyclohexanone Oxime to Caprolactam

All caprolactam manufacturers use sulfuric acid or oleum as a rearrangement medium. The rearrangement is more complete in concentrated sulfuric acid. Excess sulfur trioxide further increases the speed of the rearrangement. This step is based on chemistry that was discovered by E. Beckmann in 1886 [see Eq. 21.8)]. Commercial development of this process was carried out by BASF[46, 264].

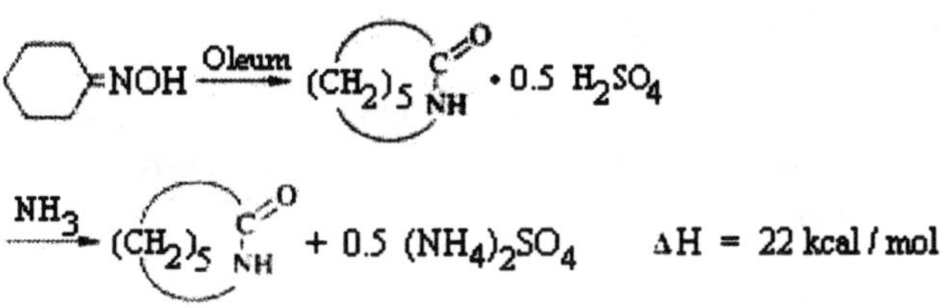

(21.8)

In the continuous process the oxime solution (which is acidified with sulfuric acid) is passed through the reaction zone that is kept at the rearrangement temperature of 90 to 120°C. The rearrangement is complete within a few minutes, and the resulting lactam sulfate solution is converted into the free lactam with ammonia in the neutralization vessel. It separates from the saturated ammonium sulfate solution as an oily layer. After extraction with benzene, toluene or chlorinated hydrocarbons and stripping with water is further purified and then distilled the selectivities amount to almost 98%[46].

21.1.2. DSM / HPO (Stamicarbon) Process

Of the processes based on modification of the conventional route, only Stamicarbon's hydroxylamine phosphate oxime (HPO) process completely avoid all production of ammonium sulfate before the Beckmann rearrangement. The steps in the HPO process are listed below and shown in Figure 21.1[46, 264]:

1. Reduction of the phosphoric acid/ammonium nitrate buffer solution with hydrogen and formation of hydroxylammonium phosphate:

$$NH_4NO_2 + 2\,H_3PO_4 + 3\,H_2 \xrightarrow{\text{catalyst}} [NH_3OH]^+\,[H_2PO_4]^- + NH_4\,H_2PO_4 + 2\,H_2O \qquad (21.9)$$

The catalyst is a palladium-on-carbon catalyst or alumina (Al_2O_3) with promoters such as germanium compounds suspended in a phosphate buffer solution[46].

2. Oxime Formation:

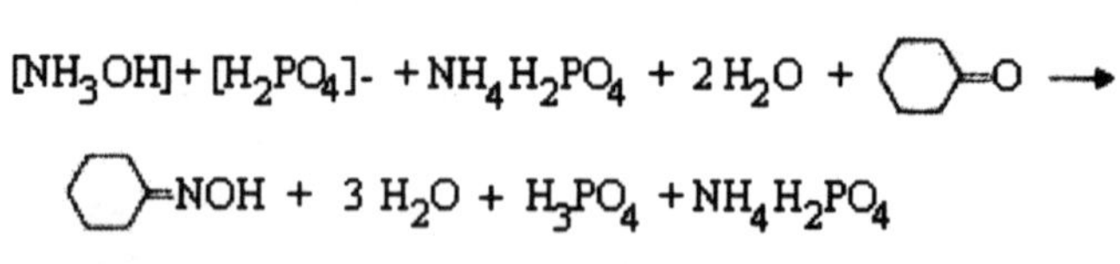

(21.10)

Since the hydroxylamine ion from the first step is unstable, the solution is reacted with a stream of cyclohexanone in the presence of toluene, which extracts the cyclohexanone oxime as it is formed. The oxime is recovered from the toluene solution by distillation[266].

3. After the oxime has been separated, 60% nitric acid is added to replace the nitrate ions that were consumed. This solution is recycled back to the first step.

$$NH_4H_2PO_4 + H_3PO_4 + 3\,H_2O + HNO_3 \rightarrow 2\,H_3PO_4 + + NH_4NO_2 + 3\,H_2O \qquad (21.11)$$

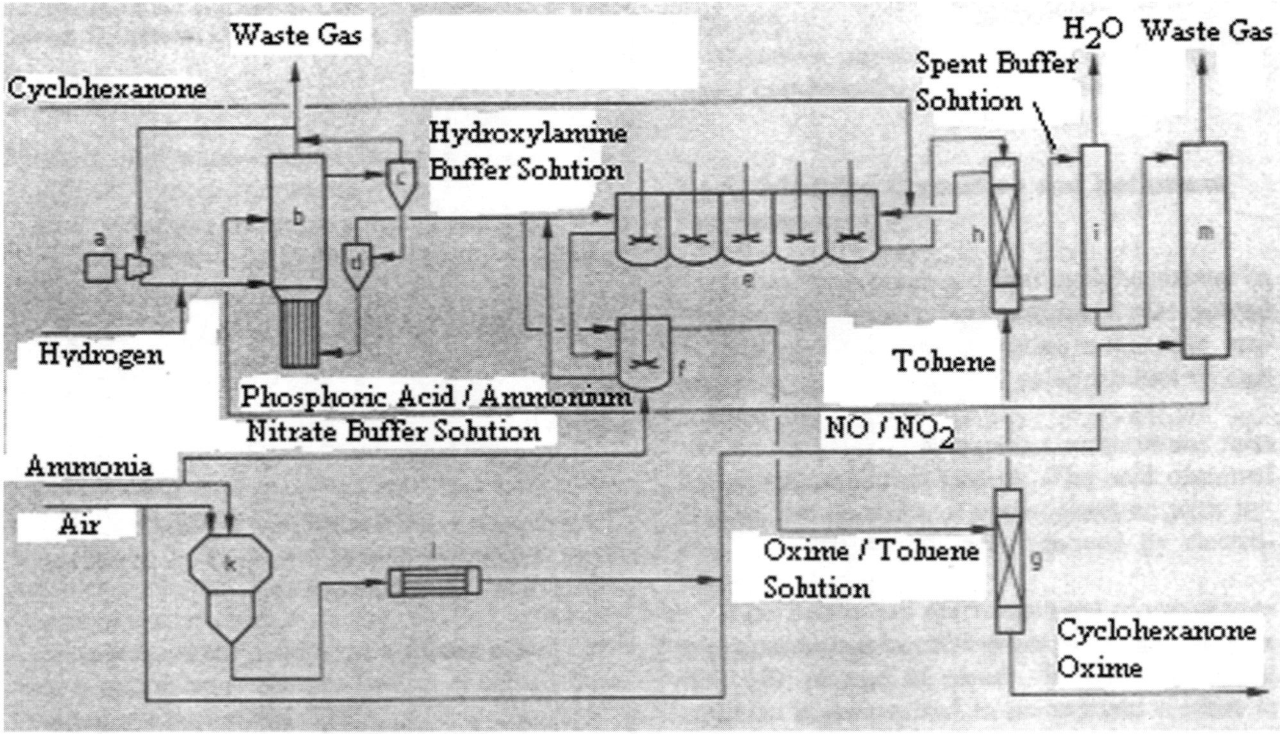

Figure 21.1. DSM / HPO hydroxylamine and cyclohexanone oxime production process.

(Reproduced by permission of Wiley-VCH)

After production of cyclohexanone oxime, the remaining DSM/HPO process steps follow those of the conventional process and include production of 1.4 pounds of ammonium sulfate per pound of caprolactam. The yield of caprolactam is at least 74% of theoretical when based on cyclohexane. One pound of caprolactam per pound is produced per pound of cyclohexane consumed. When the process is based on phenol, the yield is approximately 93% of theoretical and give 1.12 pounds of caprolactam per pound of phenol consumed[266].

21.1.3. BASF Process

The BASF process differs from the conventional process principally in the method used to prepare the hydroxylamine sulfate that is needed for the oximation step. The BASF process differs from the conventional process principally in the method used to prepare the hydroxylamine sulfate that is needed for the oximation step. The steps in the BASF process are listed below and shown in Figure 21.2[46, 264, 266]:

1. Nitric Acid Production

$$NH_3 + O_2 \xrightarrow{\text{Steam}} NO \qquad (21.12)$$

2. Nitric oxide is catalytically hydrogenated in an ammonium hydrogen sulfate solution:

$$(NH_4)HSO_4 + {}^3/_2\, H_2 + NO \xrightarrow{\text{Catalyst}} (NH_3OH)(NH_4)SO_4 \qquad (21.13)$$

The catalyst is platinum on carbon that is suspended in a dilute sulfuric acid solution.

3. Cyclohexanone oxime is formed by reacting cyclohexanone with ammonium hydroxylammonium sulfate:

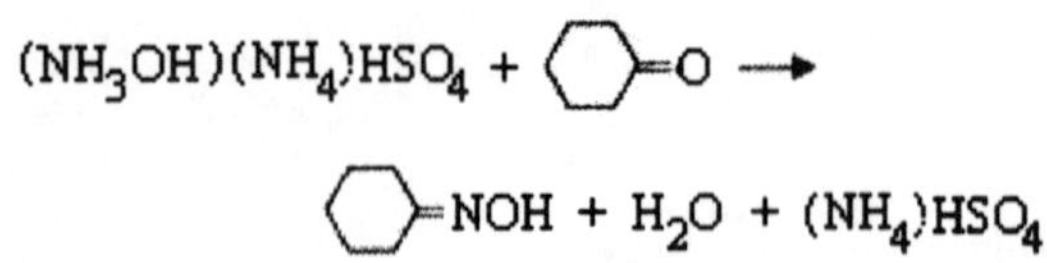

(21.14)

Oximation takes place in a special column by means of countercurrent flow of cyclohexanone and hydroxylamine solution with 97 to 98% conversion. Temperature is kept above the melting point of cyclohexanone oxime[264].

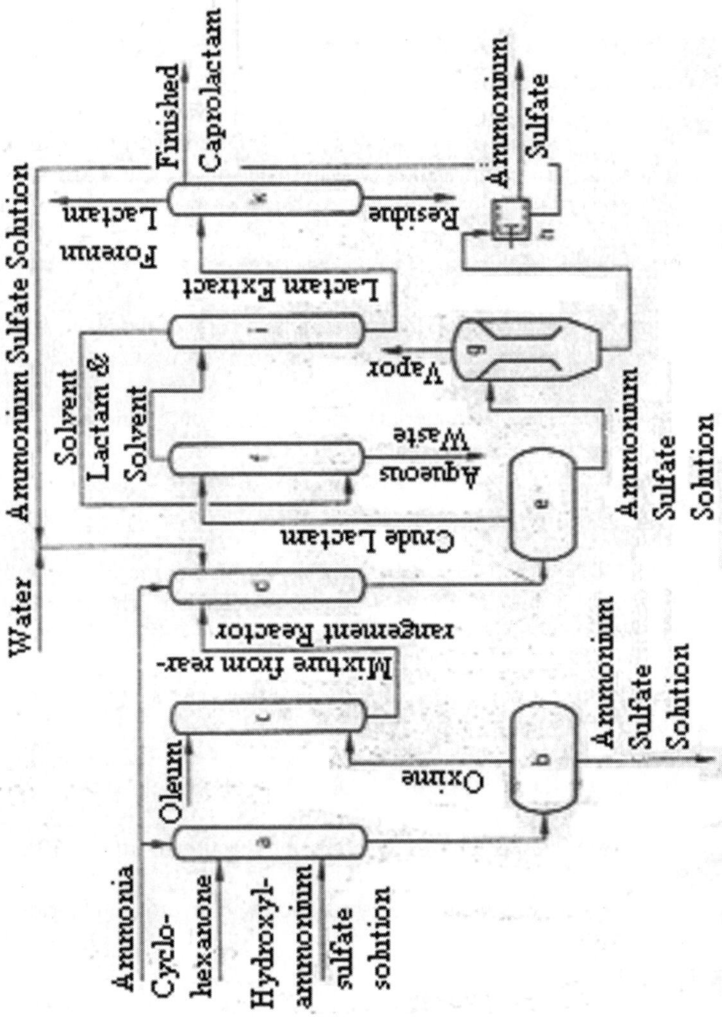

Figure 21.2. BASF caprolactam production. (Reproduced by permission of Wiley-VCH)

The above steps make 1.5 to 1.55 pounds of ammonium sulfate per pound of caprolactam produced. The entire process make 2.3 to 2.35 pounds of ammonium sulfate per pound of caprolactam. The yield of caprolactam is ~70%, which results in 0.94 pounds of caprolactam per pound of cyclohexane consumed[266].

21.1.4. Inventa-NO Reduction Process

The Inventa-NO process is similar to the BASF process in that the first source of by-product ammonium sulfate is eliminated. Like the BASF process the Inventa process prepares hydroxylamine sulfate by the reduction of nitric oxide with hydrogen over a noble metal catalyst suspended in sulfuric acid. The differences between the BASF and Inventa processes are:

1. Inventa uses plastic materials for mechanical parts of the plant, and
2. Inventa uses a palladium (rather than platinum) catalyst.

In the process the starting materials must be very pure, and an involved catalyst recovery step is necessary[46, 266].

21.1.5. PNC or Toray Process

In the 1950's Toray Industries developed a process for the Photochemical Nitrosation of Cyclohexane (PNC) to produce caprolactam. The steps in the PNC process are listed below and shown in Figure 21.3[46, 264, 266].

1. Preparation of nitrosylsulfuric acid from nitrous gases obtained from combustion of ammonia and sulfuric acid:

$$H_2SO_4 + NO + NO_2 \rightarrow 2\ NOHSO_4 + 3\ H_2O \quad (21.15)$$

The nitrous oxide (NO) can be produced by the reaction in Eq. (21.12).

2. Preparation of nitrosyl chloride by reaction with hydrogen chloride:

$$NOH_2SO_4 + HCl \rightarrow NOCl + H_2SO_4 \quad (21.16)$$

3. Photochemical reaction:

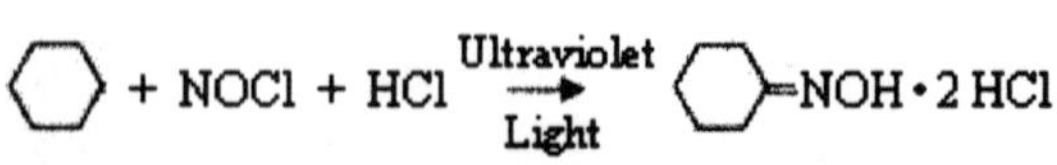

(21.17)

In the reaction of Eq. (21.17) a gas mixture consisting of HCl and NOCl is fed into cyclohexane at a temperature below 20°C. The reaction (which is initiated by Hg light) results in the formation of cyclohexanone oxime with 86% selectivity (based on benzene)[46].

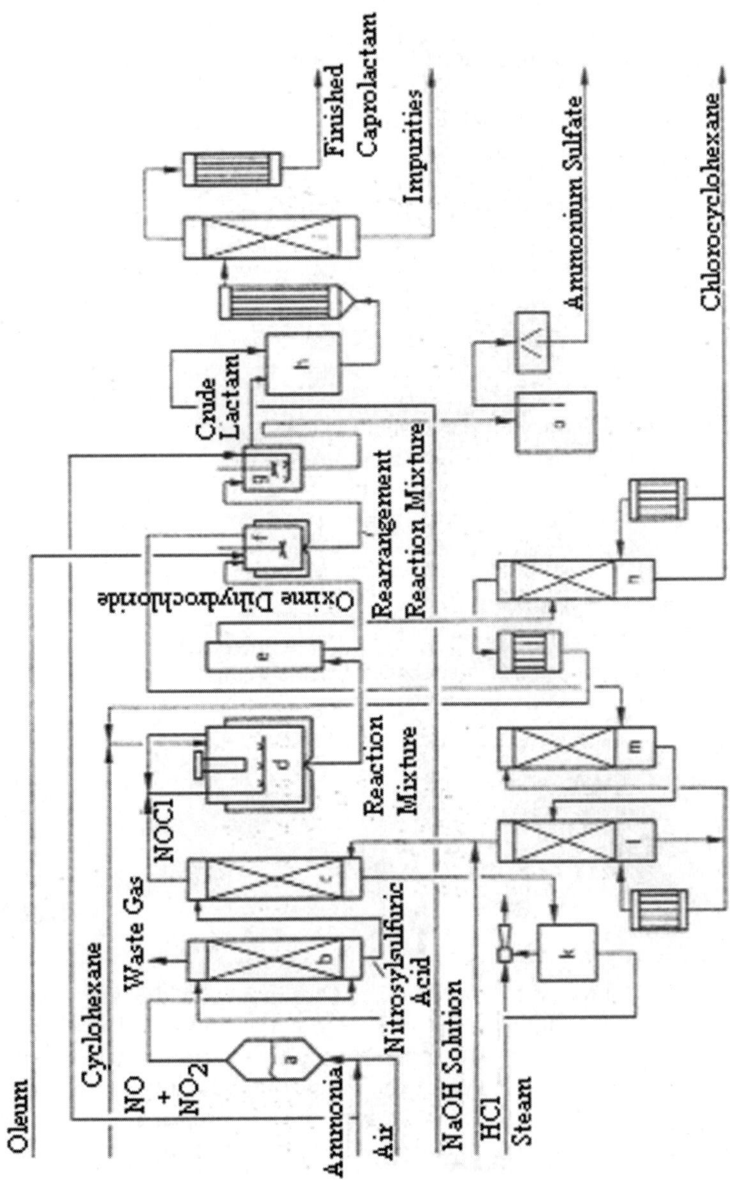

Figure 21.3. Toray PNC caprolactam production process.
(Reproduced by permission of Wiley-VCH)

The industrial PNC process is based on the development of efficient photoreactors. Toray designed an immersion lamp with a high radiation efficiency and capacity as well as long life. In order to remove the short-wave radiation below 365 nm (which contributes to tar formation on the lamps) either an absorbent is added to the cooling water, or the light source is surrounded by a glass filter[264].

The cyclohexanone in Eq. (21.17) exists in the form of oil droplets and forms a heavier lower phase in cyclohexane. This lower phase undergoes a Beckmann rearrangement with excess sulfuric acid or oleum to give caprolactam. Hydrogen chloride is displaced by the stronger acid and recycled to nitrosyl chloride production [see Eq. (21.16)]. The rearrangement reaction mixture is neutralized with ammonia water to give crude lactam and ammonium sulfate[264, 266].

Neutralization of the caprolactam-sulfuric acid mass with ammonia yields by-product ammonium sulfate equal to 1.75 pounds per pound of caprolactam. The yield of caprolactam is approximately 81% of theoretical and produces 1.09 pounds of caprolactam per pound of cyclohexane consumed[266].

As of 2003 Toray is the only commercial user of this process. They have two plants (startups in 1963 and 1971) with a total production capacity of 174,000 tonnes. Since electric power consumption is high, this process is attractive only where electric power costs are low[264, 266].

21.1.6. SNIA Viscosa Process

SNIA Viscosa developed a toluene-based process to produce caprolactam in 1960. The steps in the SNIA process are listed below and shown in Figure 21.4[46, 264, 266].

1. Catalytic oxidation of toluene with air to benzoic acid:

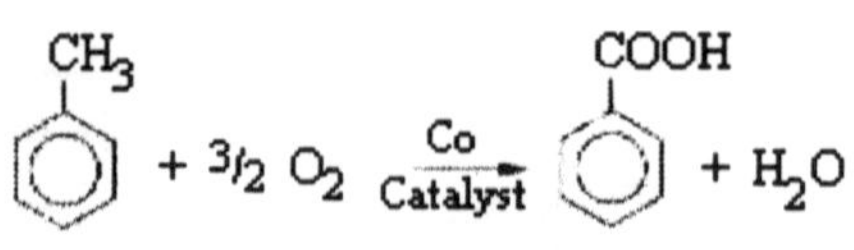

(21.18)

The oxidation of toluene with air is carried out in the liquid phase using a cobalt catalyst at 160 to 170°C and 0.8 to 1.0 MPa (8 to 10 bar). The yield is well above 90% of theoretical. The gases leaving the reactor contain mainly nitrogen with small amounts of oxygen, carbon dioxide and carbon monoxide. The are cooled to 7 to 8°C in order to recover unreacted toluene. The water from the reaction is removed in a separator drum and toluene is recycled to the reactor[264].

Most of the toluene is removed by distillation, and the remaining concentrated solution is sent to a rectification column. The lower-boiling intermediates and the remaining toluene are removed at the top and reused[264].

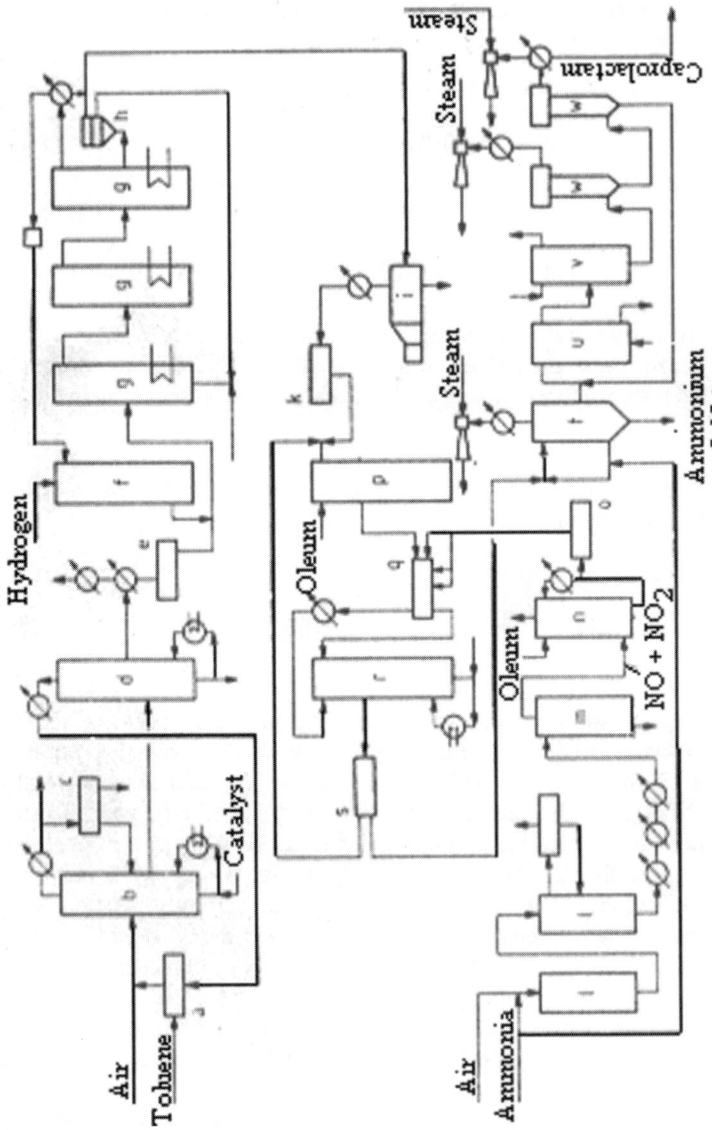

Figure 21.4. SNIA caprolactam production process.
(Reproduced by permission of Wiley-VCH)

Benzoic acid in the vapor is removed from the rectification column as a side stream. The high-boiling by-products leave the column as a residue. The benzoic acid is suitable for hydrogenation without any further purification[264].

2. Hydrogenation of benzoic acid to cyclohexanecarboxylic acid:

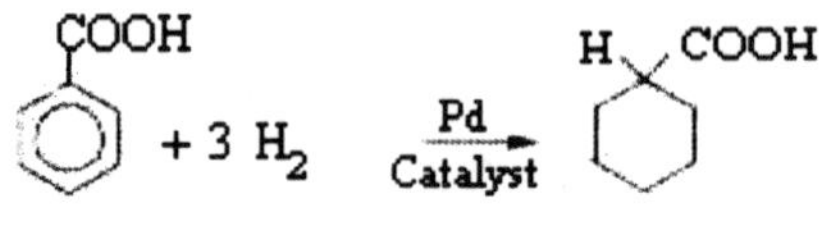

(21.19)

The reaction in Eq. (21.19) is carried out in the liquid phase in the presence of a palladium-on-graphite catalyst. A series of stirred reactors are used with a temperature of 170°C and a pressure of 1.0 to 1.17 MPa (10 to 17 bar). Conversion is 99.9%[264].

The catalyst is separated from the liquid reaction product in a centrifuge, mixed with benzoic acid and reused in hydrogenation. Cyclohexanecarboxylic acid is distilled under reduced pressure. Any catalyst still present remains in the residue[264].

3. Nitrosodecarboxylation of cyclohexane-carboxylic acid to caprolactam in the presence of oleum:

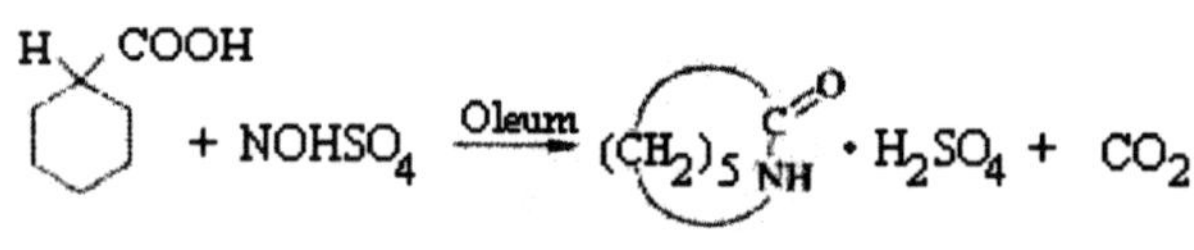

(21.20)

A 73% nitrosylsulfuric acid solution in sulfuric acid is used for the nitrosation of cyclohexanecarboxylic acid. It is obtained by absorption of nitrous gases [from ammonia combustion - see Eq. (21.12)] in concentrated sulfuric acid or oleum. Cyclohexanecarboxylic acid is premixed with oleum at ambient temperature and fed into a multistage nitrosation reactor that operates at up to 80°C. Exactly defined amounts of cyclohexanecarboxylic acid and nitrosylsulfuric acid are reacted at each stage to ensure complete reaction of the nitrosating agent. The conversion rate of cyclohexanecarboxylic acid is kept at about 50%. One mole of carbon dioxide per mole of caprolactam is set free along with small amounts of other gases such as nitric oxide, carbon monoxide, sulfur dioxide and nitrogen. To keep the reaction going and remove the heat of reaction, this process is carried out in boiling cyclohexane under atmospheric pressure. Depending on the organic acid, less than 10% of the yield is lost. During the reaction, 3.3 moles of sulfuric acid per mole of caprolactam are consumed. The acid caprolactam solution follows the Beckmann rearrangement steps shown in Eq. (21.8) to give caprolactam[46, 264].

The overall process produces 4.2 pounds of ammonium sulfate per pound of caprolactam. The yield of caprolactam in this process is approximately 72% of theoretical and makes 0.89 pounds of caprolactam per pound of toluene consumed[266].

In a newer process revision, caprolactam is extracted from the sulfuric acid solution with alkylphenols and then stripped with water. In this way the Snia Viscosa route can be a salt-free process. With a 50% conversion of cyclohexanecarboxylic acid, the selectivity to caprolactam is 90%[46].

In 2003 the Snia process was being used at two plants in Italy that have a total capacity of 100,000 tonnes per year and one plant in the former Soviet Union that has a capacity of 80,000 tonnes per year[46].

21.1.7. Recovery from Nylon 6 Waste

Solid nylon 6 waste is depolymerized in a kettle operation with the aid of a cracking catalyst and superheated steam[264].

The mixture of steam and caprolactam leaving the kettle is fed to an evaporation column, where the mixture is condensed and concentrated under atmospheric pressure. The concentrate is purified by adding an oxidizing agent that converts impurities to readily removable compounds. The caprolactam is purified through distillation[264].

The wash water obtained from extraction of nylon 6 chips contains oligomers of caprolactam as well as caprolactam. The oligomers are separated after preconcentration in a thin-film evaporator and then depolymerized in the same manner as the solid waste. The resulting aqueous caprolactam solution is purified chemically. Subsequent distillation gives water, a fraction containing low-boiling compounds and caprolactam as the principal product[264].

Oligomers can also be depolymerized in another way. The wash water from nylon 6 chip production is concentrated at a sufficiently high temperature to maintain the oligomers in solution. The caprolactam-oligomer solution is fed into a fixed- or fluidized-bed reactor. The oligomers are cracked at 275 to 350°C. A special aluminum oxide catalyst is used to give a yield of 95% relative to the oligomers in the feed. After cracking, the caprolactam is purified by conventional methods[264].

In 1999 Evergreen Nylon Recycling, a joint venture between DSM Chemicals and Honeywell, invested $US 100 million and started up a facility in Augusta, Georgia, USA, that was designed to process 200 million pounds per year (90 million kg per year) of used carpet. The plant was designed to use selective pyrolysis to produce 100 million pounds per year (45 million kg per year) of caprolactam and 100 million pounds per year of other material. The other materials include 1) calcium carbonate that comes from the chalk that is used as a filler in carpet backings (80 million pounds per year) and 2) other polymers that come from the carpet backing (20 million pounds per year). The 'other material' was initially used as feedstock in a cement kiln where the polymers added fuel value, and the chalk was a source of lime. In addition to these products, a liquid waste might be released that contains suspended dirt and other residues from the used carpet[269].

In August, 2001 production at the Evergreen Nylon Recycling plant was suspended indefinitely. Higher than expected production costs combined with poor business conditions for caprolactam led the decision to suspend operations. The plant had been operating at about 60% of its capacity, and the venture determine that further capital investment will be required to improve the recycle production costs to make the plant competitive[270].

The Polyamid 2000 plant in Premnitz, Germany is another nylon 6 carpet recycling facility. For an investment of $US 200 million in 2000, the plant was designed to process 264 million pounds per year of carpet and produce 20 million lb/yr of nylon 6 plus 25 million pounds of nylon 6,6 and other engineering plastics[271].

This plant uses novo-synthesis to remove caprolactam and the other components in nylon 6. Then the plant makes nylon 6 fiber rather than stoppping at caprolactam production[272].

In May 2002 the plant was only running at 75% of capacity (15 million lb/yr of nylon 6). One problem was that the nylon content in the carpet was lower than expected. Another problem is that landfill fees have dropped from $90/ton to $45/ton. It is estimated that landfill fees must be at least $75/ton for carpet recycling to be viable[271].

21.2. CAPROLACTAM PRODUCTION

Caprolactam production in the United States grew at 2.1% per year from 1995 through 2000. But the largest segment of caprolactam use, nylon fibers, has been characterized by no sustainable growth in the United States during the past ten years. Nylon fiber production in 2000 was 0.2% below 1990 US production. The main problem is that the US industry has not been able to compete on price with Asian imports. Nylon demand growth for carpet fibers remained positive through much of the 1990's due to economic growth. However weakness in the US economy resulted in negative growth in 2000 and 2001 in carpet and industrial fiber applications[265, 267].

Nylon resin demand had grown at a rate of 9% per year from 1989 through 1999. This growth was due to the replacement of metal parts with nylon resins in under-the-hood automotive applications and other end uses. Nylon resin production dropped 5% in 2000 and an estimated 15% in 2001 as consumer spending dropped. Poor demand for nylon throughout the world in the early years of the new century has hurt the global caprolactam market. Production was scaled back and prices have declined. The supply and demand for caprolactam are summarized in Table 21.2[265].

Table 21.2. Caprolactam Supply and Demand

	Million of Pounds						
	1995	1996	1997	1998	1999	2000	2001[b]
Capacity	1520	1540	1600	1720	1825	1925	1825
Total Prod.	1570	1595	1600	1687	1745	1700	1600
Imports	38	56	158	102	52	24	12
Exports	195	181	181	189	182	159	207
Demand[a]	1413	1470	1577	1600	1615	1565	1405

[a] Demand equals production plus imports minus exports.
[b] Forecast

(Reproduced by permission of Mannsville Chemical Products Corp.)

Caprolactam demand is projected to grow at 2.5% to 3.0% annually for several years after the US economy improves and reach 1.8 billion pounds in 2004. This growth will be due mainly to nylon engineering resins. Resin demand growth is expected to drop from 9% per year to 5% or 6% annually. Most of the growth will be due to uses in transportation and electrical/electronic devices because nylon 6 fiber production in North America is forecast to decline at 1% per year from 2001 through 2006. However global fiber demand is projected to grow at 1% to 1.5% during this time as the Asia / Pacific region provides the strongest growth. In the US apparel and industrial fiber uses are expected to decline slowly while carpet fiber demand will slowly increase with economic expansion[265, 267].

Caprolactam pricing is affected by raw material costs - principally benzene, which is used to make cyclohexane. From 1995 to 2000 the price varied from \$0.70 per pound to a high of \$0.75 per pound for molten caprolactam. Flake material in bags would cost \$0.10 to \$0.15 per pound more[265, 267].

21.3. CAPROLACTAM USES

In 2000 nylon-6 fibers (including monofilament) consumed 74% of the caprolactam for textile, industrial (i.e., tire cords, seat belts, and fishing line) and carpet yarns. Engineering resins and film consumed the other 26% of the production[265, 267].

Industrial caprolactam is 99.9% to 99.95% pure, and the main contaminant is usually water. Molten caprolactam is stored and shipped in stainless steel or aluminum containers at 75 to 80°C. The shipping containers are under a nitrogen blanket to keep the oxygen content less than 10 ppm and keep out moisture[264].

Flaked caprolactam is supplied in polyethylene bags for general purposes or in aluminum coated paper bags for anionic polymerization. Flake caprolactam should be stored in dry warehouses and protected against direct sunlight. It is advisable to keep the ambient temperature below 45°C and the relative humidity below 65%[264].

Caprolactam has a low order of toxicity. The 2001 ACGIH threshold limit value for caprolactam vapor is 5 ppm or 20 mg/m^3 for a time-weighted average, with a limit of 1 mg/m^3 for particulates[264]. In Germany the vapor limit is 25 mg/m^3.

22

OTHER NITROGEN COMPOUNDS

Several other nitrogen compounds are commercially important, and some of these compounds will be summarized in this chapter.

22.1. ADIPONITRILE AND HEXAMETHYLENEDIAMINE

Adiponitrile (ADN) is a white, odorless liquid. It is the dominant intermediate for the production of hexamethylenediamine (HMDA), a component in production of nylon 6,6.

It has a global capacity of 1.3 million tonnes per year. In the United States, production of ADN is based on the hydrocyanation of butadiene or electrochemical conversion from acrylonitrile. In Western Europe, companies produce ADN from adipic acid, butadiene and acrylonitrile. In Japan, the sole producer makes ADN from the electrodimerization of acrylonitrile. Demand for ADN is expected to be around 2% per year through at least 2010[273]. A comparison of the costs associated with two of the ADN processes are shown in Table 22.1[275].

Table 22.1. Compare Costs of ADN Production[275]

Metric Unit[a]		Hydrocyanation of Butadiene	Electrohydrodimerization of Acrylonitrile
Material	lb/$VA	1.44	0.17
Energy	kBtu/$VA	59.4	92.1
Water	gal/$VA	16.2	15.4
Toxics	lb/$VA	0.0023	0.0000
Pollutants	lb/$VA	0.81	0.008
CO_2	lb/$VA	8.85	13.2

[a] $VA = dollar value-added

The chemical formula for ADN is:

$$NC(CH_2)_4CN$$

Hexamethylenediamine (HMDA) is a colorless solid when pure, but it slowly degrades to colored products when it contacts air. It is commercially available as the anhydrous product or in aqueous solutions. In 2003 it had a global capacity that was approaching 3.0 billion pounds per year, and it was made by only one main route: hydrogenation of ADN. Commercially the most important use of HMDA is in a polycondensation reaction with adipic acid to eventually give nylon 6,6[274]. The chemical formula for HMDA is:

$$H_2N(CH_2)_6NH_2$$

22.2. ACRYLONITRILE

Acrylonitrile (ACRN) is a clear, colorless, volatile liquid with a slightly pungent odor. It is very toxic, highly flammable and unstable. It is classified as a carcinogen and a mutagen. The chemical formula for ACRN is:

$$CH_2{=}CHCN$$

22.2.1. Process Technology

The main route to ACRN is the one-step propylene ammoxidation process. In this process propylene, ammonia and air reacted in a fluidized bed reactor to produce ACRN with acetonitrile and hydrogen cyanide as by-products. New technology based on propane ammoxidation has been developed by BP, Mitsubishi (in conjunction with BOC) and Asahi Kasei with claims of a 30% production cost advantage over the propylene route[276]. However no plans have been announced to build a propane-based plant as of first quarter 2004[297].

22.2.2. Production

Demand in the United States was flat in 2000 and 2001 at about 1.68 billion pounds per year. Demand is expected to grow to 1.8 billion pounds per year by 2005[277].

Worldwide demand is expected to grow at 2% to 3% per year from 2002 through 2008 with higher growth in China. New capacity that came onstream in 2000 had not been fully absorbed by 2002. Therefore more rationalization was expected with possible plant closures of uncompetitive plants in Europe and Asia[276]. Global ACRN capacity was 5.9 million tonnes per year in 2002. World ACRN demand is shown in Table 22.2 and shows that capacity utilization is 80% to 85%[278].

Table 22.2. World Acrylonitrile Demand (1000 tonnes per year)[278]

Region	1998	1999	2000	2001	2002 Forecast
North America	759	769	793	761	758
Latin America/Mexico	255	259	267	291	294
Asia	1997	2139	2285	2298	2406
Middle East / Africa	262	254	272	261	275
Europe	1262	1205	1233	1225	1241
Totals	4535	4626	4850	4836	4974

(Reproduced by permission of Asian Chemical News, www. asianchemicalnews.com)

Global overcapacity plagued the ACRN market during the early years of the new century (see Table 22.3[292]). The overcapacity is mainly due to start up of new capacity in Asia although some European plants have been shutdown[297]. ACRN price is quite volatile. Between the first quarter of 2000 and the second quarter of 2003, the contract prices varied from 900 Euros per tonne (in 1Q00 and 1Q02) to over 1200 Euros per tonne (in 2Q03). During the same time, the spot price varied from $US 440 per tonne (in 1Q02) to almost $US 1000 per tonne (in 2Q00) [276].

Table 22.3. 2001 ACRN Capacity and Consumption (1000 tonnes per year)[292]

Region	Capacity (kt/yr)	Output (kt)	Consumption (kt)	Operating Rate (%)	Exports (kt)
North America	1,958	1,380	672	70	708
Western Europe	1,205	1,047	1,167	87	120
Japan	775	740	721	97	19
Korea	387	336	381	87	-45
China	422	380	687	90	-307
Taiwan	380	345	374	91	-29
India	30	27	122	90	-95
Thailand	0	0	96	-	-96
Totals	5,157	4,255	4,220	83	-

China's ACRN capacity is growing rapidly. It has grown from 422,000 tonnes per year in 2001, to 487,000 tonnes per year in 2002, to 571,000 tonnes per year in 2003. And it is expected to reach 900,000 tonnes per year by 2005 - an increase of over 115% in five years. However Chinese demand is expected to reach 1,100,000 tonnes per year by 2005. Therefore imports of at least 200 tonnes per year will still be needed[292, 293].

22.2.3. Uses

ACRN is used to make acrylic fibers, acrylonitrile-butadiene-styrene (ABS), and styrene-acrylonitrile (SAN). Worldwide acrylic fiber accounts for over half of total demand while ABS and SAN consume about 30% of output. Smaller applications include nitrile rubber copolymers (4%), adiponitrile (ADN) and acrylamide. Acrylic fibers are used in carpets and clothing while ABS and SAN resins are used in pipes and fittings, automobiles, furniture, and packaging[276]. In the United States the ACRN uses are distributed differently: 38% is used in ADN, 22% in ABS and SAN, 17% in acrylic fibers, 11% in acrylamide, 3% in nitrile elastomers, and 9% in miscellaneous, which includes polymers, polyols, barrier resins and carbon fibers[277].

22.3. METHYLMETHACRYLATE

Methylmethacrylate (MMA) is a clear, colorless liquid with an ester-like odor odor. It is volatile and highly flammable. Its heavy vapor can roll long distances and flash back. It will also polymerize at high temperatures. The chemical formula for MMA is:

O
||
$H_2C{=}C{-}C{-}OCH_3$
|
CH_3

22.3.1. Process Technology

About 80% of worldwide capacity is based on acetone cyanohydrin (ACH), which is made when acetone reacts with hydrogen cyanide. The ACH is converted to methacrylamide sulphate, which is treated with a methanol/water mixture and heated to form MMA and ammonium bisulfate[279].

Because of the problems with disposal of the bisulfate waste and the handling of HCN, much research has been devoted to alternative processes. The new processes range from using new feedstocks such as isobutylene / t-butyl alcohol, ethylene, isobutane or methylacetylene to techniques for recycling the HCN and / or ammonium bisulfate[279, 280]. In 1998 Asahi replaced 60,000 tonnes per year of MMA capacity based on direct oxidation of isobutylene with a new process that also starts with isobutylene. However the new "direct oxidative esterification" (DOE) process makes MMA by the simultaneous oxidation and esterification of methacrolein, which eliminates the intermediate production of methacrylic acid[298].

22.3.2. Production

Worldwide demand is expected to grow at 3% to 5% per year from 2002 through 2006 with higher growth in Asia. Poor markets have delayed some investment plans from 2002 to 2005 or 2006. Global MMA capacity was 2.3 million tonnes per year in 2002. World MMA demand was estimated at 2.1 million tonnes in 2002 so capacity utilization is 90% to 95%. Growth will depend on economic recovery, and it could be higher if the economy improves rapidly[279, 280].

European MMA prices have varied. Between the third quarter of 2001 and the fourth quarter of 2002, the contract prices varied from 1050 Euros per tonne (2Q02) to over 1300 Euros per tonne (in 4Q02). During the same time, the spot price varied from Euros 865 per tonne (in 1Q02) to over $US 1300 per tonne (in 4Q02)[279].

22.3.3. Uses

The largest use for MMA is in polymethyl methacrylate (PMMA), which is used as a substitute for glass in products such as car taillights and lighting fixtures. MMA has good optical clarity, good weather-resistance and is light weight. Another major application is in surface coatings, particularly emulsion polymers for architectural and maintenance paints. Methyl methacrylate-butadiene-styrene (MBS) resins are used as an impact modifier for clear rigid PVC and in food and pharmaceutical packaging. MMA has also partially replaced styrene in unsaturated polyester resins to give better weather resistance and longer life[279].

22.4. ACRYLAMIDE

Acrylamide is a water-soluble monomer with two reactive centers (a vinyl group - with its reactive double bond, and an amide group). Because of its high reactivity, aqueous acrylamide monomer is stabilized with dissolved cupric salts and oxygen to prevent polymerization during shipping and storage. There are some toxicity, safety and handling concerns related to dry acrylamide[281]. The chemical formula for acrylamide is:

$$\begin{array}{c} \quad\quad\;\; O \\ \quad\quad\;\; \| \\ CH_2{=}CH{-}C{-}NH_2 \end{array}$$

22.4.1. Process Technology

The principal synthetic route to making acrylamide involves the hydration of acrylonitrile (ACRN). In this process an aqueous ACRN solution reacts over a copper-oxide-chromium oxide catalyst at approximately 100°C. Several other catalyst systems have been used, and most of them contain copper - in some form. The reaction step is followed by purification and concentration to a 50% solution in a vacuum evaporator. The yield of acrylamide from ACRN is 98%[281]. The purification and concentration steps are costly and also involve the recycle of ACRN back to the reaction step. In the early part of the new century, a catalytic distillation process has been developed that converts almost 100% of the ACRN to acrylamide and allows concentration to occur in the same column where acrylamide is made. Therefore this process is less costly[282].

Nitto Chemical (now Dia-Nitrix) introduced a biosynthetic route from ACRN to acrylamide in Japan in 1985. This process uses an immobilized nitrile hydratase biocatalyst that converts the ACRN solution to acrylamide with a yield of 99.5%. This high yield allows a concentrated acrylamide solution to be made without the need for ACRN recycle or solution concentration. This process therefore has lower energy costs[282]. The initial plant capacity was 4,000 tonnes per year; it was expanded to 20,000 tonnes per year in 1991; and that is its capacity as of 2002.

22.4.2. Production

Worldwide demand is expected to grow at 3% to 5% per year from 2000 through 2005. Global acrylamide capacity was about 550,000 tonnes per year in 2002. World acrylamide demand was estimated at 400,000 tonnes in 2002 so capacity utilization is 80% to 85%. Growth in the United States was 5% per year from 1989 through 1998. It was predicted to be 4% per year from 1998 through 2003. Production in the United States was 165 million pounds in 1998, 170 million pounds in 1999, and it is forecast to be 205 million pounds in 2003[281, 283].

Acrylamide's growth depends largely on polyacrylamides (PAMs) and their primary markets in water treatment and the paper industry. Growth in the use of acrylamide for the production of N-methylolacrylamide and other cross-linking monomers is steady, but it is starting from a small base. Acrylamide and PAMs have significant growth potential outside the United States as more countries implement more thorough water treatment practices[283].

Acrylamide list prices between 1990 and 2001steadily increased from $0.70 per pound to $0.82 per pound[283].

22.4.3. Uses

Over 90% of acrylamide is used to make polyacrylamides (PAMs), and the remaining 10% is used to make N-methylolacrylamide (NMA) and other monomers. Water treatment PAMs consumed 60% of the acrylamide; PAMs for pulp and paper production consume 20% of the acrylamide; and PAMs for mineral processing consume 10% of the acrylamide[283].

Some of the specific uses of acrylamide are[283]:

- In liquid-solid separation where acrylamide polymers act as flocculants and aids in mineral processing, waste treatment and water treatment. They also help reduce sludge volumes in these applications.
- As additives in the manufacture of paper and paper board products, leather and paint industries. In the paper industry PAMs act as retention aids during wet end processing and in wet strength additives.
- In the manufacture of synthetic resins for pigment binders for textile/leather industries, and
- In enhanced oil recovery.

22.5. DIMETHYLFORMAMIDE

Dimethylformamide (DMF) is colorless liquid with an unpleasant, fishy odor. It is a solvent that dissolves a wide variety of organics. It has been called the universal solvent because of its wide organic and inorganic solvency. DMF owes this versatility to its high dielectric constant, wide liquid range and low volatility. DMF is completely miscible in water and is both chemically and thermodynamically stable[284]. The chemical formula for DMF is:

$$\begin{array}{c} \qquad\quad O \\ \qquad\quad \| \\ CH_3\text{-}N\text{-}CH \\ | \\ CH_3 \end{array}$$

22.5.1. Process Technology

Two processes are used commercially to produce DMF. In the direct or one-step process, dimethylamine and carbon monoxide react at 100°C and 200 psia in the presence of a sodium methoxide catalyst to make DMF. The homogenous catalyst is separated from the crude DMF, which is then refined to the final product[285, 286].

In the indirect process, methyl formate is isolated, and then reacted with dimethylamine to form DMF. To obtain methyl formate, two methods may be used - dehydrogenation of methanol and esterification of formic acid[285].

22.5.2. Production

Worldwide capacity in 2002 was estimated to be over 360,000 tonnes of DMF per year. China consumes about 35% of the worldwide capacity (126,000 tonnes per year, and their consumption is rising sharply at a rate of 8% to 10% per year[287].

22.5.3. Uses

DMF is used primarily in the pharmaceutical processing and acrylic fiber production industries. These uses account for about 50% of the total demand.

Some of the specific uses of DMF are[284+, 286, 288].

- As a selective solvent for acetylene, butadiene, acid gases, inorganic salts and certain constituents of petroleum.
- As a solvent for selective extraction - especially butadiene extraction and other selective gas absorptions.
- Solvent and reaction medium for the manufacture of polyacrylonitrile, polyurethanes, polyvinyl resins, dyes and pharmaceuticals.
- As a recyclable solvent in the manufacture of acrylic fiber and in polyurethane coatings.
- Aircraft fuel combustion catalyst, and
- As an intermediate in the synthesis of acetals, aldehydes, esters, formamidines, peptides and nitriles..

22.6. CALCIUM CYANAMIDE

Calcium Cyanamide ($H_2N\text{-}CN$) is used as a fertilizer, herbicide, insecticide, a steel-making additive and an ore processing material. It can also be used to make thiourea, guanidine and ferrocyanides[142]. SKW Chemicals in Germany (now part of Degussa) claims to be the world leader in production of cyanamide and cyanamide-based chemical intermediates.

To make calcium cyanamide, atmospheric nitrogen can be fixed by the Frank-Caro or Cyanamide process. This process is based on the fact that metallic carbides, particularly calcium carbide, readily react with nitrogen to form the metal cyanamide. In the first step of this process lime and carbon are heated to produce calcium carbide. Then fairly pure nitrogen reacts with the calcium carbide to form calcium cyanamide:

$$CaC_2 + N_2 \xrightarrow{1000^{\circ}C} \underset{\text{Calcium Cyanamide}}{CaNCN} + C \qquad (22.1)$$

The calcium cyanamide can be hydrolyzed with steam to yield ammonia and calcium carbonate. A. Frank and N. Caro developed this process in Germany in 1895. It was used in Germany, Norway and Italy until it was replaced by the Haber-Bosch process after World War I[1, 12]. In 2002 the Frank-Caro process is still in use in many countries including Germany, South Africa, Japan and China.

The commercially formulated product contains approximately 65% calcium cyanamide, which is 20% to 24% nitrogen. For most of the twentieth century it has been used as a fertilizer, a cotton defoliant, a herbicide and a soil insecticide.

A 1999 study in France showed that it greatly improves the quality and yield of cauliflower crops by acting as both a herbicide and an insecticide[34].

22.7. OTHER COMPOUNDS

A few other nitrogen compounds that are produced are listed below:

- Dimethylacetamide [$CH_3CON(CH_3)_2$] - an important industrial solvent for polyacrylonitrile, vinyl resins, cellulose derivatives, styrene polymers and linear polyesters.
- Isocyanates - important materials in the production of foams, resins and rubbers. (See comments in Chapter 20 about the use of Aniline to make some of these compounds.)
- Cyanoacrylates and Light-Curing Acrylic Adhesives - These types of adhesives have achieved widespread acceptance in high-speed manufacturing because they cure rapidly, offer high bond strength to many substrates and are easy to dispense[289]. (See Reference 189 for the structure of these chemicals.)
- Pyridines - The industrially significant pyridine bases are pyridine, 2-methylpyridine (2-picoline) and 2-methyl-5-ethylpyridine (MEP)[46].
  - In 1997 much of the pyridines were still isolated from coal tar, where they occur with other low-boiling derivatives at 0.1 wt.
  - Some synthetic routes to alkylpyridines are increasing in importance due to growing demand.
  - In 1995, the world capacity of synthetic pyridines was about 94,000 tonnes per year.

Numerous other nitrogen compounds exist. However this book has attempted to describe the major synthetic nitrogen products that were produced in 2003.

REFERENCES

1. Comyns, A.E., Named Processes in Chemical Technology, 2nd ed., CRC Press LLC, Boca Raton, 1999.
2. Knoblauch, K., Chem. Eng. (N.Y.), 1978, 85(25) 87.
3. Suzuki, M. in Adsorption and Ion Exchange: Fundamentals and Applications, LeVan, M.D., Ed., American Institute of Chemical Engineers, New York, 1998, 120.
4. Suzuki, M. in Adsorption and Ion Exchange: Fundamentals and Applications, LeVan, M.D., Ed., American Institute of Chemical Engineers, New York, 1998, 121-122.
5. Eur. Chem. News, 1989, 53(1391), 31.
6. Chem. Mark. Rep., 1990, 29 Oct. 5.
7. Hopkins, B.S., Smith, H.R., et. al., Chemistry and You, Lyons & Carnahan, Chicago, 1949.
8. Chem. Eng. News, p. 16 (Aug. 27, 1979).
9. Haber, F. and van Oordt, G., Z. anolrg. Allg. Chem., 1905, 43, 111.
10. Harding, A.J., Ammonia, Manufacture and Uses, Oxford University Press, London, 1959.
11. Haber, L.F., The Chemical Industry 1900-1930, Clarendon Press, Oxford, 1971, 198.
12. Vancini, C.A., Synthesis of Ammonia, translated by L. Pirt, Macmillan Press, Basingstoke, England, 1971.
13. Jennings, J.R. and Ward, S.A., in Catalyst Handbook, 2nd ed., Twigg, M.V., Ed., Wolfe Publishing, London, 1989, 384.
14. Travis, T., Chem. Ind (London), 1993, (15) 581.
15. Livingston, J.G. and Pinto, A., Chem. Eng. Prog., 1983, 79(5), 62.
16. Chem. Eng. (Rugby, Eng.), 1990, 21.
17. Hydrocarbon Process., 1991, 70(3), 132-136.
18. Isalski, W.H., Separation of Gases, Clarendon Press, Oxford, 1989, 155.
19. Nitrogen, 1996, (223), 25.
20. Chem. Eng. Prog., 1952, 48, 468.
21. Shearon, W.H., Jr. and Thompson, H.L., in Modern Chemical Processes, Vol. 3, Reinhold Publishing, New York, 1954, 16.
22. Chem. Week, 1991, 3 Apr, pp. 13.
23. Eur. Chem. News, 1992, 58(1524), pp. 42.
24. Eur. Chem. News, 1993, 60(1592), pp. 27.
25. Oil & Gas J., 1996, 94(47), 37.
26. Chem. Eng. (N.Y.), 1989, 96(7), 43.
27. Chem. Eng. (Rugby, England), 1990, (471), 21.
28. Hydrocarbon Process., 1991, 70(3), pp. 134.

29. Hill, W.H., in Chemistry of Coal Utilization, Vol. 2, Lowry, H.H., Ed., John Wiley & Sons, New York, 1945, 1029.
30. Cohen, J.M., The Life of Ludwig Mond, Methuen & Co., London, 1956, 176.
31. Scholvien, W.R., Chem. Met. Eng., 1931, 38(2) 82.
32. Spitz, P.H., Petrochemicals, The Rise of an Industry, John Wiley & Sons, New York, 1988, 84.
33. Chem. Eng. News, p. 16 (Aug. 27, 1979).
34. Terra e Vita (France), 1999, 40(22), 78-79.
35. SRI International, Chemical Economics Handbook, Nitrogen Industry Overview, 1998.
36. www.fertilizer.org/STATSIND, International Fertilizer Industry Association, June, 2001.
37. Ammonia Chemical Profile, Chemical Marketing Reporter (Nov. 29, 1999).
38. De Guzman, D., Ethanolamines Face Oversupply and Raw Material Cost Pressures, Chemical Market Reporter (June 18, 2001).
39. Personal Communication, D.L. Mills to G.R. Maxwell, BP Nitrogen Fatality, July 10, 2001.
40. SRI International, Chemical Economics Handbook, Hydrogen Product Review, 1998.
41. Hydrogen, Airco Industrial Gases, Airco, Inc., 1991.
42. Green, D.S., Chemical Processing, 2001, Jan., 74.
43. Alves, J., "Analysis and Design of Refinery Hydrogen Distribution Systems," PhD Thesis, UMIST (1999).
44. Hallale, N. and F. Liu, "Efficient Refinery Hydrogen Management for Clean Fuels Production," presented at the AIChE Annual Meeting, Los Angeles (2000).
45. Hallale, N., "Burning Bright Trends in Process Integration", Chemical Engineering Progress, 2001, 97 (7), pp. 30-41.
46. Weissermel, K and Arpe, H.-J., Industrial Organic Chemistry, 3rd Ed., VCH Publishers, New York, 1997.
47. U.S. Patent 3,827,987.
48. Wolff, D., "Producing Hydrogen On-Site", Chemical Engineering Progress, 2001, 97 (3), pp. 51-53.
49. Am. Petrol. Inst., "Selected Values of Properties of Hydrocarbons," Research Project 44, Nat. Bureau of Standards.
50. Leseur, H., Czernichowski, J., and Chapelle, J., Int. J. Hydrogen Energy, 1994, 19(2), 139.
51. Hydrocarbon Process., 1996, 75(4), 149.
52. PEP Yearbook, 1996, Process Economics Program, SRI Consulting.
53. www.efma.org/Publications, Production of Ammonia, European Fertilizer Manufacturers Association, July 2001.
54. www.mbendi.co.za, Mbendi Information Services (Pty) Ltd., South Africa, July 2001.
55. www.hydrocarbononline.com, Hydrocarbon Online News, Vert Tech LLC, February 16, 1999.
56. www.hydrocarbononline.com, Hydrocarbon Online News, Vert Tech LLC, July 17, 2001.
57. SRI International, Chemical Economics Handbook, Ammonia, 2001.
58. Tebo, M., "Honda Opens First Los Angeles-Area Hydrogen Fuel Station", PRNNewswire, American Honda Motor Co. Press Release, July 10, 2001.
59. Gronich, S., "DOE Hydrogen Program Overview", Proceedings of the 2000 U.S. DOE Hydrogen Review, May 9, 2000.
60. Muradov, N. , "Thermocatalytic CO_2-Free Production of Hydrogen from Hydrocarbon Fuels," Proceedings of the 2000 U.S. DOE Hydrogen Review, May 9, 2000.
61. Noble, T., "Microreactors: Smaller Can Be Better", Chemical Engineering Progress, 97 (7), pp. 10-11.
62. Lombard, J.F., Hydrocarbon Proc., 48 (8), 111 (Aug. 1969).
63. Lundberg, W.C., Chem. Eng. Prog., p. 81 (June 1979).
64. Abrardo, J.M., and Khurana, V., "Hydrogen Technologies to Meet Refiners' Future Needs," Hydrocarbon Processing, February 1995, pp. 43-49.
65. Chem. Eng. Prog., 69 No. (2), 67-70 (Feb. 1973).
66. Chem. Eng., 67 (19), 166-169 (Sept. 19, 1960).
67. Chem. Eng., pp. 88-89 (Dec 3, 1979).
68. Hydrocarbon Proc., pp. 145-151 (Apr. 1978).
69. "Amine Guard Systems in Hydrogen Production," AIChE, 86th National Meeting, Houston, TX, Apr. 1-5, 1979.

70. www.synetix.com, Hydrogen Plant Products, Imperial Chemical Industries PLC, August 14, 2001.
71. Foreman, J.M., "Pre-Reformer Aids Syngas Units", Hydrocarbon Processing, (69) 12, December 1990, pp. 34B-34D.
72. Farnell, P.W., "Synetix's Advanced Gas Heated Reformer", AIChE Ammonia Safety Symposium, Seattle, WA, September, 1999.
73. Carstensen, J.H., et. al., "New Developments in Shift Catalysts for Ammonia", Hydrocarbon Processing, (69) 3, March 1990, pp. 57-62.
74. Bakemeier, H., Huberich, T., et. al., in Ullmann's Encyclopedia of Industrial Chemistry, Vol. A2, 5th Ed., Reinhold Publishing, New York, 1985.
75. www.haldortopsoe.com , Synthesis Gas Technologies, Haldor Topsøe A/S, August 13, 2001.
76. www.shell.com, "Generating Energy Using Gasification", Royal Dutch/Shell Group of Companies, August 13, 2001.
77. Bajusz, I.G. and Goodwin, Jr., J.G., "Hydrogen and Temperature Effects on the Coverages and Activities of Surface Intermediates during Methanation on Ru/SiO2," J. of Catalysis 169, 157-165 (1997).
78. Hoost, T.E., "Chemical Modification of Supported-Metal Catalysis," Ph.D Dissertation, University of Pittsburgh, Pittsburgh, PA, 1991.
79. www.itd.anl.gov/highlights/9-2/ceramic.html, Argonne National Laboratory, August 13, 2001.
80. www.hydrocarbononlline.com Telepneff, M.,"Hyprotech and Synetix Integrate Catalytic Reactor Models," Hydrocarbon Online News, Vert Tech LLC, August 10, 2001.
81. www.halliburton.com/KBR/KBRTCH, Kellogg Brown & Root – a Halliburton Company, August 22, 2001.
82. www.synetix.com, Ammonia Plant Products, Imperial Chemical Industries PLC, August 22, 2001.
83. Thompson, D., "New Advances in Gold Catalysis Part II", Gold Bulletin, 1999, 32 (1), pp. 12-19.
84. Thompson, D., "Highlights from Recent Literature", Gold Bulletin, 2000, 33 (1), pp. 33.
85. www.haldortopsoe.com , Ammonia Synthesis, Haldor Topsøe A/S, August 22, 2001.
86. Fleshman, J.D., "Cost Efficient Revamps in Hydrogen Plants", Proceedings of NPRA Annual Meeting, New Orleans, USA, March 2001.
87. Fleshman, J.D., McEvoy, L.J., et. al., "New Hydrogen Plant Design Achieves Low Cost and High Efficiency", Proceedings of NPRA Annual Meeting, San Antonio, USA, 1999.
88. Nair, M.P.S., "Control Corrosion Factors in Ammonia and Urea Plants", Hydrocarbon Processing, 80 (1), pp. 85-93, January 2001.
89. www.dakota.com, The Coal Gasification Process, Dakota Gasification Company, July 16, 2001.
90. www.kvaerner.com, Compact Reformer Technology, Kvaerner Process Technology, August 27, 2001.
91. SRI International, Chemical Economics Handbook, Nitric Acid, 2001.
92. *International Critical Tables,* Vol. III, pp. 304, 305, 309, McGraw-Hill Book Co., Inc., New York,1928.
93. *International Critical Tables of Numerical Data, Physics, Chemistry, and Technology,* Vol. IV, p. 255, McGraw-Hill Book Co., Inc. New York, 1928.
94. Perry, R. H., and Chilton, C. H. (Eds.), *Chemical Engineers' Handbook,* 5th Ed., McGraw-Hill Book Co., Inc., New York, 1973.
95. Rossini, F. D., and Coworkers, National Bureau of Standards, Circular 500, 53 (1952).
96. Encarta.msn.com, "Ostwald, Wilhelm", Microsoft Corporation, 2001.
97. www.efma.org/Publications, Production of Nitric Acid, European Fertilizer Manufacturers Association, September 2001.
98. Sauchelli, "Fertilizer Nitrogen," ACS Monograph Series 161.
99. www.enviro-chem.com, Nitric Acid Plant Technology, Monsanto Enviro-Chem, 2001.
100. www.yet2.com, Nitric Acid Processes, yet2.com Inc., 2001.
101. www.chemetics.kvaerner.com, Nitric Acid Concentration, Kvaerner Pulp & Paper, 2001.
102. Powell, R., "Nitric Acid Technology – Recent Developments," Noyes Development Corp., Park Ridge, NJ, 1969.

103. www.efma.org/Publications, Production of Ammonium Nitrate, European Fertilizer Manufacturers Association, September 2001.
104. Kirk-Othmer, Encyclopedia of Chemical Technology, Vol. 17, pp. 80-107, John Wiley & Sons, Inc., New York, 1997.
105. www.agcentral.com/imcdemo, Production of Major Nitrogen Fertilizers, IMC-Agrico Company, 2001.
106. www.sasol.com/smx, Ammonium Nitrate, Sasol SMX, 1998.
107. Chemical Week, 101 (2), 75, July 8, 1967.
108. Kirk-Othmer, Encyclopedia of Chemical Technology, Supplement, pp. 597-621, John Wiley & Sons, Inc., New York, 1996.
109. Meessen, J.H. and Petersen, H., et. al., in Ullmann's Encyclopedia of Industrial Chemistry, Vol. A27, 5th Ed., Reinhold Publishing, New York, 1996.
110. www.efma.org/Publications, Production of Urea and Urea Ammonium Nitrate, European Fertilizer Manufacturers Association, September 2001.
111. www.agrium.com, Carseland Nitrogen Operations, Agrium, Canada, 3/01.
112. www.enviro-chem.com, Urea Plant Technology, Monsanto Enviro-Chem, 20001.
113. SRI International, Chemical Economics Handbook, Urea, 2001.
114. SRI International, Chemical Economics Handbook, Melamine, 1999.
115. "DSM Expands Melamine Production in Netherlands", European Paint and Resin News, 38 (12) pp. 5, Dec. 2000.
116. SRI International, Chemical Economics Handbook, Alkylamines, 2001.
117. www.kvaerner.com, Industrial Amine Process Technologies, Kvaerner Process Technology, August 27, 2001.
118. Cyclohexylamine Chemical Profile, Chemical Market Reporter (June 4, 2001).
119. Methylamines Chemical Profile, Chemical Market Reporter (April 24, 2000).
120. "Ethanolamines", European Chemical News, 74 (1953), pp.16, 16-22 April, 2001.
121. www.owlnet.rice.edu/~ceng403/ethamsep.html, Garcia, C.A., Reisdorf, M. and Palombo, L., "Separation Unit for Ruehl Corp. Ethanolamine Plant", Rice University, 10/25/97.
122. www.himtek.ru/english/technology1.shtml, Technology of Ethanolamine Production, Himtek Engineering, 2000.
123. Ethanolamine Chemical Profile, Chemical Market Reporter (June 11, 2001).
124. Ethyleneamines Chemical Profile, Chemical Market Reporter (August 9, 1999).
125. www.chemicals-india.com, Hexamine Manufacturing Process, Kothari Phytochemicals International, 10/5/01.
126. www.wrightcorp.com, Formaldehyde and Hexamine, Wright Corporation, 10/5/01.
127. www.saudiform.com, Hexamethylenetetramine (HMT), Saudi Formaldehyde Chemical Company Ltd., 1999.
128. www.hydrazine.com, About Arch Hydrazine, Arch Hydrazine, 2001.
129. Kirk-Othmer , Encyclopedia of Chemical Technology, Vol. 7, pp. 753-782, John Wiley & Sons, Inc., New York, 1993.
130. D. Hasenberg, "HCN Synthesis on Polycrystalline Platinum and Rhodium", dissertation, University of Minnesota, 1984.
131. SRI International, Chemical Economics Handbook, Hydrogen Cyanide, 2000.
132. Schirmann, J.P, in Ullmann's Encyclopedia of Industrial Chemistry, Vol. A13, pp. 177-190, Reinhold Publishing, New York, 1996.
133. Botz, M., "Cyanide Treatment Methods", Mining Environmental Management, 9 (3), pp. 28-30, May 2001.
134. "Hypergolic Propellant Hazard Response Guide, Cape Canaveral Air Force Station (CCAFS)", Draft Volume I, ICF Technology Incorporated, July 1, 1988.
135. www.hydrazinehydrate.com, Hydrazine Hydrate: Safety and Health, Bayer Corporation, 2001.
136. "Kvaerner Awarded Seal Alliance Contracts by BASF and Ineos", Hydrocarbon Processing, Int. Ed., 80 (3), pp. 41, March 2001.
137. www.epa.gov/ttn/chief/le/benzene/benz.cfe.pdf., "Emissions from Major Uses of Benzene", Section 5E, EPA-454/4-98-011, Locating & Estimating Documents, Technology Transfer Network, United States EPA, pp. 5-58 to 5-61, June 1998.
138. SRI International, Chemical Economics Handbook, Aniline, 1997.

139. www.che.LSU.edu/courses/4171/computerprojects1.htm, Aniline Process Flow Diagram, Louisiana State University, September 2001.
140. "Production and Consumption of Aniline in the World", Xiandai Hyagong, 18 (6), pp. 34-36, June 1998.
141. Aniline Chemical Profile, Chemical Market Reporter (February 8, 1999).
142. ntp-server.niehs.nih.gov/htdocs/LT-studies/TR163.html, Bioassay of Calcium Cyanamide for Possible Carcinogenicity, 1979.
143. Merritts, D.M., De Wet, A., and Menking, K., 1997, Environmental Geology - an Earth Systems Approach, W.H. Freeman & Co., pp. 452. (www.whfreeman.com/geology)
144. "Catalytic Processes in Nitric Acid Manufacture," Fertilizer Society Proceedings, London, 1978.
145. www.potashcorp.com/npk_markets/industry_overview/, Industry Overview –Nitrogen, Potash Corporation, 11/2/01.
146. McNulty, A., "Northern Business Unit: Inert Gas Lessons Learned", Internal Safety Memo, Foinaven Delivery Unit, Northern Business Unit, BP, 29 March, 2001.
147. Bright, K., "Cyanides in the South African Mining Industry", Report Prepared for DuPont by Business & Marketing Intelligence, June 1985.
148. www.ifdc.org/PDF_Files/FreeFert.statistics.PDF, International Fertilizer Development Center (IFDC), Muscle Shoals, AL, January, 2004..
149. Milmo, S., More Pressure Ahead for Europe's Fertilizers, Chemical Market Reporter (August 5, 2002).
150. Carney, M.J., Industrial Gases, www.uwec.edu/carneymj/Adobe%20slides/Chapter%203%20slides.pdf, August, 2002.
151. www.census.gov/econ/www/industry.html, U.S. Census Bureau, Industrial Products Overview, August, 2002.
152. Drnevich, F.R., et. al., Production and Use of Industrial Gases for Iron and Steelmaking, Chapter 5, www.aise.org/newpubs/ironchap%205.pdf, The AISE Steel Foundation, Pittsburgh, PA, 1999.
153. www.potashcorp.com/npk_markets/industry_overview/, Industry Overview –Nitrogen, Potash Corporation, 7/25/02.
154. www.potashcorp.com/npk_markets/potashcorp_letter/, The PotashCorp Letter, Potash Corporation of Saskatchewan Inc., February, 2002.
155. www.census.gov/econ/www/industry.html, U.S. Census Bureau, Industrial Statistics, Annual Survey of Manufactures, February 11, 2002.
156. www.epa.gov/ttn/otag, Technology Transfer Network, United States Environmental Protection Agency, 8/29/02.
157. Bradford, M.L, Grover, R. and Paul, P., "Controlling NOx Emissions - Part 2", Chemical Engineering Progress, 4, pp. 38-42, April, 2002.
158. Chemical Commodities, Chemical Market Reporter (July 22-29, 2002).
159. www.airproducts.com/gases, Gases and Equipment, Modes of Supply - Nitrogen, August 21, 2002.
160. Hydrogen, Chemical Products Synopsis, Mannsville Chemical Products Corp, Adams, NY, August, 2001.
161. www.hydrogen.org/Knowledge/w-I-energiew-eng3.html, HyWeb: Knowledge - Hydrogen in the Energy Sector, September, 2002.
162. www.ott.doe.gov/facts/archives/fotw205.shtml, Fact #205, Hydrogen Cost and Worldwide Production, Office of Transportation Technologies, February 25, 2002.
163. The H_2 Technologies Economic Advantage, H_2 Technologies, Alchemix Corporation, www.alchemix.net/hyrefining.html, September, 2002.
164. Simbeck, D., Coal - Bridge to the Hydrogen Economy, presentation at 18th Annual International Pittsburgh Coal Conference in Newcastle, NSW, Australia, www.engr.pitt.edu/pcc/2001plenary/simbeck coal to h2 revised.pdf, December, 2001.
165. Hydrogen Chemical Profile, Chemical Marketing Reporter (January 29, 2001), www.chemexpo.com/news/profile010129.cfm.
166. www.sud-chemie.com, Purification of Hydrocarbon Streams, Sud-Chemie AG, 9/11/02.
167. www.chicagobridge.com/howebaker, Howe-Bakers Engineers, Ltd., Hydrogen & Synthesis Gases, 9/9/02.

168. Kot, R., Hydrogen Attack, Detection, Assessment and Evaluation, 10th Asia-Pacific Conference on Non-Destructive Testing, Brisbane, Australia, September, 2001 (www.ndt.net/article/apcndt01/papers/1154/11154.htm).
169. www.ceteklimited.com, Ceramic Coatings for Process Tubes, Cetek Limited, 6/18/02.
170. Morris, P., Clean Fuels Means Hydrogen -Where Will It Come From And What Will It Cost?, European Fuels Week, 10-14th April, 2000. (www.airproducts.com - 9/12/2002).
171. www.halliburton.com/news/archive/2001, Halliburton KBR Signs Contract with Liaotong for Ammonia Plant Revamp, Press Release, Dallas, TX, November 26, 2001.
172. Davis, R.A., Patel, N.M., Refinery Hydrogen Management - Part 1 & Part 2, Hydrocarbon Engineering, Palladian Publications, Ltd, Farnham, England, August & September, 2001.
173. www.bellona.no/en/energy/hydrogen, Hydrogen Technologies, Bellona Report 6:2002, October, 2002.
174. Moulijn, J.A., Production Synthesis Gas, Chemical Process Technology (ST3141) Class Notes, 1Q2002, (www.dct.tudelft.nl/~wwwrace/st3141a/cpt_05.pdf).
175. Buchanan, T.L., Schoff, R.L., et. al., Hydrogen Production Facilities Plant performance and Cost Comparisons, Final Report, Prepared for United States Dept. of Energy, National Energy Technology Laboratory, Parsons Infrastructure and Technology Group Inc, Reading, PA, March 2002. (www.netl.doe.gov/coalpower/gasification/pubs/pdf/FinalCompReport.pdf)
176. Gross, M., Wolff, J., Gasification of Residue as a Source of Hydrogen for the Refining Industry in India, 2000 Gasification Technologies Conf., San Francisco, CA, USA, October 2000. (www.gasification.org/98GTC/GTC00150.pdf).
177. Beaudette, T.M., Bochow, C. Jr., Slivensky, D., Natural Gas Partial Oxidation for Chemical Processing in Longview, Texas, Gasification Technologies 2001, San Francisco, CA, October 2001. (www.gasification.org/98GTC/GTC01039.pdf)
178. www.proquip-corp.com/hydrochem/experience/plant_types.shtml, Hydro-Chem Modular Hydrogen Plants, Hydro-Chem: A Division of Pro-Quip, Holly Springs, GA, August, 2002.
179. Liebner, W., and Ulber, D., MPG - Lurgi Multi Purpose Gasification: application in "Gas-Gasification", 2000 Gasification Technologies Conference, San Francisco, CA, October 2000 (www.gasification.org/98GTC/GTC00340.pdf).
180. Higman, C. and Eppinger, M., The Zero-Residue Refiner, ACHEMA, Frankfurt, Germany, June 1994 (www.higman.de/gasification/achema94.pdf).
181. Goudarefractories.com/applications_ndustries/petrochem/reforming_and_gasification.asp, GOUDA - applications & industries, Gouda, the Netherlands, October 2002.
182. Bhandarkar, P.G., Gasification Overview Focus - India, Hydrocarbon Asia, , pp. 46-51, November/December 2001 (www.hcasia.safan.com/mag/dec01/t46.pdf).
183. Brown, J.A.G. and Griffiths, J., The Gasification of Refinery Residual Oil and Subsequent Generation of Electricity, Institution of Diesel and Gas Turbine Engineers, 2002 (www.jacobsconsultancy.com/idgte%20paper.pdf).
184. Tavoulareas, E.S., and Charpentier, J-P., Clean Coal Technologies for Developing Countries, World Bank Technical Paper No. 286, Energy Series, July 1995 (www.worldbank.org/html/fpd/em/power/EA/mitigatn/igccsubs.stm).
185. Breckenridge, W., Holiday, A., et. al., Use of SELEXOL® Process in Coke Gasification to Ammonia Project, Laurance Reid Gas Conditioning Conference, 2000 conference Proceedings, University of Oklahoma Norman, OK, March 2000. (www.uop.com/gasprocessing/TechPapers/Selexol/cokegasifiAmmonia.pdf).
186. Camps, J.A., Nitrogen Fertilizer Producers Push the Boundaries, Jacobs Consultancy, 2000. (www.jacobs-sirrine.com/aichepush.pdf).
187. Zuideveld, P.L., Overview of Shell Gasification Projects, www.icheme.org/literature/conference/gasi/Gasification%20Conf%20Papers/Session%204%20presentation-Shell.pdf, Noordwijk, Netherlands, April 2002.
188. How PEM Works, Technology Introduction, Proton Energy Systems, Wallingford, CT, www.protonenergy.com, 2002.
189. Praxair Starts-Up Industry's First On-Site Hydrogen Generating System, Praxair, Inc., News Releases - 1998, Danbury, CT, March 12, 1998. (www.praxair.com/praxair.nsf).

190. Carolan, M.F., Chen, C.M., and Rynders, S.W., ITM Syngas and ITM H2: Engineering Development of Ceramic Membrane Reactor Systems for Converting Natural Gas to Hydrogen and Synthesis Gas for Liquid Transportation Fuels, Proceedings of the 2002 U.S. DOE Hydrogen Program Review, NREL/CP-610-32405, Washington, D.C., 2002. (www.eren.doe.gov/hydrogen/pdfs/32405a19.pdf)
191. Brown, L.C., Besenbruch, G.E., et. al., High Efficiency Generation of Hydrogen Fuels Using Nuclear Energy, Hydrogen and Fuel Cells Annual Review, U.S. Department of Energy, May 2002. (www.eren.doe.gov/hydrogen/pdfs/32405.pdf).
192. Brooksbank, W.J., HydroMax®: Bridge to a Hydrogen Economy, Alchemix Corporation, Carefree, AZ, 2003. (www.alchemix.net)
193. Pressure Swing and Vacuum Swing Adsorption Systems, Air Products and Chemicals, Inc., 1996-2002. (www.airproducts.com/membranes).
194. Myers, D.B., Ariff, G.D., et. al., Cost and Performance Comparison of Stationary Hydrogen Fueling Appliances, Directed Technologies, Arlington, VA, 2002 (www.eren.doe.gov/hydrogen/pdfs/32405b2.pdf).
195. "Polymer Membrane Captures CO_2 at High Temperatures", Chemical Engineering Progress, 7, pp. 27, July, 2002.
196. Acid Gas Purification Techniques for CO_2 (carbon dioxide) Benfield Process, Air Pollution Control Technology in Japan, Exhaust Gas Treatment Equipment, Kobe Steel, LTD., Tokyo, Japan, 2002 (nett21.unep.or.jp/CTT_DATA/AIR/AIR_7/html/Air-208.html).
197. Rectisol Wash, www.linde-process-engineering.com, Linde AG, 2001.
198. Selexol Process, Gas Processing, UOP LLC, Des Plaines, IL, 2002 (www.uop.com).
199. Carbon Dioxide Capture from Power Stations - Absorption Technologies, Greenhouse Gas R&D Programme, International Energy Agency, 2002. (www.ieagreen.org.uk).
200. Amine Guard, Wet Solvent Systems, Gas Processing Technologies, UOP LLC, Des Plaines, IL, 2000. (www.uop.com).
201. Modernizing Ammonia Plants with Innovative Technical Solutions, Kellogg Brown & Root, Houston, TX, 2000 (www.mwkl.co.uk/pdf/modamplt.pdf).
202. Korotkikh, O. and Farrauto, R., Selective Catalytic Oxidation of CO in H_2 For Fuel Cell Applications, Engelhard Corporation, Iselin, New Jersey, 1999. (www.unl.edu.ar/cepac/abstract/bfarraut.htm).
203. Rhodes, A.K., "New Ammonia Process, Catalyst Proven in Canadian Plant", Oil & Gas Journal, 96 (47), pp. 37-41, November, 1996.
204. Synetix Ammonia Synthesis Catalysts Technical Manual, www.synetix.com/ammonia/pdfs/216w.pdf, Cleveland, UK, 2002.
205. www.kvaerner.com, Catalytic Rich Gas (CRG) Technology, Kvaerner Process Technology, August 27, 2001.
206. Furukawa, S.F. and Bartoo, R.K., Improved Benfield Process for Ammonia Plants, UOP Technical Papers, www.uop.com/gasprocessing/techpapers/Improve Benfield.pdf, UOP LLC, Des Plaines, Illinois, January, 1997.
207. Christensen, P.V., Revamping Ammonia Plants to Follow Feedstock Situation and Market Demands, Haldor Topsøe Ammonia Publications, Presented at FAI Symposium on "Improving Productivity of Ammonia and Urea Plants" in New Delhi, India, www.haldortopsoe.com, Lyngby, Denmark, April 2001.
208. Linde Ammonia Concept (LAC) Plant, www.linde-process-engineering.com, Linde AG, Hoellriegelskreuth, Germany, 2001.
209. Ammonia Casale Technologies, www.casale.ch, Ammonia Casale S.A., Lugano, Switzerland, 2003.
210. Ohri, I.J., Energy Conservation Measures in Ammonia Plants at IFFCO Kalol and Phulpur Units, IFFCO, India, www.fertilizer.org/ifa/publicat/pdf/tech0039.pdf, 2001.
211. Lippmann, D., Deutsch, S., Khalil, R., and Moussa, M., "The New Abu Qir III Ammonia/Urea Complex - Design and Start-Up Experiences, Paper No. S2, 44th Annual Ammonia Safety Symposium, Seattle, USA, 27-30 Dept. 1999.
212. Agarwal, P., Ammonia: The Next Step, www.cheresources.com, 2002.
213. Filippi, E. and Pinauda, L., "Casale Offers More Ammonia Revamp Options", Fertilizer International, pp. 27, July 2001.

214. Linde Ammonia Concept (LAC) - Ammonia from Hydrocarbons, Hydrocarbon Processing, 80 (3), pp. 78, March 2001.
215. Parkinson, G., "All Aboard, Ammonia", Chemical Engineering, www.che.com/mag/ce3.htm, November, 2002.
216. Qafco Preparing for Higher Exports, Gulf Industry Online.com, www.gulfindustryworldwide.com/Articles.asp?Article=1660, 11 (5), Al Hilal Publishing & Marketing Group, September-October 2002.
217. Christensen, P.V., Design and Operation of Large Capacity Ammonia Plants, 4th Conference for Development and Integration of Petrochemical Industries in Arab States -Bahrain, www. www.haldortopsoe.com, Haldor Topsøe, Lyngby, Denmark, May, 2001.
218. Larsen, J., Lippmann, D., and Hooper, C.W., A New Process for Large-Capacity Ammonia Plants, Nitrogen & Methanol, 253, pp. 41-46, September-October, 2001.
219. Ammonia Properties, www.encyclopedia.com/articles/00445Properties.html, The Columbia Electronic Encyclopedia, Sixth Edition, Columbia University Press, 2000.
220. Maurer, R. and Bartsch, U., Enhanced Plant Design for the Production of Azeotropic Nitric Acid, Prepared for Presentation at Heraeus Nitric Acid Conference, Johannesburg, South Africa, November 29-30, 2001.
221. Brink, R.W. van den, Combined Catalytic Removal of N_2O and NOx from the Tail Gases of a Nitric Acid Plant, Energy Research Centre of the Netherlands, ECN-RX-01-012, April 2001, www.ecn.nl/library/reports/2001e/rx01012.html.
222. Schwefer, M., Maurer, R. and Groves, M., Reduction of Nitrous Oxide Emissions from Nitric Acid Plants, Prepared for Presentation at Nitrogen 2000, Vienna, Austria, March, 2000.
223. Ammonia Chemical Profile, Chemical Marketing Reporter (Nov. 25, 2002).
224. Greene, R., "Turning Stranded Gas into Cold Cash", Chemical Engineering Progress, 2003, 97 (2), pp. 13.
225. Fritsch, S., Steam Reformer-Based Hydrogen Plant Optimisation, Prepared for Presentation at HYFORM 2000, Munich, Germany, September, 2000.
226. Erben, A., From Neutralisation to Granulation, Nitrogen & Methanol, No. 241, pp. 49-52, September-October, 1999.
227. Bovens, J.L. and van Hecke, F., The Carnit Ammonium Nitrate Process, Kemira SA, The International Fertiliser Society, Proceeding 320, 1992 (www.fertiliser-society.org/Proceedings/Prc320.HTM).
228. Dall'Aglio, L., Fluid Drum Granulation for Ammonium Nitrate, Kaltenbach-Thüring, The International Fertiliser Society, Proceeding 415, 1998 (www.fertiliser-society.org/Proceedings/Prc415.HTM).
229. Sato, L., A New Route for Granular Ammonium Nitrate, The International Fertiliser Society, Proceeding 296, 1990 (www.fertiliser-society.org/Proceedings/Prc296.HTM).
230. Urea Chemical Profile, Chemical Marketing Reporter (December 2, 2002).
231. DSM Melamine Product, DSM Melamine, www.dsm.com/dmm/products, Sittard, The Netherlands, 2002.
232. Bryan, K.B., Melamine by the Melamine Chemicals M-II Process, PEP Review No. 92-1-1, SRI Consulting, process-economics.com/reviews/92-2-1.htm, 1993.
233. Melamine Process Block Diagram, Eurotecnica - Melamine, www.eurotecnica.it/melamine.htm, Milan, Italy, 2003.
234. Shan, Cheng-gang, Status of Production and Demand and Market Analysis of Melamine, Modern Chemical Industry, Author's Abstract, January 2002. (www.xdhg.com.cn/article/2002/abs/e-abs-2002-01.html).
235. Market Survey on Melamine Industry in China, CCM/Market Reports, March 2002 (www.cnchemicals.com/maindocs/marketreports/maincontent/Melamine.htm).
236. OMV Lifts Melamine Sales by Targeting New Markets, Chemical Marketing Reporter (August 19, 2002).
237. Nitroprocess Technologies, Platecxis, Engelhard-CLAL, Paris, France, 2002 (www.engelhard-clal.fr).
238. Kirk-Othmer , Encyclopedia of Chemical Technology, Vol. 17, pp. 444-467, John Wiley & Sons, Inc., New York, 1997.
239. Ammonium Phosphates Chemical Profile, Chemical Marketing Reporter (Nov. 18, 2002).

240. Ammonium Nitrate, Chemical Products Synopsis, Mannsville Chemical Products Corp, Adams, NY, October 2001.
241. Hagen, K., The Future of Ammonium Sulphate - Co and By-Production, The International Fertiliser Society, Proceeding 295, 1990 (www.fertiliser-society.org/Proceedings/Prc295.HTM).
242. Lauriente, D. H., CEH Marketing Research Report, Ammonium Sulfate, SRI International, July 1997.
243. Ammonium Sulfate, Chemical Products Synopsis, Mannsville Chemical Products Corp, Adams, NY, November 2000.
244. Kirk-Othmer , Encyclopedia of Chemical Technology, 4th Ed., Vol. 10, pp. 452-453, John Wiley & Sons, Inc., New York, 1993.
245. Heilen, G., Mercker, H.J., et. al., Aliphatic Amines, in Ullmann's Encyclopedia of Industrial Chemistry, Vol. A2, 5th Ed., Reinhold Publishing, New York, 1996.
246. The Nitto Chemical Methylamines Process, PEP Review No. 89-3-4, SRI Consulting, process-economics.com/reviews/89-3-4.htm, February 1991.
247. Methylamines Chemical Profile, Chemical Marketing Reporter (August 4, 2003).
248. Morpholine Health and Safety Guide, IPCS International Programme on Chemical Safety, Health and Safety Guide No. 92, World Health Organization Geneva, Switzerland, 1995 (www.inchem.org/documents/hsg/hsg/hsg092.htm).
249. Ethanolamines Chemical Profile, Chemical Marketing Reporter (June 11, 2001).
250. Davenport, R.E., CEH Product Review, Chelating Agents, SRI International, March 2000.
251. Chelating Agents, Chemical of the Week, scifun.chem.wisc.edu/chemweek/Chelating Agents/ChelatingAgents.html, October 2001.
252. Hexamine, Chemical Products Synopsis, Mannsville Chemical Products Corp, Adams, NY, October 2001.
253. Kirk-Othmer , Encyclopedia of Chemical Technology, Vol. 13, pp. 560-606, John Wiley & Sons, Inc., New York, 1997.
254. Hydrogen Cyanide Technology Manual, DuPont Chemicals Solution Enterprise, Memphis, Tennessee, May 1994.
255. Aniline, Chemical Products Synopsis, Mannsville Chemical Products Corp, Adams, NY, July, 2001.
256. Markets and Economics: Aniline, Chemical Week, 161 (43), pg. 41, November 17, 1999.
257. Aniline: Supply Demand Balance and Stable Price in China, China Chemical Reporter, 11 (25), pg. 13-15, September 5, 2000.
258. Camera, E., CEH Data Summary, Nitrobenzene, SRI International, June 1997.
259. Hydrogen Data, Air Liquide, Paris, France, 2001 (www.airliquide.com/en/business/products/hydrogen).
260. Shell Sulfinol Process, Shell Global Solutions, Shell Oil Products Company, Houston, Texas, 2003 (www.sulfinol.com).
261. Hydrogen Cyanide Chemical Profile, Chemical Marketing Reporter (Nov. 26, 2001).
262. Aniline Chemical Profile, Chemical Marketing Reporter (February 4, 2002).
263. Nitrobenzene Chemical Profile, Chemical Marketing Reporter (April 15, 2002).
264. Ritz, J., Kieczka, H. and Moran, W.C., Caprolactam, in Ullmann's Encyclopedia of Industrial Chemistry, Vol. A5, pp. 31-50, Reinhold Publishing, New York, 1996.
265. Caprolactam, Chemical Products Synopsis, Mannsville Chemical Products Corp, Adams, NY, August, 2001.
266. Ring, K.L, CEH Marketing Research Report, Caprolactam, SRI International, April 2000.
267. Caprolactam Chemical Profile, Chemical Marketing Reporter (April 16, 2001).
268. Chiyoda and DSM Alliance in Caprolactam Projects, Chiyoda Corporation News Release, Yokohama, Japan, April 26, 2001.
269. True Closed-Loop Recycling for Nylon, Environmental Building News, BuildingGreen.com, Vol. 8 (9), Brattleboro, VT, September 1999.
270. Production at Evergreen Nylon Recycling Suspended, DSM Corporate Communications, Heerlen, The Netherlands, August 30, 2001. (www.dsm.com/newsarchive)
271. Schut, J.H., Big German Plant May Relieve U.S. Bottleneck in Recycling Carpet Nylon, Plastics Technology Online, New York, NY, May 2002. (www.plasticstechnology.com/articles)

272. Production Process, Polyamid 2000, Premnitz, Germany, October 2000. (www.polyamid2000.com)
273. Adipic Acid / Adiponitrile, Product Focus, Chemical Week, 161 (9), pg. 31, March 10, 1999.
274. Stahl, W.F., CEH Product Review, Hexamethylenediamine (HMDA), SRI International, November 1996.
275. Schwarz, J., Beloff, B. and Beaver, E. "Use Sustainability Metrics to Guide Decision-Making", Chemical Engineering Progress, 2002, 98 (7), pp. 58-63.
276. "Acrylonitrile", European Chemical News, pp.16, 29 July-4 August, 2002.
277. Acrylonitrile Chemical Profile, Chemical Marketing Reporter (August 19-26, 2001).
278. "Acrylonitrile", Asian Chemical News, pp.10-13, 11 March, 2002.
279. "Methyl Methacrylate", European Chemical News, 77 (2029), pp.15, November 11-17, 2002.
280. Methyl Methacrylate, Product Focus, Chemical Week, 164 (43), pg. 32, October 30, 2002.
281. Will., R. and Toki, G.., CEH Product Review, Acrylamide, SRI International, March 2002.
282. 'Acrylamide, PERP Program - New Report Alert, Chem Systems', December 2002. (www.chemsystems.com/newsletters)
283. Acrylamide Chemical Profile, Chemical Marketing Reporter (March 15, 1999).
284. Dimethylformamide from Supercritical Carbon Dioxide, Design Projects, University of Notre Dame, South Bend, Indiana, June 2003. (www.nd.edu/~enviro)
285. Dimethylformamide Producing Companies, China Report: Chemical Product Trends, SRI Consulting, 2003. (www.cnonline.com/cni_home)
286. Dimethylformamide, Methylamine and Derivatives Process Technologies, Davy Process Technology, www.davyprotech.com, 2003.
287. Mitsubishi Rayon Reaches Agreement on Joint Venture to Manufacture and Sell Petrochemical Products in China, Chemchannels.com, Tokyo, Japan, December 19, 2002. (www.chemchannels.com)
288. Dimethylformamide, Solvents, UCB Chemicals - Methylamines & Derivatives, www.chemicals.ucb-group.com, 2003.
289. Courtney, P.J., "Light-Curing Cyanoacrylates", Adhesives Age, March 2002, pg. s8-s11.
290. www.chemetics.kvaerner.com, Mononitrobenzene, Kvaerner Pulp & Paper, 2003.
291. Davey, W.L.E. and Filippi, E., MEGAMMONIA - the Mega-Ammonia Process for the new Century, Nitrogen 2003, Warsaw Poland, February 26, 2003.
292. Demand Growth of Acrylonitrile Faster than Output Increase, China Chemical Reporter, 14 (30), pg. 18, October 26, 2003.
293. Acrylonitrile Supply Short of Demand, China Chemical Week, 464, pg. 7, November 7, 2003.
294. Norsk Hydro A.S., Oslo, Norway, "Method for Reduction of Nitrogen Oxide", EP 0 359 286 B1, (1993).
295. Zapp, K.H.., Ammonium Compounds, in Ullmann's Encyclopedia of Industrial Chemistry, , Wiley-VCH Verlag GmbH & Co. , Weinheim, Germany, June 15, 2000. (www.mrw.interscience.wiley.com/ueic/articles/a02_243)
296. Kolmeijer, N.W., Phosphate Fertilizers, in Ullmann's Encyclopedia of Industrial Chemistry, Wiley-VCH Verlag GmbH & Co. , Weinheim, Germany, June 15, 2000. (www.mrw.interscience.wiley.com/ueic/articles/a19_421)
297. "Acrylonitrile", European Chemical News, pp.16, 29 July-4 August, 2002.
298. The Commercialization of a New Production Process for MMA Monomer, Asahi Chemical Industry Co., Ltd. Press Release, March 10, 1997 (www.asahi-kasei.co.jp/asahi/en/news/1996/e970310.html).

INDEX